AF157376

Tucholsky   Wagner   Zola   Scott   Sydow   Freud   Schlegel
Turgenev   Wallace   Fonatne
Twain   Walther von der Vogelweide   Fouqué   Friedrich II. von Preußen
Weber   Freiligrath   Frey
Kant   Ernst
Fechner   Fichte   Weiße Rose   von Fallersleben   Frommel
Hölderlin   Richthofen
Engels   Fielding   Eichendorff   Tacitus   Dumas
Fehrs   Faber   Flaubert
Eliasberg   Ebner Eschenbach
Feuerbach   Maximilian I. von Habsburg   Fock   Eliot   Zweig
Ewald   Vergil
Goethe   London
Mendelssohn   Balzac   Shakespeare   Elisabeth von Österreich
Lichtenberg   Rathenau   Dostojewski   Ganghofer
Trackl   Stevenson   Doyle   Gjellerup
Tolstoi   Hambruch
Mommsen   Lenz   Droste-Hülshoff
Thoma   Hanrieder
Dach   Verne   von Arnim   Hägele   Hauff   Humboldt
Reuter   Rousseau   Hagen   Hauptmann   Gautier
Karrillon   Garschin   Defoe   Baudelaire
Damaschke   Descartes   Hebbel
Hegel   Kussmaul   Herder
Wolfram von Eschenbach   Schopenhauer
Dickens   Rilke   George
Bronner   Darwin   Melville   Grimm Jerome
Campe   Horváth   Aristoteles   Bebel   Proust
Bismarck   Vigny   Voltaire   Federer   Herodot
Gengenbach   Barlach   Heine
Storm   Casanova   Tersteegen   Grillparzer   Georgy
Chamberlain   Lessing   Langbein   Gilm
Brentano   Lafontaine   Gryphius
Strachwitz   Claudius   Schiller   Kralik   Iffland   Sokrates
Katharina II. von Rußland   Bellamy   Schilling
Gerstäcker   Raabe   Gibbon   Tschechow
Löns   Hesse   Hoffmann   Gogol   Wilde   Vulpius
Luther   Heym   Hofmannsthal   Klee   Hölty   Morgenstern   Gleim
Roth   Heyse   Klopstock   Kleist   Goedicke
Luxemburg   Puschkin   Homer   Mörike   Musil
Machiavelli   La Roche   Horaz
Kierkegaard   Kraft   Kraus
Navarra   Aurel   Musset   Kind   Moltke
Nestroy   Marie de France   Lamprecht   Kirchhoff   Hugo
Laotse   Ipsen   Liebknecht
Nietzsche   Nansen   Ringelnatz
Marx   Lassalle   Gorki   Klett   Leibniz
von Ossietzky   May   Irving
vom Stein   Lawrence
Petalozzi   Knigge
Platon   Kafka
Sachs   Pückler   Michelangelo   Kock
Poe   Liebermann   Korolenko
de Sade   Praetorius   Mistral   Zetkin

Wild Flowers An Aid to Knowledge of Our Wild Flowers and Their Insect Visitors

Neltje Blanchan

Imprint

This book is part of the TREDITION CLASSICS series.

Author: Neltje Blanchan
Cover design: toepferschumann, Berlin (Germany)

Publisher: tredition GmbH, Hamburg (Germany)
ISBN: 978-3-8491-9252-5

www.tredition.com
www.tredition.de

y tramp a whole day without finding a single native ladies' slip-
. What of the sundew that not only catches insects, but secretes
tric juice to digest them? What of the bladderwort, in whose
ated traps tiny crustaceans are imprisoned, or the pitcher plant,
t makes soup of its guests? Why are gnats and flies seen about
tain flowers, bees, butterflies, moths or humming birds about
ers, each visitor choosing the restaurant most to his liking? With
at infinite pains the wants of each guest are catered to! How
entlessly are pilferers punished! The endless devices of the more
nbitious flowers to save their species from degeneracy by close
breeding through fertilization with their own pollen, alone prove
e operation of Mind through them. How plants travel, how they
nd seeds abroad in the world to found new colonies, might be
udied with profit by Anglo-Saxon expansionists. Do vice and vir-
te exist side by side in the vegetable world also? Yes, and every
nner is branded as surely as was Cain. The dodder, Indian pipe,
roomrape and beech-drops wear the floral equivalent of the
triped suit and the shaved head. Although claiming most respecta-
le and exalted kinsfolk, they are degenerates not far above the
ungi. In short, this is a universe that we live in; and all that share
he One Life are one in essence, for natural law is spiritual law.
'Through Nature to God," flowers show a way to the scientist lacking faith.

Although it has been stated by evolutionists for many years that
in order to know the flowers, their insect relationships must first be
understood, it is believed that "Nature's Garden" is the first American work to explain them in any considerable number of species.
Dr. Asa Gray, William Hamilton Gibson, Clarence Moores Weed,
and Miss Maud Going in their delightful books or lectures have
shown the interdependence of a score or more of different blossoms
and their insect visitors. Hidden away in the proceedings of scientific societies' technical papers are the invaluable observations of
such men as Dr. William Trelease of Wisconsin and Professor
Charles Robertson of Illinois. To the latter especially, I am glad to
acknowledge my indebtedness. Sprengel, Darwin, Muller, Delpino,
and Lubbock, among others, have given the world classical volumes
on European flora only, but showing a vast array of facts which the
theory of adaptation to insects alone correlates and explains. That

PREFACE

Surely a foreword of explanation is called for from
the temerity to offer a surfeited public still another b
flowers. Inasmuch as science has proved that almost ev
in the world is everything it is because of its necessit
insect friends or to repel its foes - its form, mechanism, c
ings, odor, time of opening and closing, and its season o
being the result of natural selection by that special in
which each depends more or less absolutely for help in
ing its species - it seems fully time that the vitally impo
interesting relationship existing between our common wi
and their winged benefactors should be presented in a
book.

Is it enough to know merely the name of the flower yo
the meadow? The blossom has an inner meaning, hopes a
that inspire its brief existence, a scheme of salvation for its s
the struggle for survival that it has been slowly perfecti
some insect's help through the ages. It is not a passive thin
admired by human eyes, nor does it waste its sweetness on
sert air. It is a sentient being, impelled to act intelligently t
the same strong desires that animate us, and endowed with
powers differing only in degree, but not in kind, from those
animal creation. Desire ever creates form.

Do you doubt it? Then study the mechanism of one of our
mon orchids or milkweeds that are adjusted with such marv
delicacy to the length of a bee's tongue or of a butterfly's leg;
why so many flowers have sticky calices or protective hairs;
the skunk cabbage, purple trillium, and carrion flower emit a
odor while other flowers, especially the white or pale yellow n
bloomers, charm with their delicious breath; see if you cannot
cover why the immigrant daisy already whitens our fields w
descendants as numerous as the sands of the seashore, whereas y

the results of illumining researches should be so slow in enlightening the popular mind can be due only to the technical, scientific language used in setting them forth, language as foreign to the average reader as Chinese, and not to be deciphered by the average student either, without the help of a glossary. These writings, as well as the vast array of popular books - too many for individual mention - have been freely consulted after studies made afield.

To Sprengel belongs the glory of first exalting flowers above the level of botanical specimens. After studying the wild geranium he became convinced, as he wrote in 1787, that "the wise Author of Nature has not made even a single hair without a definite design. A hundred years before, one, Nehemias Grew, had said that it was necessary for pollen to reach the stigma of a flower in order that it might set fertile seed, and Linnaeus bad to come to his rescue with conclusive evidence to convince a doubting world that he was right. Sprengel made the next step forward, but his writings lay neglected over seventy years because he advanced the then incredible and only partially true statement that a flower is fertilized by insects which carry its pollen from its anthers to its stigma. In spite of his discoveries that the hairs within the wild geranium protect its nectar from rain for the insect benefactor's benefit; that most flowers which secrete nectar have what he termed "honey guides" - spots of bright color, heavy veining, or some such pathfinder for the visitor on the petals; that sometimes the male flowers, the staminate ones, are separated from the seed-bearing or pistillate ones on distinct plants, he left it to Darwin to show that cross-fertilization by insects, the transfer of pollen from one blossom to another - not from anthers to stigma of the same flower - is the great end to which so much marvelous floral mechanism is adapted. The wind is a wasteful, uncertain pollen distributor. Insects transfer it more economically, especially the more highly organized and industrious ones. In a few instances hummingbirds, as well, unwittingly do the flower's bidding while they feast now here, now there. In spite of Sprengel's most patient and scientific research, that shed great light on the theory of natural selection a half century before Darwin advanced it, he never knew that flowers are nearly always sterile to pollen of another species when carried to them on the bodies of insect visitors, or that cross-pollenized blossoms defeat the self-pollinated ones in the struggle for survival. These facts Darwin proved in endless experiments.

Because bees depend absolutely upon flowers, not only for their own food but for that of future generations for whom they labor; because they are the most diligent of all visitors, and are rarely diverted from one species of flower to another while on their rounds collecting, as they must, both nectar and pollen, it follows they are the most important fertilizing agents. It is estimated that, should they perish, more than half the flowers in the world would be exterminated with them! Australian farmers imported clover from Europe, and although they had luxuriant fields of it, no seed was set for next year's planting, because they had failed to import the bumblebee. After his arrival, their loss was speedily made good.

Ages before men cultivated gardens, they had tiny helpers they knew not of. Gardeners win all the glory of producing a Lawson pink or a new chrysanthemum; but only for a few seasons do they select, hybridize, according to their own rules of taste. They take up the work where insects left it off after countless centuries of toil. Thus it is to the night-flying moth, long of tongue, keen of scent, that we are indebted for the deep, white, fragrant Easter lily, for example, and not to the florist; albeit the moth is in his turn indebted to the lily for the length of his tongue and his keen nerves: neither could have advanced without the other. What long vistas through the ages of creation does not this interdependence of flowers and insects open!

Over five hundred flowers in this book have been classified according to color, because it is believed that the novice, with no knowledge of botany whatever, can most readily identify the specimen found afield by this method, which has the added advantage of being the simple one adopted by the higher insects ages before books were written. Technicalities have been avoided in the text wherever possible, not to discourage the beginner from entering upon one of the most enjoyable and elevating branches of Nature study. The scientific names and classification follow that method adopted by the International Botanical Congress which has now superseded all others; nevertheless the titles employed by Gray, with which older botanists in this country are familiar, are also indicated where they differ from the new nomenclature.

NELTJE BLANCHAN, New York, March, 1900

TABLE OF CONTENTS

"Let us content ourselves no longer with being mere 'botanists' - historians of structural facts. The flowers are not mere comely or curious vegetable creations, with colors, odors, petals, stamens and innumerable technical attributes. The wonted insight alike of scientist, philosopher, theologian, and dreamer is now repudiated in the new revelation. Beauty is not 'its own excuse for being,' nor was fragrance ever 'wasted on the desert air.' The seer has at last heard and interpreted the voice in the wilderness. The flower is no longer a simple passive victim in the busy bee's sweet pillage, but rather a conscious being, with hopes, aspirations and companionships. The insect is its counterpart. Its fragrance is but a perfumed whisper of welcome, its color is as the wooing blush and rosy lip, its portals are decked for his coming, and its sweet hospitalities humored to his tarrying; and as it speeds its parting affinity, rests content that its life's consummation has been fulfilled." - William Hamilton Gibson.

"I often think, when working over my plants, of what Linnaeus once said of the unfolding of a blossom: 'I saw God in His glory passing near me, and bowed my head in worship.' The scientific aspect of the same thought has been put into words by Tennyson:

'Flower in the crannied wall
I pluck you out of the crannies,
I hold you here, root and all in my hand
Little flower, - but if I could understand
What you are, root and all, and all in all,
I should know what God and man is.'

No deeper thought was ever uttered by poet. For in this world of plants, which, with its magician, chlorophyll, conjuring with sunbeams, is ceaselessly at work bringing life out of death, - in this quiet vegetable world we may find the elementary principles of all life in almost visible operation." - JOHN FISKE in "Through Nature to God."

FROM BLUE TO PURPLE FLOWERS

"If blue is the favorite color of bees, and if bees have so much to do with the origin of flowers, how is it that there are so few blue ones? I believe the explanation to be that all blue flowers have descended from ancestors in which the flowers were green; or, to speak more precisely, in which the leaves surrounding the stamens and pistil were green; and that they have passed through stages of white or yellow, and generally red, before becoming blue." - Sir John Lubbock in "Ants, Bees, and Wasps."

VIRGINIA or COMMON DAY-FLOWER
(Commelina Virginica) Spiderwort family

Flowers - Blue, 1 in. broad or less, irregular, grouped at end of stem, and upheld by long leaf-like bracts. Calyx of 3 unequal sepals; 3 petals, 1 inconspicuous, 2 showy, rounded. Perfect stamens 3; the anther of 1 incurved stamen largest; 3 insignificant and sterile stamens; 1 pistil. Stem: Fleshy, smooth, branched, mucilaginous. Leaves: Lance-shaped, 3 to 5 in. long, sheathing the stem at base; upper leaves in a spathe-like bract folding like a hood about flowers. Fruit: A 3-celled capsule, seed in each cell. Preferred Habitat - Moist, shady ground. Flowering Season - June - September. Distribution - Southern New York to Illinois and Michigan, Nebraska, Texas, and through tropical America to Paraguay. - Britton and Browne.

Delightful Linnaeus, who dearly loved his little joke, himself confesses to have named the day-flowers after three brothers Commelyn, Dutch botanists, because two of them - commemorated in the two showy blue petals of the blossom - published their works; the third, lacking application and ambition, amounted to nothing, like the inconspicuous whitish third petal! Happily Kaspar Commelyn died in 1731, before the joke was perpetrated in "Species Plantarum."

In the morning we find the day-flower open and alert-looking, owing to the sharp, erect bracts that give it support; after noon, or as soon as it has been fertilized by the female bees, that are its chief benefactors while collecting its abundant pollen, the lovely petals roll up, never to open again, and quickly wilt into a wet, shapeless mass, which, if we touch it, leaves a sticky blue fluid on our fingertips.

The SLENDER DAY-FLOWER (C. erecta), the next of kin, a more fragile-looking, smaller-flowered, and narrower-leafed species, blooms from August to October, from Pennsylvania southward to tropical America and westward to Texas.

SPIDERWORT; WIDOW'S or JOB'S TEARS
(Tradescantia Virginiana) Spiderwort family

Flowers - Purplish blue, rarely white, showy, ephemeral, 1 to 2 in. broad; usually several flowers, but more drooping buds, clustered and seated between long blade-like bracts at end of stern. Calyx of 3 sepals, much longer than capsule. Corolla of 3 regular petals; 6 fertile stamens, bearded; anthers orange; 1 pistil. Stem: 8 in. to 3 ft. tall, fleshy, erect, mucilaginous, leafy. Leaves: Opposite, long, blade-like, keeled, clasping, or sheathing stem at base. Fruit: 3-celled capsule. Preferred Habitat - Rich, moist woods, thickets, gardens. Flowering Season - May-August. Distribution - New York and Virginia westward to South Dakota and Arkansas.

As so very many of our blue flowers are merely naturalized immigrants from Europe, it is well to know we have sent to England at least one native that was considered fit to adorn the grounds of Hampton Court. John Tradescant, gardener to Charles I, for whom the plant and its kin were named, had seeds sent him by a relative in the Virginia colony; and before long the deep azure blossoms with their golden anthers were seen in gardens on both sides of the Atlantic - another one of the many instances where the possibilities of our wild flowers under cultivation had to be first pointed out to us by Europeans.

Like its relative the dayflower, the spiderwort opens for part of a day only. In the morning it is wide awake and pert; early in the afternoon its petals have begun to retreat within the calyx, until presently they become "dissolved in tears," like Job or the traditional widow. What was flower only a few hours ago is now a fluid jelly that trickles at the touch. Tomorrow fresh buds will open, and a continuous succession of bloom may be relied upon for a long season. Since its stigma is widely separated from the anthers and surpasses them, it is probable the flower cannot fertilize itself, but is wholly dependent on the female bees and other insects that come to it for pollen. Note the hairs on the stamens provided as footholds for the bees.

The plant is a cousin of the "Wandering Jew" (T. repens), so commonly grown either in water or earth in American sitting-rooms. In a shady lane within New York city limits, where a few stems were thrown out one spring about five years ago, the entire bank is now covered with the vine, that has rooted by its hairy joints, and, in spite of frosts and blizzards, continues to bear its true-blue flowers throughout the summer.

PICKEREL WEED
(Pontederia cordata) Pickerel-weed family

Flowers - Bright purplish blue, including filaments, anthers, and style; crowded in a dense spike; quickly fading; unpleasantly odorous. Perianth tubular, 2-lipped, parted into 6 irregular lobes, free from ovary; middle lobe of upper lip with 2 yellow spots at base within. Stamens 6, placed at unequal distances on tube, 3 opposite each lip. Pistil 1, the stigma minutely toothed. Stem: Erect, stout, fleshy, to 4 ft. tall, not often over 2 ft. above water line. Leaves: Several bract-like, sheathing stem at base; leaf only, midway on flower-stalk, thick, polished, triangular, or arrow-shaped, 4 to 8 in. long, 2 to 6 in. across base. Preferred Habitat - Shallow water of ponds and streams. Flowering Season - June-October. Distribution - Eastern half of United States and Canada.

Grace of habit and the bright beauty of its long blue spikes of ragged flowers above rich, glossy leaves give a charm to this vigorous wader. Backwoodsmen will tell you that pickerels lay their eggs among the leaves; but so they do among the sedges, arums, wild rice, and various aquatic plants, like many another fish. Bees and flies, that congregate about the blossoms to feed, may sometimes fly too low, and so give a plausible reason for the pickerel's choice of haunt. Each blossom lasts but a single day; the upper portion, withering, leaves the base of the perianth to harden about the ovary and protect the solitary seed. But as the gradually lengthened spike keeps up an uninterrupted succession of bloom for months, more than ample provision is made for the perpetuation of the race - a necessity to any plant that refuses to thrive unless it stands in water. Ponds and streams have an unpleasant habit of drying up in summer, and often the pickerel weed looks as brown as a bulrush where it is stranded in the baked mud in August. When seed falls on such ground, if indeed it germinates at all, the young plant naturally withers away.

In the Bulletin of the Torrey Botanical Club, Mr. W. H. Leggett, who made a careful study of the flower, tells that three forms occur, not on the same, but on different plants, being even more distinctly trimorphic than the purple Loosestrife. As these flowers set no seed without insects' aid, the provisions made to secure the greatest benefit from their visits are marvelous. Of the three kinds of blossoms, one raises its stigma on a long style reaching to the top of the flower; a second form lifts its stigma only halfway up, and the third keeps its stigma in the bottom of the tube. Now, there are two sets of stamens, three in each set bearing pollen grains of different size and value. Whenever the stigma is high, the two sets of stamens keep out of its way by occupying the lowest and middle positions, or just where the stigmas occur in the two other forms; or, let us say, whenever the stigma is in one of the three positions, the different sets of stamens occupy the other two. In a long series of experiments on flowers occurring in two and three forms - dimorphic and trimorphic - Darwin proved that perfect fertility can be obtained only when the stigma in each form is pollenized with grains carried from the stamens of a corresponding height. For example, a bee on entering the flower must get his abdomen dusted with pollen from the long stamens, his chest covered from the middle-length stamens, and his tongue and chin from the set in the bottom of the tube nearest the nectary. When he flies off to visit another flower, these parts of his body coming in contact with the stigmas that occupy precisely the position where the stamens were in other individuals, he necessarily brushes off each lot of pollen just where it will do the most good. Pollen brought from high stamens, for example, to a low stigma, even should it reach it, which is scarcely likely, takes little or no effect. Thus cross-fertilization is absolutely essential, and in three-formed flowers there are two chances to one of securing it.

WILD HYACINTH, SCILLA or SQUILL. QUAMASH
(Quamasia kyacinthina; Scilla Fraseri of Gray) Lily family

Flowers - Several or many, pale violet blue, or rarely white, in a long, loose raceme; perianth of 6 equal, narrowly oblong, widely spreading divisions, the thread-like filaments inserted at their bases; style thread-like, with 3-lobed stigma. Scape: 1 to 2 ft. high, from egg-shaped, nearly black bulb, 1 to 1 1/2 in. long. Leaves: Grasslike, shorter than flowering scape, from the base. Fruit: A 3-angled, oval capsule containing shining black seeds. Preferred Habitat - Meadows, prairies, and along banks of streams. Flowering Season - April-May. Distribution - Pennsylvania and Ohio westward to Minnesota, south to Alabama and Texas.

Coming with the crocuses, before the snow is off the ground, and remaining long after their regal gold and purple chalices have withered, the Siberian scillas sold by seedsmen here deserve a place in every garden, for their porcelain-blue color is rare as it is charming; the early date when they bloom makes them especially welcome; and, once planted and left undisturbed, the bulbs increase rapidly, without injury from overcrowding. Evidently they need little encouragement to run wild. Nevertheless they are not wild scillas, however commonly they may be miscalled so. Certainly ladies' tresses, known as wild hyacinth in parts of New England, has even less right to the name.

Our true native wild hyacinth, or scilla, is quite a different flower, not so pure a blue as the Siberian scilla, and paler; yet in the middle West, where it abounds, there are few lovelier sights in spring than a colony of these blossoms directed obliquely upward from slender, swaying scapes among the lush grass. Their upward slant brings the stigma in immediate contact with an incoming visitor's pollen-laden body. As the stamens diverge with the spreading of the divisions of the perianth, to which they are attached, the stigma receives pollen brought from another flower, before the visitor dusts himself anew in searching for refreshment, thus effecting cross-pollination. Ants, bees, wasps, flies, butterflies, and beetles may be seen about the wild hyacinth, which is obviously best adapted to the bees. The smallest insects that visit it may possibly defeat Nature's plan and obtain nectar without fertilizing the flower, owing to the wide passage between stamens and stigma. In about an hour, one May morning, Professor Charles Robertson captured over six hundred insects, representing thirty-eight distinct species, on a patch of wild hyacinths in Illinois.

The bulb of a MEDITERRANEAN SCILLA (S. maritima) furnishes the sourish-sweet syrup of squills used in medicine for bronchial troubles.

The GRAPE HYACINTH (Muscari botrycides), also known as Baby's Breath, because of its delicate faint fragrance, escapes from gardens at slight encouragement to grow wild in the roadsides and meadows from Massachusetts to Virginia and westward to Ohio. Its tiny, deep-blue, globular flowers, stiffly set around a fleshy scape that rises between erect, blade-like, channeled leaves, appear spring after spring wherever the small bulbs have been planted. On the east end of Long Island there are certain meadows literally blued with the little runaways.

PURPLE TRILLIUM, ILL-SCENTED WAKE-ROBIN or BIRTH-ROOT
(Trillium erectum) Lily-of-the-Valley family

Flowers - Solitary, dark, dull purple, or purplish red; rarely greenish, white, or pinkish; on erect or slightly inclined footstalk. Calyx of 3 spreading sepals, 1 to 1 1/2 in. long, or about length of 3 pointed, oval petals; stamens 6; anthers longer than filaments; pistil spreading into 3 short, recurved stigmas. Stem: Stout, 8 to i6 in. high, from tuber-like rootstock. Leaves: In a whorl of 3; broadly ovate, abruptly pointed, netted-veined. Fruit: A 6-angled, ovate, reddish berry. Preferred Habitat - Rich, moist woods. Flowering Season - April-June. Distribution - Nova Scotia westward to Manitoba, southward to North Carolina and Missouri.

Some weeks after the jubilant, alert robins have returned from the South, the purple trillium unfurls its unattractive, carrion-scented flower. In the variable colors found in different regions, one can almost trace its evolution from green, white, and red to purple, which, we are told, is the course all flowers must follow to attain to blue. The white and pink forms, however attractive to the eye, are never more agreeable to the nose than the reddish-purple ones. Bees and butterflies, with delicate appreciation of color and fragrance, let the blossom alone, since it secretes no nectar; and one would naturally infer either that it can fertilize itself without insect aid - a theory which closer study of its organs goes far to disprove - or that the carrion-scent, so repellent to us, is in itself an attraction to certain insects needful for cross-pollination. Which are they? Beetles have been observed crawling over the flower, but without effecting any methodical result. One inclines to accept Mr. Clarence M. Weed's theory of special adaptation to the common green flesh-flies (Lucilia carnicina), which would naturally be attracted to a flower resembling in color and odor a raw beefsteak of uncertain age. These little creatures, seen in every butcher shop throughout the summer, the flower furnishes with a free lunch of pollen in consideration of the transportation of a few grains to another blossom. Absence of the usual floral attractions gives, the carrion flies a practical monopoly of the pollen food, which no doubt tastes as it smells.

The SESSILE-FLOWERED WAKE-ROBIN (T. sessile), whose dark purple, purplish-red, or greenish blossom, narrower of sepal and petals than the preceding, is seated in a whorl of three egg-shaped, sometimes blotched, leaves, possesses a rather pleasant odor; nevertheless it seems. to have no great attraction for insects. The stigmas, which are very large, almost touch the anthers surrounding them; therefore the beetles which one frequently sees crawling over them to feed on the pollen so jar them, no doubt, as to self-fertilize the flower; but it is scarcely probable these slow crawlers often transfer the grains from one blossom to another. A degraded flower like this has little need of color and perfume, one would suppose; yet it may be even now slowly perfecting its way toward an ideal of which we see a part only complete. In deep, rich, moist woods and thickets the. sessile trillium blooms in April or May, from Pennsylvania, Ohio, and Minnesota southward nearly to the Gulf.

LARGER BLUE FLAG; BLUE IRIS; FLEUR-DE-LIS; FLOWER-DE-LUCE
 (Iris versicolor) Iris family

Flowers - Several, 2 to 3 in. long, violet-blue variegated with yellow, green, or white, and purple veined. Six divisions of the perianth: 3 outer ones spreading, recurved; 1 of them bearded, much longer and wider than the 3 erect inner divisions; all united into a short tube. Three stamens under 3 overhanging petal-like divisions of the style, notched at end; under each notch is a thin plate, smooth on one side, rough and moist (stigma) on side turned away from anther. Stem: 2 to 3 ft. high, stout, straight, almost circular, sometimes branching above. Leaves: Erect, sword-shaped, shorter than stem, somewhat hoary, from 1/2 to 1 in. wide, folded, and in a compact flat cluster at base; bracts usually longer than stem of flower. Fruit: Oblong capsule, not prominently 3-lobed, and with 2 rows of round, flat seeds closely packed in each cell. Rootstock: Creeping, horizontal, fleshy. Preferred Habitat - Marshes, wet meadows.

Flowering Season - May-July. Distribution - Newfoundland and Manitoba to Arkansas and Florida.

"The fleur-de-lys, which is the flower of chivalry," says Ruskin, "has a sword for its leaf and a lily for its heart." When that young and pious Crusader, Louis VII, adopted it for the emblem of his house, spelling was scarcely an exact science, and the fleur-de-Louis soon became corrupted into its present form. Doubtless the royal flower was the white iris, and as li is the Celtic for white, there is room for another theory as to the origin of the name. It is our far more regal looking, but truly democratic blossom, jostling its fellows in the marshes, that is indeed "born in the purple."

When Napoleon wished to pose as the true successor of those ancient French kings whose territory included the half of Europe - ignoring every Louis who ever sat on the throne, for their very name and emblem had become odious to the people - he discarded the fleur-de-lis, to replace it with golden bees, the symbol in armory for industry and perseverance. It is said some relics of gold and fine stones, somewhat resembling an insect in shape, had been found in the tomb of Clovis's father, and on the supposition that these had been bees, Napoleon appropriated them for the imperial badge. Henceforth "Napoleonic bees" appeared on his coronation robe and wherever a heraldic emblem could be employed.

But even in the meadows of France Napoleon need not have looked far from the fleurs-de-lis growing there to find bees. Indeed, this gorgeous flower is thought by scientists to be all that it is for the bees' benefit, which, of course, is its own also. Abundant moisture, from which to manufacture nectar - a prime necessity with most irises - certainly is for our blue flag. The large showy blossom cannot but attract the passing bee, whose favorite color (according to Sir John Lubbock) it waves. The bee alights on the convenient, spreading platform, and, guided by the dark veining and golden lines leading to the nectar, sips the delectable fluid shortly to be changed to honey. Now, as he raises his head and withdraws it from the nectary, he must rub it against the pollen-laden anther above, and some of the pollen necessarily falls on the visitor. As the sticky side of the plate (stigma), just under the petal-like division of the style, faces away from the anther, which is below it in any case, the flower is marvelously guarded against fertilization from its own pollen. The bee, flying off to another iris, must first brush past the projecting lip of the over-arching style, and leave on the stigmatic outer surface of the plate some of the pollen brought from the first flower, before reaching the nectary. Thus cross-fertilization is effected; and Darwin has shown how necessary this is to insure the most vigorous and beautiful offspring. Without this wonderful adaptation of the flower to the requirements of its insect friends, and of the insect to the needs of the flower, both must perish; the former from hunger, the latter because unable to perpetuate its race. And yet man has greedily appropriated all the beauties of the floral kingdom as designed for his sole delight

The name iris, meaning a deified rainbow, which was given this group of plants by the ancients, shows a fine appreciation of their superb coloring, their ethereal texture, and the evanescent beauty of the blossom.

In spite of the name given to another species, the SOUTHERN BLUE FLAG (I. hexagona) is really the larger one; its leaves, which are bright green, and never hoary, often equaling the stem in its height of from two to three feet. The handsome solitary flower, similar to that of the larger blue flag, nevertheless has its broad outer divisions fully an inch larger, and is seated in the axils at the top of the circular stem. The oblong, cylindric, six-angled capsule also contains two rows of seeds in each cavity. From South Carolina and Florida to Kentucky, Missouri, and Texas one finds this iris blooming in the swamps during April and May.

The SLENDER BLUE FLAG (I. prismatica; I. Virginica of Gray), found growing from New Brunswick to North Carolina, but mainly near the coast, and often in the same oozy ground with the larger blue flag, may be known by its grass-like leaves, two or three of which usually branch out from the slender flexuous stem; by its solitary or two blue flowers, variegated with white and veined with yellow, that rear themselves on slender foot-stems; and by the sharply three-angled, narrow, oblong capsule, in which but one row of seeds is borne in each cavity. This is the most graceful member of a rather stiffly stately family.

POINTED BLUE-EYED GRASS; EYE-BRIGHT; BLUE STAR
(Sisyrinchium angustifolium) Iris family

Flowers - From blue to purple, with a yellow center; a Western variety, white; usually several buds at the end of stem, between 2 erect unequal bracts; about 1/2 in. across; perianth of 6 spreading divisions, each pointed with a bristle from a notch; stamens 3, the filaments united to above the middle; pistil 1, its tip 3-cleft. Stem: 3 to 14 in. tall, pale hoary green, flat, rigid, 2-edged. Leaves: Grass-like, pale, rigid, mostly from base. Fruit: 3-celled capsule, nearly globose. Preferred Habitat - Moist fields and meadows. Flowering Season - May-August. Distribution - Newfoundland to British Columbia, from eastern slope of Rocky Mountains to Atlantic, south to Virginia and Kansas.

Only for a day, and that must be a bright one, will this "little sister of the stately blue flag" open its eyes, to close them in indignation on being picked; nor will any coaxing but the sunshine's induce it to open them again in water, immediately after. The dainty flower, growing in dense tufts, makes up in numbers what it lacks in size and lasting power, flecking our meadows with purplish ultramarine blue in a sunny June morning. Later in the day, apparently there are no blossoms there, for all are tightly closed, never to bloom again. New buds will unfold to tinge the field on the morrow.

Usually three buds nod from between a pair of bracts, the lower one of which may be twice the length of the upper one but only one flower opens at a time. Slight variations in this plant have been considered sufficient to differentiate several species formerly included by Gray and other American botanists under the name of S. Bermudiana.

LARGE or EARLY, PURPLE-FRINGED ORCHIS
(Habenaria grandiflora; H. fimbriata of Gray) Orchid family

Flowers - Pink-purple and pale lilac, sometimes nearly white; fragrant, alternate, clustered in thick, dense spikes from 3 to 15 in. long. Upper sepal and toothed petals erect; the lip of deepest shade,

1/2 in. long, fan-shaped, 3-parted, fringed half its length, and prolonged at base into slender, long spur; stamen united with style into short column; 2 anther sacs slightly divergent, the hollow between them glutinous, stigmatic. Stem. 1 to 5 ft. high, angled, twisted. Leaves: Oval, large, sheathing the stem below; smaller, lance-shaped ones higher up; bracts above. Root: Thick, fibrous. Preferred Habitat - Rich, moist meadows, muddy places, woods. Flowering Season - June-August. Distribution - New Brunswick to Ontario; southward to North Carolina, westward to Michigan.

Because of the singular and exquisitely unerring adaptations of orchids as a family to their insect visitors, no group of plants has greater interest for the botanist since Darwin interpreted their marvelous mechanism, and Gray, his instant disciple, revealed the hidden purposes of our native American species, no less wonderfully constructed than the most costly exotic in a millionaire's hothouse.

A glance at the spur of this orchid, one of the handsomest and most striking of its clan, and the heavy perfume of the flower, would seem to indicate that only a moth with a long proboscis could reach the nectar secreted at the base of the thread-like passage. Butterflies, attracted by the conspicuous color, sometimes hover about the showy spikes of bloom, but it is probable that, to secure a sip, all but possibly the very largest of them must go to the smaller purple-fringed orchis, whose shorter spur holds out a certain prospect of reward; for, in these two cases, as in so many others, the flower's welcome for an insect is in exact proportion to the length of its visitor's tongue. Doubtless it is one of the smaller sphinx moths, such as we see at dusk working about the evening primrose and other flowers deep of chalice, and heavily perfumed to guide visitors to their feast, that is the great purple-fringed orchid's benefactor, since the length of its tongue is perfectly adapted to its needs. Attracted by the showy, broad lower petal, his wings ever in rapid motion, the moth proceeds to unroll his proboscis and drain the cup, that is frequently an inch and a half deep. Thrusting in his head, either one or both of his large, projecting eyes are pressed against the sticky button-shaped disks to which the pollen masses are attached by a stalk, and as he raises his head to depart, feeling that he is caught, he gives a little jerk that detaches them, and away he flies with these still fastened to his eyes.

Even while he is flying to another flower, that is to say, in half a minute, the stalks of the pollen masses bend downward from the perpendicular and slightly toward the center, or just far enough to require the moth, in thrusting his proboscis into the nectary, to strike the glutinous, sticky stigma. Now, withdrawing his head, either or both of the golden clubs he brought in with him will be left on the precise spot where they will fertilize the flower. Sometimes, but rarely, we catch a butterfly or moth from the smaller or larger purple orchids with a pollen mass attached to his tongue, instead of to his eyes; this is when he does not make his entrance from the exact center - as in these flowers he is not obliged to do - and in order to reach the nectary his tongue necessarily brushes against one of the sticky anther sacs. The performance may be successfully imitated by thrusting some blunt point about the size of a moth's head, a dull pencil or a knitting-needle, into the flower as an insect would enter. Withdraw the pencil, and one or both of the pollen masses will be found sticking to it, and already automatically changing their attitude. In the case of the large, round-leaved orchis, whose greenish-white flowers are fertilized in a similar manner by the sphinx moth, the anther sacs converge, like little horns; and their change of attitude while they are being carried to fertilize another flower is quite as exquisitely exact.

Usually in wetter ground than we find its more beautiful big sister growing in, most frequently in swamps and bogs, the SMALLER PURPLE-FRINGED ORCHIS (H. psycodes) lifts its perfumed lilac spires. Thither go the butterflies and long-lipped bees to feast in July and August. Inasmuch as without their aid the orchid must perish from its inability to set fertile seed, no wonder it woos its benefactors with a showy mass of color, charming fringes, sweet perfume, and copious draughts of nectar, and makes their visits of the utmost value to itself by the ingenious mechanism described above. Here is no waste of pollen; that is snugly packed in little bundles, ready to be carried off, but placed where they cannot come in contact with the adjoining stigma, since every orchid, almost without exception, refuses to be deteriorated through self-fertilization.

>From New Jersey and Illinois southward, particularly in mountainous regions, if not among the mountains themselves, the

FRINGELESS PURPLE ORCHIS (H. perarnoena) may be found blooming in moist meadows through July and August. Moisture, from which to manufacture the nectar that orchids rely upon so largely to entice insects to work for them, is naturally a prime necessity; yet Sprengel attempted to prove that many orchids are gaudy shams and produce no nectar, but exist by an organized system of deception. "Scheinsaftblumen" he called them. From the number of butterflies seen hovering about this fringeless orchis and its more attractive kin, it is small wonder their nectaries are soon exhausted and they are accused of being gay deceivers. Sprengel's much-quoted theory would credit moths, butterflies, and even the highly intelligent bees with scant sense; but Darwin, who thoroughly tested it, forever exonerated these insects from imputed stupidity and the flowers from gross dishonesty. He found that many European orchids secrete their nectar between the outer and inner walls of the tube, which a bumblebee can easily pierce, but where Sprengel never thought to look for it. The large lip of this orchis is not fringed, but has a fine picotee edge. The showy violet-purple, long-spurred flowers are alternately set on a stem that is doing its best if it reach a height of two and a half feet.

WATER-SHIELD or WATER TARGET
  (Brasenia purpurea; B. peltata of Gray) Water-lily family

  Flowers - Small, dull purplish, about 1/2 in. across, on stout footstalks from axils of upper leaves; 3 narrow sepals and petals; stamens 12 to 18; pistils 4 to 18, forming 1 to 3-seeded pods. Stem: From submerged rootstock; slender, branching, several feet long, covered with clear jelly, as are footstalks and lower leaf surfaces. Leaves: On long petioles attached to center of underside of leaf, floating or rising, oval to roundish, 2 to 4 in. long, 1 1/2 to 2 in. wide. Preferred Habitat - Still, rather deep water of ponds and slow streams. Flowering Season - All summer. Distribution - Parts of Asia, Africa, and Australia, Nova Scotia to Cuba, and westward from California to Puget Sound.

  Of this pretty water plant Dr. Abbott says, in "Wasteland Wanderings": "I gathered a number of floating, delicate leaves, and endeavored to secure the entire stem also; but this was too difficult a task for an August afternoon. The under side of the stem and leaf are purplish brown and were covered with translucent jelly, embedded in which were millions of what I took to be insects' eggs. They certainly had that appearance. I was far more interested to find that, usually, beneath each leaf there was hiding a little pike. The largest was not two inches in length. When disturbed, they swam a few inches, and seemed wholly 'at sea' if there was not another leaf near by to afford them shelter."

EUROPEAN or COMMON GARDEN COLUMBINE
 (Aquilegia vulgaris) Crowfoot family

Flowers - Showy, blue, purple, or white, 1 1/2 to 2 in. broad, or about as broad as long; spurs stout and strongly incurved.
General characteristics of plant resembling wild columbine.
Preferred Habitat - Escaped from gardens to woods and fields in Eastern and Middle States. Native of Europe.
Flowering Season - May-July.

A heavier, less graceful flower than either the wild red and yellow columbine or the exquisite, long-spurred, blue and white species (A. coerulea) of the Rocky Mountain region; nevertheless this European immigrant, now making itself at home here, is a charming addition to our flora. How are insects to reach the well of nectar secreted in the tip of its incurved, hooked spur? Certain of the long-lipped bees, large bumblebees, whose tongues have developed as rapidly as the flower, are able to drain it. Hummingbirds, partial to red flowers, fertilize the wild columbine, but let this one alone. Muller watched a female bumblebee making several vain attempts to sip this blue one. Soon the brilliant idea of biting a hole through each spur flashed through her little brain, and the first experiment proving delightfully successful, she proceeded to bite holes through other flowers without first trying to suck them. Apparently she satisfied her feminine conscience with the reflection that the flower which made dining so difficult for its benefactors deserved no better treatment.

FIELD or BRANCHED LARKSPUR; KNIGHT'S-SPUR; LARK-HEEL
(Delphinium Consoilda) Crowfoot family

Flowers - Blue to pinkish and whitish, 1 to 1 1/2 in. long, hung on slender stems and scattered along spreading branches; 5 petal-like sepals, the rear one prolonged into long, slender, curving spur; 2 petals, united. Stem: 1 to 2 1/2 ft. high. Leaves: Divided into very finely cut linear segments. Fruit: Erect, smooth pod tipped with a short beak; open on one side. Preferred Habitat - Roadsides and fields. Flowering Season - June-August. Distribution - Naturalized from Europe; from New Jersey southward, occasionally escaped from gardens farther north.

Keats should certainly have extolled the larkspurs in his sonnet on blue. No more beautiful group of plants contributes to the charm of gardens, woods, and roadsides, where some have escaped cultivation and become naturalized, than the delphinium, that take their name from a fancied resemblance to a dolphin (delphin), given them by Linnaeus in one of his wild flights of imagination. Having lost the power to fertilize themselves, according to Muller, they are pollenized by both bees and butterflies, insects whose tongues have kept pace with the development of certain flowers, such as the larkspur, columbine, and violet, that they may reach into the deep recesses of the spurs where the nectar is hidden from all but benefactors.

The TALL WILD LARKSPUR (D. urceolatum; D. exaltatum of Gray) waves long, crowded, downy wands of intense purplish blue in the rich woods of Western Pennsylvania, southward to the Carolinas and Alabama, and westward to Nebraska. Its spur is nearly straight, not to increase the difficulty a bee must have in pressing his lips through the upper and lower petals to reach the nectar at the end of it. First, the stamens successively raise themselves in the passage back of the petals to dust his head; then, when each has shed its pollen and bent down again, the pistil takes its turn in occupying the place, so that a pollen-laden bee, coming to visit the blossom from an earlier flower; can scarcely help fertilizing it. It is said there are but two insects in Europe with lips long enough to reach the bottom of the long horn of plenty hung by the BEE LARK-SPUR (D. elatum), that we know only in gardens here. Its yellowish bearded lower petals readily deceive one into thinking a bee has just alighted there.

>From April to June the DWARF LARKSPUR or STAGGER-WEED (D. tricorne), which, however, may sometimes grow three feet high, lifts a loose raceme of blue, rarely white, flowers an inch or more long, at the end of a stout stem rising from a tuberous root. Its slightly ascending spur, its three widely spreading seed vessels, and the deeply cut leaf of from five to seven divisions are distinguishing characteristics. From Western Pennsylvania and Georgia to Arkansas and Minnesota it is found in rather stiff soil. Butterflies, which prefer erect flowers, have some difficulty to cling while they drain the almost upright spurs, especially the Papilios, which usually suck with their wings in motion. But the bees, to which the delphinium are best adapted, although butterflies visit them quite as frequently, find a convenient landing place prepared for them, and fertilize the flower while they sip with ease.

More slender, downy, and dwarf of stem than the preceding is the CAROLINA LARKSPUR (D. Carolinianum), whose blue flowers, varying
to white, and its very finely cleft leaves, may be found in the South, on prairies in the North and West, and in the Rocky Mountain region.

LIVER-LEAF; HEPATICA; LIVERWORT; ROUND-LOBED or KIDNEY
LIVER-LEAF; NOBLE LIVER-WORT; SQUIRREL CUP
  (Hepalica Hepatica; H. triloba of Gray) Crowfoot family

   Flowers - Blue, lavender, purple, pinkish, or white; occasionally, not always, fragrant; 6 to 12 petal-like, colored sepals (not petals, as they appear to be), oval or oblong; numerous stamens, all bearing anthers; pistils numerous 3 small, sessile leaves, forming an involucre directly under flower, simulate a calyx, for which they might be mistaken. Stems: Spreading from the root, 4 to 6 in. high, a solitary flower or leaf borne at end of each furry stem. Leaves: 3-lobed and rounded, leathery, evergreen; sometimes mottled with, or entirely, reddish purple; spreading on ground, rusty at blooming time, the new leaves appearing after the flowers. Fruit: Usually as many as pistils, dry, 1-seeded, oblong, sharply pointed, never opening. Preferred Habitat - Woods; light soil on hillsides. Flowering Season - December-May. Distribution - Canada to Northern Florida, Manitoba to Iowa and Missouri. Most common East.

Even under the snow itself bravely blooms the delicate hepatica, wrapped in fuzzy furs as if to protect its stems and nodding buds from cold. After the plebeian skunk cabbage, that ought scarcely to be reckoned among true flowers - and William Hamilton Gibson claimed even before it - it is the first blossom to appear. Winter sunshine, warming the hillsides and edges of woods, opens its eyes,

> "Blue as the heaven it gates at,
> Startling the loiterer in the naked groves
> With unexpected beauty; for the time
> Of blossoms and green leaves is yet afar."

"There are many things left for May," says John Burroughs, "but nothing fairer, if as fair, as the first flower, the hepatica. I find I have never admired this little firstling half enough. When at the maturity of its charms, it is certainly the gem of the woods. What an individuality it has! No two clusters alike; all shades and sizes…. A solitary blue-purple one, fully expanded and rising over the brown leaves or the green moss, its cluster of minute anthers showing like a group of pale stars on its little firmament, is enough to arrest and hold the dullest eye. Then,…there are individual hepaticas, or individual families among them, that are sweet scented. The gift seems as capricious as the gift of genius in families. You cannot tell which the fragrant ones are till you try them. Sometimes it is the large white ones, sometimes the large purple ones, sometimes the small pink ones. The odor is faint and recalls that of the sweet violets. A correspondent, who seems to have carefully observed these fragrant hepaticas, writes me that this gift of odor is constant in the same plant; that the plant which bears sweet-scented flowers this year will bear them next."

It is not evident that insect aid is necessary to transfer the tiny, hairy spiral ejected from each cell of the antherid, after it has burst from ripeness, to the canal of the flask-shaped organ at whose base the germ-cell is located. Perfect flowers can fertilize themselves. But pollen-feeding flies, and female hive bees which collect it, and the earliest butterflies trifle about the blossoms when the first warm days come. Whether they are rewarded by finding nectar or not is still a mooted question. Possibly the papillae which cover the receptacle secrete nectar, for almost without exception the insect visitors thrust their proboscides down between the spreading filaments as if certain of a sip. None merely feed on the pollen except the flies and the hive bee.

The SHARP-LOBED LIVER-LEAF (Hepatica acuta) differs chiefly from the preceding in having the ends of the lobes of its leaves and the tips of the three leaflets that form its involucre quite sharply pointed. Its range, while perhaps not actually more westerly, appears so, since it is rare in the East, where its cousin is so abundant; and common in the West, where the round-lobed liver-leaf is scarce. It blooms in March and April. Professor Halsted has noted that this species bears staminate flowers on one plant and pistillate flowers on another; whereas the Hepatica Hepatica usually bears flowers of both sexes above the same root. The blossoms, which close at night to keep warm, and open in the morning, remain on the beautiful plant for a long time to accommodate the bees and flies that, in this case, are essential to the perpetuation of the species.

PURPLE VIRGIN'S BOWER
(Atragene Americana) Crowfoot family

Flowers - Showy, purplish blue, about 3 in. across; 4 sepals, broadly expanded, thin, translucent, strongly veined, very large, simulating petals; petals small, spoon-shaped; stamens very numerous ; styles long, persistent, plumed throughout. Stem: Trailing or partly climbing with the help of leafstalks and leaflets. Leaves: Opposite, compounded of 3 egg-shaped, pointed leaflets on slender petioles. Preferred Habitat - - Rocky woodlands. Flowering Season - May-June. Distribution - Hudson Bay westward, south to Minnesota and Virginia.

The day on which one finds this rare and beautiful flower in some rocky ravine high among the hills or mountains becomes memorable to the budding botanist. At an elevation of three thousand feet in the Catskills it trails its way over the rocks, fallen trees, and undergrowth of the forest, suggesting some of the handsome Japanese species introduced by Sieboldt and Fortune to Occidental gardens. No one who sees this broadly expanded blossom could confuse it either with the thick and bell-shaped purple LEATHER-FLOWER (C. Viorna), so exquisitely feathery in fruit, that grows in rich, moist soil from Pennsylvania southward and westward; or with the far more graceful and deliciously fragrant purple MARSH CLEMATIS (C. crispa) of our Southern States. The latter, though bell-shaped also, has thin, recurved sepals, and its persistent styles are silky, not feathery at seed-time.

ORPINE; LIVE-FOREVER; MIDSUMMER-MEN; LIVE-LONG; PUDDING-BAG
PLANT; GARDEN STONECROP; WITCHES' MONEY
  (Sedum Telephium) Orpine family

Flowers - Dull purplish, very pale or bright reddish purple in close, round, terminal clusters, each flower 1/3 in. or less across, 5-parted, the petals twice as long as the sepals; 10 stamens, alternate ones attached to petals; pistils 4 or 5. Stem: 2 ft. high or less, erect, simple, in tufts, very smooth, pale green, juicy, leafy. Leaves: Alternate, oval, slightly scalloped, thick, fleshy, smooth, juicy, pale gray green, with stout midrib, seated on stalk. Preferred Habitat - Fields, waysides, rocky soil, originally escaped from gardens. Flowering Season - June- September. Distribution - Quebec westward, south to Michigan and Maryland.

Children know the live-forever, not so well by the variable flower - for it is a niggardly bloomer - as by the thick leaf that they delight to hold in the mouth until, having loosened the membrane, they are able to inflate it like a paper bag. Sometimes dull, sometimes bright, the flower clusters never fail to attract many insects to their feast, which is accessible even to those of short tongues. Each blossom is perfect in itself, i.e., it contains both stamens and pistils; but to guard against self-fertilization it ripens its anthers and sheds its pollen on the insects that carry it away to older flowers before its own stigmas mature and become susceptible to imported pollen. After the seed-cases take on color, they might be mistaken for blossoms.

As if the plant did not already possess enough popular names, it needs must share with the European goldenrod and our common mullein the title of Aaron's rod. Sedere, to sit, the root of the generic name, applies with rare appropriateness to this entire group that we usually find seated on garden walls, rocks, or, in Europe, even on the roofs of old buildings. Rooting freely from the joints, our plant forms thrifty tufts where there is little apparent nourishment; yet its endurance through prolonged drought is remarkable. Long after the farmer's scythe, sweeping over the roadside, has laid it low, it thrives on the juices stored up in fleshy leaves and stem until it proves its title to the most lusty of all folk names.

PURPLE or WATER AVENS
 (Geum rivale) Rose family

 Flowers - Purple, with some orange chrome, 1 in. broad or less, terminal, solitary, nodding; calyx 5-lobed, purplish, spreading; 5 petals, abruptly narrowed into claws, forming a cup-shaped corolla; stamens and pistils of indefinite number; the styles, jointed and bent in middle, persistent, feathery below. Stem: 1 to 2 ft. high, erect, simple or nearly so, hairy, from thickish rootstock. Leaves: Chiefly from root, on footstems; lower leaves irregularly parted; the side segments usually few and small; the 1 to 3 terminal segments sharply, irregularly lobed; the few distant stem leaves 3-foliate or simple, mostly seated on stem. Fruit: A dry, hairy head stalked in calyx. Preferred Habitat - Swamps and low, wet ground. Flowering Season - May-July. Distribution - Newfoundland far westward, south to

Colorado, eastward to Missouri and Pennsylvania, also northern parts of Old World.

Mischievous bumblebees, thrusting their long tongues between the sepals and petals of these unopened flowers, steal nectar without conferring any favor in return. Later, when they behave properly and put their heads inside to feast at the disk on which the stamens are inserted, they dutifully carry pollen from old flowers to the early maturing stigmas of younger ones. Self-fertilization must occur, however, if the bees have not removed all the pollen when a blossom closes. When the purple avens opens in Europe, the bees desert even the primrose to feast upon its abundant nectar. Since water is the prime necessity in the manufacture of this sweet, and since insects that feed upon it have so much to do with the multiplication of flowers, it is not surprising that the swamp, which has been called "nature's sanctuary," should have its altars so exquisitely decked. This blossom hangs its head, partly to protect its precious nectar from rain, and partly to make pilfering well nigh impossible to the unwelcome crawling insect that may have braved the forbidding hairy stems.

WILD LUPINE; OLD MAID'S BONNETS; WILD PEA; SUN DIAL
(Lupinus perennis) Pea family

Flowers - Vivid blue, very rarely pink or white, butterfly-shaped corolla consisting of standard, wings, and keel; about 1/2 in. long, borne in a long raceme at end of stern; calyx 2-lipped, deeply toothed. Stem: Erect, branching, leafy, to 2 ft. high. Leaves: Palmnate, compounded of from 7 to 11 (usually 8) leaflets. Fruit: A broad, flat, very hairy pod, 1 1/2 in. long, and containing 4 or 5 seeds. Preferred Habitat - Dry, sandy places, banks, and hillsides. Flowering Season - May-June. Distribution - United States east of Mississippi, and eastern Canada.

Farmers once thought that this plant preyed upon the fertility of their soil, as we see in the derivation of its name, from lupus, a wolf; whereas the lupine contents itself with sterile waste land no one should grudge it - steep gravelly banks, railroad tracks, exposed sunny hills, where even it must often burn out under fierce sunshine did not its root penetrate to surprising depths. It spreads far and wide in thrifty colonies, reflecting the vivid color of June skies, until, as Thoreau says, "the earth is blued with it."

What is the advantage gained in the pea-shaped blossom? As usual, the insect that fertilizes the flower best knows the answer. The corolla has five petals, the upper one called the standard, chiefly a flaunted advertisement; two side wings, or platforms, to alight on; and a keel like a miniature boat, formed by the two lower petals, whose edges meet. In this the pistil, stamens, and nectar are concealed and protected. The pressure of a bee's weight as he alights on the wings, light as it must be, is nevertheless sufficient to depress and open the keel, which is elastically affected by their motion, and so to expose the pollen just where the long-lipped bee must rub off some against his underside as he sucks the nectar. He actually seems to pump the pollen that has fallen into the forward part of the keel upon himself, as he moves about. As soon as he leaves the flower, the elastic wings resume their former position, thus closing the keel to prevent waste of pollen. Take a sweet pea from the garden, press down its wings with the thumb and forefinger to imitate the action of the bee on them; note how the keel opens to display its treasures, and resumes its customary shape when the pressure is removed.

The lupine is another of those interesting plants which go to sleep at night. Some members of the genus erect one half of the leaf and droop the other half until it becomes a vertical instead of the horizontal star it is by day. Frequently the leaflets rotate as much as 90 degrees on their own axes. Some lupines fold their leaflets, not at night only, but during the day also there is more or less movement in the leaves. Sun dial, a popular name for the wild lupine, has reference to this peculiarity. The leaf of our species shuts downward around its stem, umbrella fashion, or the leaflets are erected to prevent the chilling which comes to horizontal surfaces by radiation, some scientists think. "That the sleep movements of leaves are in some manner of high importance to the plants which exhibit them," says Darwin, "few will dispute who have observed how complex they sometimes are."

CANADIAN or SHOWY TICK-TREFOIL
 (Meibomia Canadensis; Desmodium Canadense of Gray) Pea family

Flowers - Pinkish or bluish purple, butterfly-shaped, about 1/2 in. long, borne in dense, terminal, elongated racemes. Stem; Erect, hairy, leafy, 2 to 8 ft. high. Leaves: Compounded of 3 oblong leaflets, the central one largest; upper leaves nearly seated on stem; bracts, conspicuous before flowering, early falling off. Fruit: A flat pod, about 1 in. long, jointed, and covered with minute hooked bristles, the lower edge of pod scalloped; almost seated in calyx. Preferred Habitat - Thickets, woods, riverbanks, bogs. Flowering Season - July-September. Distribution - New Brunswick to Northwest Territory, south to North Carolina, westward to Indian Territory and Dakota.

As one travels hundreds or even thousands of miles in a comfortable railway carriage and sees the same flowers growing throughout the length and breadth of the area, one cannot but wonder however the plants manage to make the journey. We know some creep along the ground, or under it, a tortoise pace, but a winning one; that some send their offspring flying away from home, like dandelions and thistles; and many others with wings and darts are blown by the wind. Berries have their seeds dropped afar by birds. Aquatic plants and those that grow beside running water travel by river and flood. European species reach our shores among the ballast. Darwin raised over sixty wild plants from seed carried in a pellet of mud taken from the leg of a partridge. So on and so on. The imagination delights to picture these floral vagabonds, each with its own clever method of getting a fresh start in the world. But by none of these methods just mentioned do the tick-trefoils spread abroad. Theirs is indeed a by hook or by crook system. The scalloped, jointed pod, where the seeds lie concealed, has minute crooked bristles, which catch in the clothing of man or beast, so that every herd of sheep, every dog, every man, woman, or child who passes through a patch of trefoils gives them a lift. After a walk through the woods and lanes of late summer and autumn, one's clothes reveal scores of tramps that have stolen a ride in the hope of being picked off and dropped amid better conditions in which to rear a family.

Only the largest bees can easily "explode" the showy tick-trefoil. A bumblebee alights upon a flower, thrusts his head under the base of the standard petal, and forces apart the wing petals with his legs, in order to dislodge them from the standard. This motion causes the keel, also connected with the standard, to snap down violently, thus releasing the column within and sending upward an explosion of pollen on the under surface of the bee. Here we see the wing petals acting as triggers to discharge the flower. Depress them and up flies the fertilizing dust - once. The little gun will not "go off" twice. No nectar rewards the visitor, which usually is a pollen-collecting bee. The highly intelligent and important bumblebee has the advantage over his smaller kin in being able to discharge the pollen from both large and smaller flowers.

The NAKED-FLOWERED TICK-TREFOIL (M. nudiflora; D. nudiflorum of Gray) lifts narrow, few-flowered panicles of rose-purple blooms during July and August. The flowers are much smaller than those of the showy trefoil; however, when seen in masses, they form conspicuous patches of color in dry woods. Note that there is a flower stalk which is usually leafless and also a leaf-bearing stem rising from the base of the plant, the latter with its leaves all crowded at the top, if you would distinguish this very common species from its multitudinous kin. The trefoliate leaves are pale beneath. The two or three jointed pod rises far above the calyx on its own stalk, as in the next species.

The POINTED-LEAVED TICK-TREFOIL (M. grandifiora; D. acuminatum of Gray) has for its distinguishing feature a cluster of leaves high up on the same stem from which rises a stalk bearing a quantity of purple flowers that are large by comparison only. The leaves have leaflets from two to six inches long, rounded on the sides, but acutely pointed, and with scattered hairs above and below. This trefoil is found blooming in dry or rocky woods, throughout a wide range, from June to September.

Lying outstretched for two to six feet on the dry ground of open woods and copses east of the Mississippi, the PROSTRATE TICK-TREFOIL (M. Michauxii; D. rotundifoliurn of Gray) can certainly be named by its soft hairiness, the almost perfect roundness of its trefoliate leaves, its rather loose racemes of deep purple flowers that spring both from the leaf axils and from the ends of the sometimes branching stem; and by its three to five jointed pod, which is deeply scalloped on its lower edge and somewhat indented above, as well.

BLUE, TUFTED, or COW VETCH or TARE; CAT PEAS; TINEGRASS
 (Vicia Cracca) Pea family

   Flowers - Blue, later purple; 1/2 in. long, growing downward in 1-sided spike, 15 to 40 flowered; calyx oblique, small, with unequal teeth; corolla butterfly-shaped, consisting of standard, wings, and keel, all oblong; the first clawed, the second oblique, and adhering to the shorter keel; 10 stamens, 1 detached from other 9. Stem: Slender, weak, climbing or trailing, downy, 2 to 4 ft. long. Leaves: Tendril bearing, divided into 18 to 24 thin, narrow, oblong leaflets. Fruit: A smooth pod 1 in. long or less, 5 to 8 seeded. Preferred Habitat - Dry soil, fields, wastelands. Flowering Season - June-August. Distribution - United States from New Jersey, Kentucky, and Iowa northward and northwestward. Europe and Asia.

   Dry fields blued with the bright blossoms of the tufted vetch, and roadsides and thickets where the angular vine sends forth vivid patches of color, resound with the music of happy bees. Although the parts of the flower fit closely together, they are elastic, and opening with the energetic visitor's weight and movement give ready access to the nectary. On his departure they resume their original position, to protect both nectar and pollen from rain and pilferers whose bodies are not perfectly adapted to further the flower's cross-fertilization. The common bumblebee (Bombus terrestris) plays a mean trick, all too frequently, when he bites a hole at the base of the blossom, not only gaining easy access to the sweets for himself, but opening the way for others less intelligent than he, but quite ready to profit by his mischief, and so defeat nature's plan. Dr. Ogle observed that the same bee always acts in the same manner, one sucking the nectar legitimately, another always biting a hole to obtain it surreptitiously, the natural inference, of course, being that some bees, like small boys, are naturally depraved.

In cultivated fields and waste places farther south and westward to the Pacific Coast roams the COMMON or PEBBLE VETCH OR TARE (V. saliva), another domesticated weed that has come to us from Europe, where it is extensively grown for fodder. Let no reproach fall on these innocent plants that bear an opprobrious name: the tare of Scripture is altogether different, the bearded darnel of Mediterranean regions, whose leaves deceive one by simulating those of wheat, and whose smaller seeds, instead of nourishing man, poison him. Only one or two light blue-purple flowers grow in the axils of the leaves of our common vetch. The leaf, compounded of from eight to fourteen leaflets, indented at the top, has a long terminal tendril, whose little sharp tip assists the awkward vine, like a grappling hook.

The AMERICAN VETCH or TARE or PEA VINE (V. Americana) boasts slightly larger bluish-purple flowers than the blue vetch, but fewer of them; from three to nine only forming its loose raceme. In moist soil throughout a very broad northerly and westerly range it climbs and trails its graceful way, with the help of the tendrils on the tips of leaves compounded of from eight to fourteen oblong, blunt, and veiny leaflets.

BEACH, SEA, SEASIDE, or EVERLASTING PEA
(Lathyrus maritimus) Pea family

Flowers - Purple, butterfly-shaped, consisting of standard petal, wings, and keel; 1 in. long or less, clustered in short raceme at end of slender footstalk from leaf axils; calyx 5-toothed; stamens 10 (9 and 1); style curved, flattened, bearded on inner side. Stem: to 2 ft. long, stout, reclining, spreading, leafy. Leaves: Compounded of 3 to 6 pairs of oblong leaflets somewhat larger than halberd-shaped stipules at base of leaf; branched tendrils at end of it. Fruit: A flat, 2-valved, veiny pod, continuous between the seeds. Preferred Habitat - Beaches of Atlantic and Pacific Oceans, also of Great Lakes. Flowering Season - May-August. Sometimes blooming again in autumn. Distribution - New Jersey to Arctic Circle; also Northern Europe and Asia.

Sturdy clumps of the beach pea, growing beyond reach of the tide in the dunes and sandy wastelands back of the beach, afford the bee the last restaurant where he may regale himself without fear of drowning. From some members of the pea family, as from the wild lupine, for example, his weight, as he moves about, actually pumps the pollen that has fallen into the forward part of the blossom's keel onto his body, that he may transfer it to another flower. In some other members his weight so depresses the keel that the stamens are forced out to dust him over, the flower resuming its original position to protect its nectar and the remaining pollen just as soon as the pressure is removed. Other peas, again, burst at his pressure, and discharge their pollen on him. Now, in the beach pea, and similarly in the vetches, the style is hairy on its inner side, to brush out the pollen on the visitor who sets the automatic sweeper in motion as he alights and moves about. So perfectly have many members of this interesting family adapted their structure to the requirements of insects, and so implicitly do they rely on their automatic mechanism, that they have actually lost the power to fertilize themselves.

In moist or wet ground throughout a northern range from ocean to ocean, the MARSH VETCHLING (Lathyrus palustris) bears its purple, butterfly-shaped flowers, that are the merest trifle over half the size of those of the beach pea. From two to six of these little blossoms are alternately set along the end of the stalk. The leaflets, which are narrowly oblong, and acute at the apex, stand up opposite each other in pairs (from two to four) along the main leafstalk, that splits at the end to form hooked tendrils.

BUTTERFLY or BLUE PEA
(Clitoria Mariana) Pea family

Flowers - Bright lavender blue, showy, about 2 in. long; from 1 to 3 borne on a short peduncle. Calyx tubular, 5-toothed; corolla butterfly-shaped, consisting of very large, erect standard petal, notched at rounded apex; 2 oblong, curved wings, and shorter, acute keel; 10 stamens; style incurved, and hairy along inner side. Stem: Smooth, ascending or partly twining, 1 to 3 ft. high. Leaves: Compounded of 3 oblong leaflets, paler beneath, each on short stalk. Fruit: A few-seeded, acutely pointed pod about 1 in. long. Preferred Habitat - Dry soil. Flowering Season - June-July. Distribution - New Jersey to Florida, westward to Missouri, Texas, and Mexico.

A beautiful blossom, flaunting a large banner out of all proportion to the size of its other parts, that it may arrest the attention of its benefactors the bees. According to Henderson, the plant, which is found in our Southern States and over the Mexican border, grows also in the Khasia Mountains of India, but in no intervening place. Several members of the tropic-loving genus, that produce large, highly colored flowers, have been introduced to American hothouses; but the blue butterfly pea is our only native representative. The genus is thought to take its name from kleio, to shut up, in reference to the habit these peas have of seeding long before the flower drops off.

WILD or HOG PEANUT
(Falcata comosa; Amphicarpaea monoica of Gray) Pea family

Flowers - Numerous small, showy ones, borne in drooping clusters from axils of upper leaves; lilac, pale purplish, or rarely white, butterfly-shaped, consisting of standard petal partly enfolding wings and keel. Calyx tubular, 4 or 5 toothed; 10 stamens (9 and 1); 1 pistil. (Also solitary fertile flowers, lacking petals, on thread-like, creeping branches from lower axils or underground). Stem: Twining wiry brownish-hairy, to 8 ft. long. Leaves: Compounded of 3 thin leaflets, egg-shaped at base, acutely pointed at tip. Fruit: Hairy pod 1 in. long. Also 1-seeded, pale, rounded, underground peanut. Preferred Habitat - Moist thickets, shady roadsides. Flowering Season - August-September. Distribution - New Brunswick westward to Nebraska, south to Gulf of Mexico.

Amphicarpaea ("seed at both ends"), the Greek name by which this graceful vine was formerly known, emphasizes its most interesting feature, that, nevertheless, seems to many a foolish duplication of energy on Nature's part. Why should the same plant bear two kinds of blossoms and seeds? Among the foliage of low shrubbery and plants in shady lanes and woodside thickets, we see the delicate, drooping clusters of lilac blossoms hanging where bees can readily discover them and, in pilfering their sweets, transfer their pollen from flower to flower. But in case of failure to intercross these blossoms that are dependent upon insect help to set fertile seed, what then? Must the plant run the risk of extinction? Self-fertilization may be an evil, but failure to produce seed at all is surely the greatest one. To guard against such a calamity, insignificant looking flowers that have no petals to open for the enticing of insects, but which fertilize themselves with their own pollen, produce abundant seed close to the ground or under it.Then what need of the showy blossoms hanging in the thicket above? Close inbreeding in the vegetable world, as in the animal, ultimately produces degenerate offspring; and although the showy lilac blossoms of the wild peanut yield comparatively few cross-fertilized seeds, these are quite sufficient to enable the vine to maintain those desired features which are the inheritance from ancestors that struggled in their day and generation after perfection. No plant dares depend upon its cleistogamous or blind flowers alone for offspring; and in the sixty or more genera containing these curious growths, that usually look like buds arrested in development, every plant that bears them bears also showy flowers dependent upon cross-pollination by insect aid.

The boy who

"Drives home the cows from the pasture
Up through the long shady lane"

knows how reluctantly they leave the feast afforded by the wild peanut. Hogs, rooting about in the moist soil where it grows, unearth the hairy pods that should produce next year's vines; hence the poor excuse for branding a charming plant with a repellent folk-name,

VIOLETS
(Viola) Violet family

Lacking perfume only to be a perfectly satisfying flower, the COMMON, PURPLE, MEADOW, or HOODED BLUE VIOLET (V. obliqua; V. cucullata of Gray) has nevertheless established itself in the hearts of the people from the Arctic to the Gulf as no sweet-scented, showy, hothouse exotic has ever done. Royal in color as in lavish profusion, it blossoms everywhere - in woods, waysides, meadows, and marshes, but always in finer form in cool, shady dells; with longer flowering scapes in meadow bogs; and with longer leaves than wide in swampy woodlands. The heart-shaped, saw-edged leaves, folded toward the center when newly put forth, and the five-petalled, bluish-purple, golden-hearted blossom are too familiar for more detailed description. From the three-cornered stars of the elastic capsules, the seeds are scattered abroad.

Beards on the spurred lower petal and the two side petals give the bees a foothold when they turn head downward, as some must, to suck nectar. This attitude enables them to receive the pollen dusted on their abdomens, when they jar the flower, at a point nearest their pollen-collecting hairs. It is also an economical advantage to the flower which can sift the pollen downward on the bee instead of exposing it to the pollen-eating interlopers. Among the latter may be classed the bumblebees and butterflies whose long lips and tongues pilfer ad libitum. "For the proper visitors of the bearded violets," says Professor Robertson, "we must look to the small bees, among which the Osmias are the most important."

When science was younger and hair splitting an uncommon in-dulgence of botanists, the EARLY BLUE VIOLET (Viola palmata) was thought to be simply a variety of the common purple violet, whose heart-shaped leaves frequently show a tendency to divide into lobes. But the early blue violet, however roundish or heart-shaped its early leaves may be, has the later ones variously divided into from three to thirteen lobes, often almost as much cut on the sides as the leaves of the bird's-foot violet. In dry soil, chiefly in the woods, this violet may be found from Southern Canada westward to Minnesota, and south to northern boundaries of the Gulf States. Only its side petals are bearded to form footrests for the insects that search for the deeply secreted nectar. Many butterflies visit this flower. On entering it a bee must first touch the stigma before any fresh golden pollen is released from the anther cone, and cross-fertilization naturally results.

In shale and sandy soil, even in the gravel of hillsides, one finds the narrowly divided, finely cut leaves and the bicolored beardless blossom of the BIRD'S-FOOT VIOLET (V. pedata), pale bluish purple on the lower petals, dark purple on one or two upper ones, and with a heart of gold. The large, velvety, pansy-like blossom and the unusual foliage which rises in rather dense tufts are sufficient to distinguish the plant from its numerous kin. This species produces no cleistogamous or blind flowers. Frequently the bird's-foot violet blooms a second time, in autumn, a delightful eccentricity of this family. The spur of its lower petal is long and very slender, and, as might be expected, the longest-tongued bees and butterflies are its most frequent visitors. These receive the pollen on the base of the proboscis.

The WOOLLY BLUE VIOLET (V. sororia), whose stems and younger leaves, at least, are covered with hairs, and whose purplish-blue flowers are more or less bearded within, prefers a shady but dry situation; whereas its next of kin, the ARROW-LEAVED VIOLET (V. sagittata), delights in moist but open meadows and marshes. The latter's long, arrow, or halberd-shaped leaves, usually entire above the middle, but slightly lobed below it, may rear themselves nine inches high in favorable soil, or in dry uplands perhaps only two inches. The flowering scapes grow as tall as the leaves. All but the lower petal of the large, deep, dark, purplish-blue flower are bearded. This species produces an abundance of late cleistogamous flowers on erect stems. These peculiar greenish flowers without petals, that are so often mistaken for buds or seed vessels; that never open, but without insect aid ripen quantities of fertile seed, are usually borne, if not actually under ground, then not far above it, on nearly all violet plants. It will be observed that all species which bear blind flowers rely somewhat on showy, cross-fertilized blossoms also to counteract degeneracy from close inbreeding.

The OVATE-LEAVED VIOLET (V. ovata), formerly reckoned as a mere variety of the former species, is now accorded a distinct rank. Not all the blossoms, but an occasional clump, has a faint perfume like sweet clover. The leaf is elongated, but rather too round to be halberd-shaped; the stems are hairy; and the flowers, which closely resemble those of the arrow-leaved violet, are earlier; making these two species, which are popularly mistaken for one, among the earliest and commonest of their clan. The dry soil of upland woods and thickets is the ovate-leaved violet's preferred habitat.

In course of time the lovely ENGLISH, MARCH, or SWEET VIOLET, (V. odorata), which has escaped from gardens, and which is now rapidly increasing with the help of seed and runners on the Atlantic and the Pacific coasts, may be established among our wild flowers. No blossom figures so prominently in European literature. In France, it has even entered the political field since Napoleon's day. Yale University has adopted the violet for its own especial flower, although it is the corn-flower, or bachelor's button (Centaurea cyanus) that is the true Yale blue. Sprengel, who made a most elaborate study of the violet, condensed the result of his research into the following questions and answers, which are given here because much that he says applies to our own native species, which have been too little studied in the modern scientific spirit:

"1. Why is the flower situated on a long stalk which is upright, but curved downwards at the free end? In order that it may hang down; which, firstly, prevents rain from obtaining access to the nectar; and, secondly, places the stamens in such a position that the pollen falls into the open space between the pistil and the free ends of the stamens. If the flower were upright, the pollen would fall into the space between the base of the stamen and the base of the pistil, and would not come in contact with the bee.

"2. Why does the pollen differ from that of most other insect-fertilized flowers? In most of such flowers the insects themselves remove the pollen from the anthers, and it is therefore important that the pollen should not easily be detached and carried away by the wind. In the present case, on the contrary, it is desirable that it should be looser and dryer, so that it may easily fall into the space between the stamens and the pistil. If it remained attached to the anther, it would not be touched by the bee, and the flower would remain unfertilized.

"3. Why is the base of the style so thin? In order that the bee may be more easily able to bend the style.

"4. Why is the base of the style bent? For the same reason. The result of the curvature is that the pistil is much more easily bent than would be the case if the style were straight.

"5. Finally, why does the membranous termination of the upper filament overlap the corresponding portions of the two middle stamens? Because this enables the bee to move the pistil, and thereby to set free the pollen more easily than would be the case under the reverse arrangement."

In high altitudes of New England, Colorado. and northward, where the soil is wet and cold, the pale lilac, slightly bearded petals, streaked with darker veins, of the MARSH VIOLET (V. palustris), with its almost round leaves, may be found from May to June. All through the White Mountains one finds it abundant.

A peculiarity of the DOG or RUNNING VIOLET (V. Labradorica) is that its small, heart-shaped leaves are set along the branching stem, and its pale purple blossoms rise from their angles, pansy fashion. From March to May it blooms throughout its wide range in wet, shady places. Its English prototype, called by the same invidious name, was given the prefix "dog," because the word, which is always intended to express contempt in the British mind, is applied in this case for the flower's lack of fragrance. When a bee visits this violet, his head coming in contact with the stigma jars it, thus opening the little pollen box, whose contents must fall out on his head and be carried away and rubbed off where it will fertilize the next violet visited.

SEA LAVENDER; MARSH ROSEMARY; CANKER-ROOT; INK-ROOT (Limonium Carolinianum; Statice Limonium of Gray) Plumbago family

Flowers - Very tiny, pale, dull lavender, erect, set along upper side of branches. Calyx 5-toothed, tubular, plaited; corolla of 5 petals opposite as many stamens; 1 pistil with 5 thread-like styles. Scape: 1 to 2 ft. high, slender, leafless, much branched above. Leaves: All from thick, fleshy rootstock, narrowly oblong, tapering into margined petioles, thick, the edges slightly waved, not toothed; midrib prominent. Preferred Habitat - Salt meadows and marshes.

Flowering Season - July-October. Distribution - Atlantic coast from Labrador to Florida, westward along the Gulf to Texas; also in Europe.

Seen in masses, from a little distance, this tiny flower looks like blue-gray mist blown in over the meadows from the sea, and on closer view each plant suggests sea-spray itself. Thrifty housewives along the coast dry it for winter bouquets, partly for ornament and partly because there is an old wives' tradition that it keeps away moths. Statice, from the Greek verb to stop, hence an astringent, was the generic name formerly applied to the plants, with whose roots these same old women believed they cured canker sores.

FRINGED GENTIAN
(Gentiana crinita) Gentian family

Flowers - Deep, bright blue, rarely white, several or many, about 2 in. high, stiffly erect, and solitary at ends of very long foot-stalk. Calyx of 4 unequal, acutely pointed lobes. Corolla funnel form, its four lobes spreading, rounded, fringed around ends, but scarcely on sides. Four stamens inserted on corolla tube; 1 pistil with 2 stigmas. Stem: 1 to 3 ft. high, usually branched, leafy. Leaves: Opposite, upper ones acute at tip, broadening to heart-shaped base, seated on stem. Fruit: A spindle-shaped, 2-valved capsule, containing numerous scaly, hairy seeds. Preferred Habitat - Low, moist meadows and woods. Flowering Season - September-November. Distribution - Quebec, southward to Georgia, and westward beyond the Mississippi.

"Thou waitest late, and com'st alone
When woods are bare and birds have flown,
And frosts and shortening days portend
The aged year is near his end.

"Then doth thy sweet and quiet eye
Look through its fringes to the sky,
Blue - blue - as if that sky let fail

A flower from its cerulean wall."

When we come upon a bed of gentians on some sparkling October day, we can but repeat Bryant's thoughts and express them prosaically who attempt description. In dark weather this sunshine lover remains shut, to protect its nectar and pollen from possible showers. An elusive plant is this gentian, which by no means always reappears in the same places year after year, for it is an annual whose seeds alone perpetuate it. Seating themselves on the winds when autumn gales shake them from out of the home wall, these little hairy scales ride afar, and those that are so fortunate as to strike into soft, moist soil at the end of the journey, germinate. Because this flower is so rarely beautiful that few can resist the temptation of picking it, it is becoming sadly rare near large settlements.

The special importance of producing a quantity of fertile seed has led the gentians to adopt proterandry - one of the commonest, because most successful, methods of insuring it. The anthers, coming to maturity early, shed their pollen on the bumblebees that have been first attracted by their favorite color and the enticing fringes before they crawl half way down the tube where they can reach the nectar secreted in the walls. After the pollen has been carried from the early flowers, and the stamens begin to wither, up rises the pistil to be fertilized with pollen brought from a newly opened blossom by the bee or butterfly. The late development of the pistil accounts for the error often stated, that some gentians have none. No doubt the fringe, which most scientists regard simply as an additional attraction for winged insects, serves a double purpose in entangling the feet of ants and other crawlers that would climb over the edge to pilfer sweets clearly intended for the bumblebee alone.

Fifteen species of gentian have been gathered during a half-hour walk in Switzerland, where the pastures are spread with sheets of blue. Indeed, one can little realize the beauty of these heavenly flowers who has not seen them among the Alps.

The FIVE-FLOWERED or STIFF GENTIAN, or AGUE-WEED (Gentiana quinquefolia; G. quinqueflora of Gray) has its five-parted, small, picotee-edged blue flowers arranged in clusters, not exceeding seven, at the ends of the branches or seated in the leaf-axils. The slender, branching, ridged stem may rise only two inches in dry soil; or perhaps two feet in rich, moist, rocky ground, where it grows to perfection, especially in mountainous regions. From Canada to Florida and westward to Missouri is its range, and beginning to bloom in August southward, it may not be found until September in the Catskills, and in October it is still in its glory in Ontario. The colorless, bitter juice of many of the gentian tribe has long been valued as a tonic in medicine. Evidently the butterflies that pilfer this "ague-weed," and the bees that are its legitimate feasters, find something more delectable in its blue walls.

A deep, intense blue is the CLOSED, BLIND, or BOTTLE GENTIAN (G. Andrewsii), more truly the color of the "male bluebird's back," to which Thoreau likened the paler fringed gentian. Rarely some degenerate plant bears white flowers. As it is a perennial, we are likely to find it in its old haunts year after year; nevertheless its winged seeds sail far abroad to seek pastures new. This gentian also shows a preference for moist soil. Gray thought that it expanded slightly, and for a short time only in sunshine, but added that, although it is proterandrous, i.e. it matures and sheds its pollen before its stigma is susceptible to any, he believed it finally fertilized itself by the lobes of the stigma curling backward until they touched the anthers. But Gray was doubtless mistaken. Several authorities have recently proved that the flower is adapted to bumblebees. It offers them the last feast of the season, for although it comes into bloom in August southward, farther northward - and it extends from Quebec to the Northwest Territory - it lasts through October.

Now, how can a bumblebee enter this inhospitable-looking flower? If he did but know it, it keeps closed for his special benefit, having no fringes or hairs to entangle the feet of crawling pilferers, and no better way of protecting its nectar from rain and marauding butterflies that are not adapted to its needs. But he is a powerful fellow. Watch him alight on a cluster of blossoms, select the younger, nectar-bearing ones, that are distinctly marked white against a light-blue background at the mouth of the corolla for his special guidance. Old flowers from which the nectar has been removed turn deep reddish purple, and the white pathfinders become indistinct. With some difficulty, it is true, the bumblebee (B. Americanorum) thrusts his tongue through the valve of the chosen flower where the five plaited lobes overlap one another; then he pushes with all his might until his head having passed the entrance most of his body follows, leaving only his hind legs and the tip of his abdomen sticking out as he makes the circuit. He has much sense as well as muscle, and does not risk imprisonment in what must prove a tomb by a total and unnecessary disappearance within the bottle. Presently he backs out, brushes the pollen from his head and thorax into his baskets, and is off to fertilize an older, stigmatic flower with the few grains of quickening dust that must remain on his velvety head.

WILD BLUE PHLOX
(Phlox divaricata) Phlox family

Flowers - Pale lilac blue, slightly fragrant, borne on sticky pedicels, in loose, spreading clusters. Calyx with 5 long, sharp teeth. Corolla of 5 flat lobes, indented like the top of a heart, and united into a slender tube; 5 unequal, straight, short stamens in corolla tube; 1 pistil with 3 stigmas. Stem: to 2 ft. high, finely coated with sticky hairs above, erect or spreading, and producing leafy shoots from base. Leaves: Of flowering stem - opposite, oblong, tapering to a point; of sterile shoots - oblong or egg-shaped, not pointed, 1 to 2 in. long. Preferred habitat - Moist, rocky woods. Flowering Season - April-June. Distribution - Eastern Canada to Florida, Minnesota to Arkansas.

The merest novice can have no difficulty in naming the flower whose wild and cultivated relations abound throughout North America, the almost exclusive home of the genus, although it is to European horticulturists, as usual the first to see the possibilities in our native flowers, that we owe the gay hybrids in our gardens. Mr. Drummond, a collector from the Botanical Society of Glasgow, early in the thirties sent home the seeds of a species from Texas, which became the ancestor of the gorgeous annuals, the Drummond phloxes of commerce today; and although he died of fever in Cuba before the plants became generally known, not even his kinsman, the author of "Natural Law in the Spiritual World," has done more to immortalize the family name.

While the wild blue phlox is sometimes cultivated, it is the GAR-DEN PHLOX (P. paniculata), common in woods and thickets from Pennsylvania to Illinois and southward, that under a gardener's care bears the large terminal clusters of purple, magenta, crimson, pink, and white flowers abundant in old-fashioned, hardy borders. From these it has escaped so freely in many sections of the North and East as to be counted among the local wildflowers. Unless the young offshoots are separated from the parent and given a nook of their own, the flower quickly reverts to the original type. European cultivators claim that the most brilliant colors are obtained by crossing annual with perennial phloxes.

WILD SWEET WILLIAM (P. maculata), another perennial much sought by cultivators, loves the moisture of low woods and the neighborhood of streams in the Middle and Western States when it is free to choose its habitat; but it, too, has so freely escaped from gardens farther north into dry and dusty roadsides, that anyone who has passed the ruins of Hawthorne's little red cottage at Lenox, for example, and seen the way his wife's clump of white phlox under his study window has spread to cover an acre of hillside, would suppose it to be luxuriating in its favorite locality. This variety of the species (var. Candida) lacks the purplish flecks on stem and lower leaves responsible for the specific name of the type. Pinkish purple or pink blossoms are borne in a rather narrow, elongated panicle on the typical Sweet William.

Most members of the phlox family resort to the trick of coating the upper stem and the peduncles immediately below the flowers with a sticky secretion in which crawling insects, intent on pilfering sweets, meet their death, just as birds are caught on limed twigs. Butterflies, for whom phloxes have narrowed their tubes to the exclusion of most other insects, are their benefactors; but long-tongued bees and flies often seek their nectar. Indeed, the number of strictly butterfly-flowers is surprisingly small.

VIRGINIA COWSLIP; TREE or SMOOTH LUNGWORT; BLUE-BELLS
  (Mertensia Virginica) Borage family

Flowers - Pinkish in bud, afterward purplish blue, fading to light blue; about 1 in. long, tubular, funnel form, the tube of corolla not crested; spreading or hanging on slender pedicels in showy, loose clusters at end of smooth stem from 1 to 2 ft. high; stamens 5, inserted on corolla; 1 pistil; ovary of 4 divisions. Leaves: Large, entire, alternate, veiny, oblong or obovate, the upper ones seated on stem; lower very large ones diminishing toward base into long petioles; at first rich, dark purple, afterward pale bluish gray. Fruit: 4 seed-like little nuts, leathery, wrinkled when mature. Preferred Habitat - Alluvial ground, low meadows, and along streams. Flowering Season - March-May. Distribution - Southern Canada to South Carolina and Kansas, west to Nebraska; most abundant in middle West.

Not to be outdone by its cousins the heliotrope and the forget-me-not, this lovely and far more showy spring flower has found its way into the rockwork and sheltered, moist nooks of many gardens, especially in England, where Mr. W. Robinson, who has appealed for its wider cultivation in that perennially charming book, "The English Flower Garden," says of the Mertensias: "There is something about them more beautiful in form of foliage and stem, and in the graceful way in which they rise to panicles of blue, than in almost any other family…. Handsomest of all is the Virginia cowslip." And yet Robinson never saw the alluvial meadows in the Ohio Valley blued with lovely masses of the plant in April.

A great variety of insects visit this blossom, which, being tubular, conducts them straight to the ample feast; but not until they have deposited some pollen brought from another flower on the stigma in their way. The anthers are too widely separated from the stigma to make self-fertilization likely. Occasionally one finds the cowslips perforated by clever bumblebees. As only the females, which are able to sip far deeper cups, are flying when they bloom, they must be either too mischievous or too lazy to drain them in the legitimate manner. Butterflies have only to stand on a flower, not to enter it, in order to sip nectar from the four glands that secrete it abundantly.

FORGET-ME-NOT; MOUSE-EAR; SCORPION GRASS; SNAKE GRASS; LOVE ME

(Myosotis Palustris) Borage family

Flowers - Pure blue, pinkish, or white, with yellow eye; flat, 5-lobed, borne in many-flowered, long, often 1-sided racemes. Calyx 5-cleft; the lobes narrow, spreading, erect, and open in fruit; 5 stamens inserted on corolla tube; style threadlike; ovary 4-celled. Stem: Low, branching, leafy, slender, hairy, partially reclining. Leaves: (Myosotis = mouse-ear) oblong, alternate, seated on stem, hairy. Fruit: Nutlets, angled and keeled on inner side. Preferred Habitat - Escaped from gardens to brooksides, marshes, and low meadows. Flowering Season - May-July. Distribution - Native of Europe and Asia, now rapidly spreading from Nova Scotia southward to New Jersey, Pennsylvania, and beyond.

How rare a color blue must have been originally among our flora is evident from the majority of blue and purple flowers that, although now abundant here and so perfectly at home, are really quite recent immigrants from Europe and Asia. But our dryer, hotter climate never brings to the perfection attained in England

"The sweet forget-me-nots
That grow for happy lovers."

Tennyson thus ignores the melancholy association of the flower in the popular legend which tells how a lover, when trying to gather some of these blossoms for his sweetheart, fell into a deep pool, and threw a bunch on the bank, calling out, as he sank forever from her sight, "Forget me not." Another dismal myth sends its hero forth seeking hidden treasure caves in a mountain, under the guidance of a fairy. He fills his pockets with gold, but not heeding the fairy's warning to "forget not the best" - i.e., the myosotis - he is crushed by the closing together of the mountain. Happiest of all is the folk-tale of the Persians; as told by their poet Shiraz: "It was in the golden morning of the early world, when an angel sat weeping outside the closed gates of Paradise. He had fallen from his high estate through loving a daughter of earth, nor was he permitted to enter again until she whom he loved had planted the flowers of the forget-me-not in every corner of the world. He returned to earth and assisted her, and together they went hand in hand. When their task was ended, they entered Paradise together, for the fair woman, without tasting the bitterness of death, became immortal like the angel whose love her beauty had won when she sat by the river twining forget-me-nots in her hair."

It was the golden ring around the forget-me-not's center that first led Sprengel to believe the conspicuous markings at the entrance of many flowers served as pathfinders to insects. This golden circle also shelters the nectar from rain, and indicates to the fly or bee just where it must probe between stigma and anthers to touch them with opposite sides of its tongue. Since it may probe from any point of the circle, it is quite likely that the side of the tongue that touched a pollen-laden anther in one flower will touch the stigma in the next one visited, and so cross-fertilize it. But forget-me-nots are not wholly dependent on insects. When these fail, a fully mature flower is still able to set fertile seed by shedding its own pollen directly on the stigma.

The SMALLER FORGET-ME-NOT (M. laxa), formerly accounted a mere variety of palustris, but now defined as a distinct species, is a native, and therefore may serve to show how its European relative here will deteriorate in the dryer atmosphere of the New World. Its tiny turquoise flowers, borne on long stems from a very loose raceme, gleam above wet, muddy places from Newfoundland and Eastern Canada to Virginia and Tennessee.

Even smaller still are the blue or white flowers of the FIELD FORGET-ME-NOT, SCORPION GRASS, or MOUSE-EAR (M. arvenis), whose stems and leaves are covered with bristly hairs. It blooms from August to July in dry places, even on hillsides, an unusual locality in which to find a member of this moisture-loving clan. All the flowers remain long in bloom, continually forming new buds on a lengthening stem, and leaving behind little empty green calices.

VIPER'S BUGLOSS; BLUE-WEED; VIPER'S HERB or GRASS;
SNAKE-FLOWER;
BLUE-THISTLE

(Echium vulgare) Borage family

Flowers - Bright blue, afterward reddish purple, pink in the bud, numerous, clustered on short, 1-sided, curved spikes rolled up at first, and straightening out as flowers expand. Calyx deeply 5-cleft; corolla 1 in. long or less, funnel form, the 5 lobes unequal, acute; 5 stamens inserted on corolla tube, the filaments spreading below, and united above into slender appendage, the anthers forming a cone. 1 pistil with 2 stigmas. Stem: 1 to 2 1/2 ft. high; bristly-hairy, erect, spotted. Leaves: Hairy, rough, oblong to lance-shaped, alternate, seated on stem, except at base of plant. Preferred Habitat - Dry fields, waste places; roadsides. Flowering Season - June-July. Distribution - New Brunswick to Virginia, westward to Nebraska; Europe and Asia.

In England, from whose gardens this plant escaped long ago, a war of extermination that has been waged against the vigorous, beautiful weed by the farmers has at last driven it to the extremity of the island, where a few stragglers about Penzance testify to the vanquishing of what must once have been a mighty army. From England a few refugees reached here in i683, no one knows how; but they proved to be the vanguard of an aggressive and victorious host that quickly overran our open, hospitable country, as if to give vent to revenge for long years of persecution at the hands of Europeans. "It is a fact that all our more pernicious weeds, like our vermin, are of Old-World origin," says.John Burroughs. "…Perhaps the most notable thing about them, when compared with our native species, is their persistence, not to say pugnacity. They fight for the soil; they plant colonies here and there, and will not be rooted out. Our native weeds are for the most part shy and harmless, and retreat before civilization…. We have hardly a weed we can call our own."

Years ago, when simple folk believed God had marked plants with some sign to indicate the special use for which each was intended, they regarded the spotted stem of the bugloss, and its seeds shaped like a serpent's head, as certain indications that the herb would cure snake bites. Indeed, the genus takes its name from Echis, the Greek for viper.

Because it is showy and offers accessible nectar, a great variety of insects visit the blue-weed; Muller alone observed sixty-seven species about it. We need no longer wonder at its fertility. Of the five stamens one remains in the tube, while the other four project and form a convenient alighting place for visitors, which necessarily dust their under sides with pollen as they enter; for the red anthers were already ripe when the flower opened. Then, however, the short, immature pistil was kept below. After the stamens have shed their pollen and there can be no longer danger of self-fertilization, it gradually elongates itself beyond the point occupied by them, and divides into two little horns whose stigmatic surfaces an incoming pollen-laden insect cannot well fail to strike against. Cross-pollination is so thoroughly secured in this case that the plant has completely lost the power of fertilizing itself. Unwelcome visitors like ants, which would pilfer nectar without rendering any useful service in return, are warded off by the bristly, hairy foliage. Several kinds of female bees seek the bugloss exclusively for food for their larvae as well as for themselves, sweeping up the abundant pollen with their abdominal brushes as they feast without effort.

BLUE VERVAIN; WILD HYSSOP; SIMPLER'S JOY
 (Verbena hastala) Vervain family

Flowers - Very small, purplish blue, in numerous slender, erect, compact spikes. Calyx 5-toothed; corolla tubular, unequally 5-lobed; 2 pairs of stamens; 1 pistil. Stem: 3 to 7 ft. high, rough, branched above, leafy, 4-sided. Leaves: Opposite, stemmed, lance-shaped, saw-edged, rough; lower ones lobed at base. Preferred Habitat - Moist meadows, roadsides, waste places. Flowering Season - June-September. Distribution - United States and Canada in almost every part.

Seeds below, a circle of insignificant purple-blue flowers in the center, and buds at the top of the vervain's slender spires do not produce a striking effect, yet this common plant certainly does not lack beauty. John Burroughs, ever ready to say a kindly, appreciative word for any weed, speaks of its drooping, knotted threads, that "make a pretty etching upon the winter snow." Bees, the vervain's benefactors, are usually seen clinging to the blooming spikes, and apparently sleep on them. Borrowing the name of simpler's joy from its European sister, the flower has also appropriated much of the tradition and folk-lore centered about that plant which herb-gatherers, or simplers, truly delighted to see, since none was once more salable.

EUROPEAN VERVAIN (V. officinalis) HERB-OF-THE-CROSS, BERBINE, HOLY-HERB, ENCHANTER'S PLANT, JUNO'S TEARS, PIGEON-GRASS, LIGHTNING PLANT, SIMPLER'S JOY, and so on through a long list of popular names for the most part testifying to the plant's virtue as a love-philter, bridal token, and general cure-all, has now become naturalized from the Old World on the Atlantic and Pacific Slopes; and is rapidly appropriating waste arid cultivated ground until, in many places, it is truly troublesome. In general habit like the blue vervain, its flowers are more purplish than blue, and are scattered, not crowded, along the spikes. The leaves are deeply, but less acutely, cut.

Ages before Christians ascribed healing virtues to the vervain - found growing on Mount Calvary, and therefore possessing every sort of miraculous power, according to the logic of simple peasant folk - the Druids had counted it among their sacred plants. "When the dog-star arose from unsunned spots" the priests gathered it. Did not Shakespeare's witches learn some of their uncanny rites from these reverend men of old? One is impressed with the striking similarity of many customs recorded of both. Two of the most frequently used ingredients in witches' cauldrons were the vervain and the rue. "The former probably derived its notoriety from the fact of its being sacred to Thor, an honor which marked it out, like other lightning plants, as peculiarly adapted for occult uses," says Mr. Thiselton Dyer in his "Folk-lore of Plants." "Although vervain, therefore, as the enchanter's plant, was gathered by witches to do mischief in their incantations, yet, as Aubrey says, it 'hinders witches from their will,' a circumstance to which Drayton further refers when he speaks of the vervain as "gainst witchcraft much avayling.'" Now we understand why the children of Shakespeare's time hung vervain and dill with a horseshoe over the door.

In his eighth Eclogue, Virgil refers to vervain as a charm to recover lost love. Doubtless this was the verbena, the herba sacra employed in ancient Roman sacrifices, according to Pliny. In his day the bridal wreath was of verbena, gathered by the bride herself.

NARROW-LEAVED VERVAIN (V. angustifolia), like the blue vervain, has a densely crowded spike of tiny purple or blue flowers that quickly give place to seeds, but usually there is only one spike at the end of a branch. The leaves are narrow, lance-shaped, acute, saw-edged, rough. From Massachusetts and Florida westward to Minnesota and Arkansas one finds the plant blooming in dry fields from June to August, after the parsimonious manner of the vervain tribe.

It is curious that the vervain, or verbena, employed by brides for centuries as the emblem of chastity, should be one of the notorious botanical examples of a willful hybrid. Generally, the individuals of distinct species do not interbreed; but verbenas are often difficult to name correctly in every case because of their susceptibility to each other's pollen - the reason why the garden verbena may so easily be made to blossom forth into whatever hue the gardener wills. His plants have been obtained, for the most part, from the large-flowered verbena, the beautiful purple, blue, or white species of our Western States (V. Canadensis) crossed with brilliant-hued species imported from South America.

MAD-DOG SKULLCAP or HELMET-FLOWER; MAD-WEED; HOODWORT
(Scutellaria lateriflora) Mint family

Flowers - Blue, varying to whitish; several or many, 1/4 in. long, growing in axils of upper leaves or in 1-sided spike-like racemes. Calyx 2-lipped, the upper lip with a helmet-like protuberance; corolla 2-lipped; the lower, 3-lobed lip spreading; the middle lobe larger than the side ones. Stamens, 4, in pairs, under the upper lip; upper pair the shorter; one pistil, the style unequally cleft in two. Stem: Square, smooth, leafy, branched, 8 in. to 2 ft. high. Leaves: Opposite, oblong to lance-shaped, thin, toothed, on slender pedicles, 1 to 3 in. long, growing gradually smaller toward top of stem. Fruit: 4 nutlets. Preferred Habitat - Wet, shady ground. Flowering Season - July-September. Distribution - Uneven throughout United States and the British Possessions.

By the helmet-like appendage on the upper lip of the calyx, which to the imaginative mind of Linnaeus suggested Scutellum (a little dish), which children delight to spring open for a view of the four tiny seeds attached at the base when in fruit, one knows this to be a member of the skullcap tribe, a widely scattered genus of blue and violet two-lipped flowers, some small to the point of insignificance, like the present species, others showy enough for the garden, but all rich in nectar, and eagerly sought by bees. The wide middle lobe of the lower lip forms a convenient platform on which to alight; the stamens in the roof of a newly opened blossom dust the back of the visitor as he explores the nectary; and as the stamens of an older flower wither when they have shed their pollen, and the style then rises to occupy their position, it follows that, in flying from the top of one spike of flowers to the bottom of another, where the older ones are, the visitor, for whom the whole scheme of color, form, and arrangement was planned, deposits on the sticky top of the style some of the pollen he has brought with him and so cross-fertilizes the flower. When the seeds begin to form and the now useless corolla drops off, the helmet-like appendage on the top of the calyx enlarges and meets the lower lip, so enclosing and protecting the tiny nutlets. After their maturity, either the mouth gapes from dryness, or the appendage drops off altogether, from the same cause, to release the seeds. Old herb doctors, who professed to cure hydrophobia with this species, are responsible for its English misnomer.

Perhaps the most beautiful member of the genus is the SHOWY SKULLCAP (S. serrata), whose blue corolla, an inch long, has its narrow upper lip shorter than the spreading lower one. The flowers are set opposite each other at the end of the smooth stem, which rises from one to two feet high in the woods throughout a southerly and westerly range. As several other skullcaps have distinctly saw-edged leaves, this plant might have been given a more distinctive adjective, thinks one who did not have the naming of 200,000 species!

Above dry, sandy soil from New York and Michigan southward the HAIRY SKULLCAP (S. pilosa) lifts short racemes of blue flowers that are only half an inch long, and whose lower lip and lobes at either side are shorter than the arched upper lip. Most parts of the plant are covered with down, the lower stem being especially hairy; and this fact determines the species when connected with its rather distant pairs of indented, veiny leaves, ranging from oblong to egg-shaped, and furnished with petioles which grow gradually shorter toward the top, where pairs of bracts, seated on the stem, part to let the flowers spring from their axils.

The LARGER or HYSSOP SKULLCAP (S. integrifolia) rarely has a dent in its rounded oblong leaves ,which, like the stem, are covered with fine down. Its lovely, bright blue flowers, an inch long, the lips of about equal length, are grouped opposite each other at the top of a stem that never lifts them higher than two feet; and so their beauty is often concealed in the tall grass of roadsides and meadows and the undergrowth of woods and thickets, where they bloom from May to August, from southern New England to the Gulf of Mexico, westward to Texas.

This tribe of plants is almost exclusively North American, but the hardy MARSH SKULLCAP or HOODED WILLOW-HERB (S. galericulata), at least, roams over Europe, and Asia also, with the help of runners, as well as seeds that, sinking into the soft earth of swamps and the borders of brooks, find growth easy. The blue flowers which grow singly in the axils of the upper leaves are quite as long as those of the larger and the showy skullcaps; the oblong, lance-shaped leaves, which are mostly seated on the branching stem, opposite each other, have low teeth. Why do leaves vary as they do, especially in closely allied species? "The causes which have led to the different forms of leaves have been, so far as I know," says Sir John Lubbock, "explained in very few cases: those of the shapes and structure of seeds are tolerably obvious in some species, but in the majority they are still entirely unexplained; and, even as regards the blossoms themselves, in spite of the numerous and conscientious labors of so many eminent naturalists, there is as yet no single species thoroughly known to us."

GROUND IVY or JOY; GILL-OVER-THE-GROUND; FIELD BALM; CREEPING CHARLIE
(Glecoma hederacea; Nepeta Glechoma of Gray) Mint family

Flowers - Light bluish purple, dotted with small specks of reddish violet; growing singly or in clusters along stem, seated in leaf axils; calyx hairy, with 5 sharp teeth; corolla tubular, over 1/2 in. long, 2-lipped, the upper lip 2-lobed, lower lip with 3 spreading lobes, middle one largest; 4 stamens in pairs under upper lip; the anther sacs spreading; pistil with 2-lobed style. Stem: Trailing, rooting at intervals, sometimes 18 in. long, leafy, the branches ascending. Leaves: From 1/2 to 1 1/2 in. across; smooth, rounded, kidney-shaped, scallop-edged. Preferred Habitat - Waste places, shady ground. Flowering Season - March-May. Distribution - Eastern half of Canada and the United States, from Georgia and Kansas northward.

Besides the larger flowers, containing both stamens and pistils, borne on this little immigrant, smaller female flowers, containing a pistil only, occur just as they do in thyme, mint, marjoram, and doubtless other members of the great family to which all belong. Muller attempted to prove that these small flowers, being the least showy, are the last to be visited by insects, which, having previously dusted themselves with pollen from the stamens of the larger flowers when they first open, are in a condition to make cross-fertilization certain. So much for the small flower's method of making insects serve its end; the larger flowers have another way. At first they are male; that is, the pistil is as yet undeveloped and the four stamens are mature, ready to shed pollen on any insect alighting on the lip. Later, when the stamens are past maturity, the pistil elongates itself and is ready for the reception of pollen brought from younger flowers. Many blossoms are male on the first day of opening, and female later, to protect themselves against self-fertilization.

In Europe, where the aromatic leaves of this little creeper were long ago used for fermenting and clarifying beer, it is known by such names as ale-hoof and gill ale-gill, it is said, being derived from the old French word, guiller, to ferment or make merry. Having trailed across Europe, the persistent hardy plant is now creeping its way over our continent, much to the disgust of cattle, which show unmistakable dislike for a single leaf caught up in a mouthful of herbage.

Very closely allied to the ground ivy is the CATMINT or CATNIP (Nepela Cataria) ,whose pale-purple, or nearly white flowers, dark-spotted, may be most easily named by crushing the coarsely toothed leaves in one's hand. It is curious how cats will seek out this hoary-hairy plant in the waste places where it grows and become half-crazed with delight over its aromatic odor.

SELF-HEAL; HEAL-ALL; BLUE CURLS; HEART-OF-THE-EARTH; BRUNELLA
(Prunella vulgaris) Mint family

Flowers - Purple and violet, in dense spikes, somewhat resembling a clover head; from 1/2 to 1 in. long in flower, becoming 4 times the length in fruit. Corolla tubular, irregularly 2-lipped, the upper lip darker and hood-like; the lower one 3-lobed, spreading, the middle and largest lobe fringed; 4 twin-like stamens ascending under upper lip; filaments ofthe lower and longer pair 2-toothed at summit, one of the teeth bearing an anther, the other tooth sterile; style thread-like, shorter than stamens, and terminating in a 2-cleft stigma. Calyx 2-parted, half the length of corolla, its teeth often hairy on edges. Stem: 2 in. to 2 ft. high, erect or reclining, simple or branched. Leaves: Opposite, oblong. Fruit: 4 nutlets, round and smooth. Preferred Habitat - Fields, roadsides, waste places. Flowering Season - May-October. Distribution - North America, Europe, Asia.

This humble, rusty green plant, weakly lopping over the surrounding grass, so that often only its insignificant purple, clover-like flower heads are visible, is another of those immigrants from the old countries which, having proved fittest in the fiercer struggle for existence there, has soon after its introduction here exceeded most of our more favored native flowers in numbers. Everywhere we find the heal-all, sometimes dusty and stunted by the roadside, sometimes truly beautiful in its fresh purple, violet, and white when perfectly developed under happy conditions. In England, where most flowers are deeper hued than with us, the heal-all is rich purple. What is the secret of this flower's successful march across three continents? As usual, the chief reason is to be found in the facility it offers insects to secure food; and the quantity of fertile seed it is therefore able to ripen as the result of their visits is its reward. Also, its flowering season is unusually long, and it is a tireless bloomer. It is finical in no respect; its sprawling stems root easily at the joints, and it is very hardy.

Several species of bumblebees enter the flower, which being set in dense clusters enables them to suck the nectar from each with the minimum loss of time, the smaller bee spending about two seconds to each. After allowing for the fraction of time it takes him to sweep his eyes and the top of his head with his forelegs to free them from the pollen which must inevitably be shaken from the stamen in the arch of the corolla as he dives deeply after the nectar in the bottom of the throat, and to pass the pollen, just as honeybees do, with the most amazing quickness, from the forelegs to the middle ones, and thence to the hairy "basket" on the hind ones - after making all allowances for such delays, this small worker is able to fertilize all the flowers in the fullest cluster in half a minute! When the contents of the baskets of two different species of bumblebees caught on this blossom were examined under the microscope, the pollen in one case proved to be heal-all, with some from the goldenrod, and a few grains of a third kind not identified; and in the other case; heal-all pollen and a small proportion of some unknown kind. Bees that are evidently out for both nectar and pollen on the same trip have been detected visiting white and yellow flowers on their way from one heal-all cluster to another; and this fact, together with the presence of more than one kind of pollen in the basket, shows that the generally accepted statement that bees confine themselves to flowers of one kind or color during a trip is not always according to fact.

The older name of the plant, Brunella, and the significant one, altered by Linnaeus into the softer sound it now bears, is doubtless derived from the German word, braune, the quinsy. Quaint old Parkinson reads: "This is generally called prunella and brunella from the Germans who called it brunellen, because it cureth that disease which they call die bruen, common to soldiers in campe, but especially in garrison, which is an inflammation of the mouth, throat, and tongue." Among the old herbalists who pretended to cure every ill that flesh is heir to with it, it was variously known as carpenter's herb, sicklewort, hook-heal, slough-heal, and brown-wort.

AMERICAN or MOCK PENNYROYAL; TICKWEED; SQUAW MINT
(Hedeoma pulegioides) Mint family

Flowers - Very small, bluish purple, clustered in axils of upper leaves. Calyx tubular, unequally 5-cleft; teeth of upper lip triangular, hairy in throat. Corolla 2-lipped, upper lip erect, notched; lower one 3-cleft, spreading; 2 anther-bearing stamens under upper lip; 2 sterile but apparent; 1 pistil with 2-cleft style. Stem: Low, erect, branched, square, hairy, 6 to 18 in. high. Leaves: Small, opposite, ovate to oblong, scantily toothed, strongly aromatic, pungent. Preferred Habitat - Dry fields, open woodland. Flowering Season - July-September. Distribution - Cape Breton Island westward to Nebraska, south to Florida.

However insignificant its flower, this common little plant unmistakably proclaims its presence throughout the neighborhood. So powerful is the pungent aroma of its leaves that dog doctors sprinkle them about freely in the kennels to kill fleas, a pest by no means exterminated in Southern Europe, however, where the true pennyroyal of commerce (Mentha Pulegium) is native. Herb gatherers who collect our pennyroyal, that is so similar to the European species it is similarly employed in medicine, say they can scent it from a greater distance than any other plant.

BASTARD PENNYROYAL, which, like the Self-heal, is sometimes called BLUE CURLS (Trichostema dichotomum), chooses dry fields, but preferably sandy ones, where we find its abundant, tiny blue flowers, that later change to purple, from July to October. Its balsam-like odor is not agreeable, neither has the plant beauty to recommend it; yet where it grows, from Maine to Florida, and west to Texas, it is likely to be so common we cannot well pass it unnoticed. The low, stiff, slender, much-branched, and rather clammy stem bears opposite, oblong, smooth-edged leaves narrowed into petioles. One, two, or three flowers, borne at the tips of the branches, soon fall off, leaving the 5-cleft calyx to cradle four exposed nutlets.

>From the five-lobed tubular corolla protrude four very long, curling, blue or violet stamens - hair stamens the Greek generic title signifies - and the pretty popular name of blue curls also has reference to these conspicuous filaments that are spirally coiled in the bud.

In general habit like the two preceding plants, the FALSE PENNYROYAL (Isanthus brachiatus) nevertheless prefers that its sandy home should be near streams. From Quebec to Georgia, westward to Minnesota and Texas, it blooms in midsummer, lifting its small, tubular, pale blue flowers from the axils of pointed, opposite leaves. An unusual characteristic in one of the mint tribe is that the five sharp lobes of its bell-shaped calyx, and the five rounded, spreading lobes of the corolla, are of equal length, hence its Greek name signifying an equal flower.

WILD or CREEPING THYME
  (Thymus Serpyllum) Mint family

   Flowers - Very small purple or pink purple, fragrant, clustered at ends of branches or in leaf axils. Hairy calyx and corolla 2-lipped, the latter with lower lip 3-cleft; stamens 4; style 2-cleft. Leaves: Oblong, opposite, aromatic. Stem: 4 to 12 in. long) creeping, woody, branched, forming dense cushions. Preferred Habitat - Roadsides, dry banks, and waste places. Flowering Season - June-September. Distribution - Naturalized from Europe. Nova Scotia to Middle States.

   "I know a bank where the wild thyme blows,
     Where oxlips and the nodding violet grows;
     Quite over-canopied with luscious woodbine,
     With sweet musk-roses, and with eglantine."

- A Midsummer Night's Dream.

According to Danish tradition, anyone waiting by an elder-bush on Midsummer Night at twelve o'clock will see the king of fairyland and all his retinue pass by and disport themselves in favorite haunts, among others the mounds of fragrant wild thyme. How well Shakespeare knew his folklore!

Thyme is said to have been one of the three plants which made the Virgin Mary's bed. Indeed, the European peasants have as many myths as there are quotations from the poets about this classic plant. Its very name denotes that it was used as an incense in Greek temples. No doubt it was the Common Thyme (T. vulgaris), an erect, tall plant cultivated in gardens here as a savory, that Horace says the Romans used so extensively for bee culture.

Dense cushions of creeping thyme usually contain two forms of blossoms on separate plants - hermaphrodite (male and female which are much the commoner; and pistillate, or only female, flowers, in which the stamens develop no pollen. The latter are more fertile; none can fertilize itself. But blossoms so rich in nectar naturally attract quantities of insects - bees and butterflies chiefly. A newly opened hermaphrodite flower, male on the first day, dusts its visitors as they pass the ripe stamens. This pollen they carry to a flower two days old, which, having reached the female stage, receives it on the mature two-cleft stigma, now erect and tall, whereas the stamens are past maturity.

GARDEN, SPEAR, or MACKEREL MINT
 (Mentha spicata; M. viridis of Gray) Mint family

Flowers - Small, pale bluish, or pinkish purple, in whorls, forming terminal, interrupted, narrow spikes, 2 to 4 in. long in fruit, the central one surpassing lateral ones. Calyx bell-shaped, toothed; corolla tubular, 4-cleft. Stamens 4; style 2-cleft. Stem: Smooth, 1 to 1 1/2 ft. high, branched. Leaves: Opposite, narrowly oblong, acute, saw-edged, aromatic. Preferred Habitat - Moist soil. Flowering Season - July-September. Distribution - Eastern half of Canada and United States. Also Europe and Asia.

The poets tell us that Proserpine, Pluto's wife, in a fit of jealousy changed a hated rival into the mint plant, whose name Mentha, in its Latin form, or Minthe, the Greek equivalent, is still that of the metamorphosed beauty, a daughter of Cocytus, who was also Pluto's wife. Proserpine certainly contrived to keep her rival's memory fragrant. But how she must delight in seeing her under the chopping-knife and served up as sauce!

It is a curious fact that among the Labiates, or two-lipped blossoms to which thymes and mints belong, there very frequently occur species bearing flowers that are male on the first day (staminate) and female, or pistillate, on the second day, and also smaller female flowers on distinct plants. Muller believed this plan was devised to attract insects, first by the more showy hermaphrodite flower, that they might carry its pollen to the less conspicuous female flower, which they would naturally visit last; but this interesting theory has yet to be proved. Nineteen species of flies, to which the mints are specially adapted, have been taken in the act of transferring pollen. Ten varieties of the lower hymenoptera (bees, wasps, and others) commonly resort to the fragrant spikes of bloom.

PEPPERMINT (M. piiterita), similar in manner of growth to the preceding, is another importation from Europe now thoroughly at home here in wet soil. The volatile oil obtained by distilling its leaves has long been an important item of trade in Wayne County, New York. One has only to crush the leaves in one's hand to name the flower.

Our native WILD MINT (M. Canadensis), common along brooksides and in moist soil from New Brunswick to Virginia and far westward, has its whorls of small purplish flowers seated in the leaf axils. Its odor is like pennyroyal. The true PENNYROYAL, not to be confused with our spurious woodland annual, is M. Pulegium, a native of Europe, whence a number of its less valuable relatives, all perennials, have traveled to become naturalized Americans.

In dry open woods and thickets and by the roadside, from late August throughout September, we find blooming the aromatic fragrant STONE MINT, SWEET HORSE-MINT, or AMERICAN DIT-

TANY (Cunila origanoides; C. Mariana of Gray). Its small pink-purple, lilac, or whitish flowers, that are only about half as long as the protruding pair of stamens, are borne in loose terminal clusters at the ends of the stiff, branched, slender, sometimes reddish, stem. A pair of rudimentary, useless stamens remain within the two-lipped tube; the exserted pair, affording the most convenient alighting place for the visiting flies, dust their undersides with pollen the first day the flower opens; on the next, the stigma will be ready to receive pollen carried from young flowers.

NIGHTSHADE; BLUE BINDWEED; FELONWORT; BITTER-SWEET; SCARLET or
SNAKE BERRY; POISON-FLOWER; WOODY NIGHTSHADE
(Solanum Dulcamara) Potato family

Flowers - Blue, purple, or, rarely, white with greenish spots on each lobe; about 1/2 in. broad, clustered in slender, drooping cymes. Calyx 5-lobed, oblong, persistent on the berry; corolla deeply, sharply 5-cleft, wheel-shaped, or points curved backward; 5 stamens inserted on throat, yellow, protruding, the anthers united to form a cone; stigma small. Stem: Climbing or straggling, woody below, branched, 2 to 8 ft. long. Leaves: Alternate, 2 to 4 in. long, 1 to 2 1/2 in. wide, pointed at the apex, usually heart-shaped at base; some with 2 distinct leaflets below on the petiole, others have leaflets united with leaf like lower lobes or wings. Fruit: A bright red, oval berry. Preferred Habitat - Moist thickets, fence rows. Flowering Season - May-September. Distribution - United States east of Kansas, north of New Jersey. Canada, Europe, and Asia.

More beautiful than the graceful flowers are the drooping cymes of bright berries, turning from green to yellow, then to orange and scarlet, in the tangled thicket by the shady roadside in autumn, when the unpretending, shrubby vine, that has crowded its way through the rank midsummer vegetation, becomes a joy to the eye. Another bittersweet, so-called, festoons the hedgerows with yellow berries which, bursting, show their scarlet-coated seeds. Rose hips and mountain-ash berries, among many other conspicuous bits of color, arrest attention, but not for us were they designed. Now the birds are migrating, and, hungry with their long flight, they gladly stop to feed upon fare so attractive. Hard, indigestible seeds traverse the alimentary canal without alteration and are deposited many miles from the parent that bore them. Nature's methods for widely distributing plants cannot but stir the dullest imagination.

The purple pendent flowers of this nightshade secrete no nectar, therefore many insects let them alone; but it is now believed that no part of the plant is poisonous. Certainly one that claims the potato, tomato, and eggplant among its kin has no right to be dangerous. The BLACK, GARDEN, or DEADLY NIGHTSHADE, also called MOREL (S. nigrum), bears jet-black berries that are alleged to be fatal. Nevertheless, female bumblebees, to which its white flowers are specially adapted, visit them to draw out pollen from the chinks of the anthers with their jaws, just as they do in the case of the wild, sensitive plant, and with no more disastrous result. It has been well said that the nightshades are a blessing both to the sick and to the doctors. The present species takes its name from dulcis, sweet, and amaras, bitter, referring to the taste of the juice; the generic name is derived from solamen, solace or consolation, referring to the relief afforded by the narcotic properties of some of these plants.

BLUE or WILD TOADFLAX; BLUE LINARIA
  (Linaria Canadensis) Figwort family

Flowers - Pale blue to purple, small, irregular, in slender spikes. Calyx 5-pointed; corolla 2-lipped, with curved spur longer than its tube, which is nearly closed by a white, 2-ridged projection or palate; the upper lip erect, 2-lobed; lower lip 3-lobed, spreading. Stamens 4, in pairs, in throat; 1 pistil. Stem: Slender, weak, of sterile shoots, prostrate; flowering stem, ascending or erect, 4 in. to 2 ft. high. Leaves: Small, linear, alternately scattered along stem, or oblong in pairs or threes on leafy sterile shoots. Preferred Habitat - Dry soil, gravel, or sand. Flowering Season - May-October. Distribution - North, Central, and South Americas.

Sometimes lying prostrate in the dust, sometimes erect, the linaria's delicate spikes of bloom wear an air of injured innocence, yet the plant, weak as it looks, has managed to spread over three Americas from ocean to ocean. More beautiful than the rather scrawny flowers are the tufts of cool green foliage made by the sterile shoots that take complete possession of a wide area around the parent plants.

Unlike its relative butter-and-eggs, the corolla of this toadflax is so contracted that bees cannot enter it; but by inserting their long tongues, they nevertheless manage to drain it. Small, short-tongued bees contrive to reach only a little nectar. The palate, so valuable to the other linaria, has in this one lost its function; and the larger flies, taking advantage of the flower's weakness, pilfer both sweets and pollen. Butterflies, to which a slender spurred flower is especially attractive, visit this one in great numbers, and as they cannot regale themselves without touching the anthers and stigma, they may be regarded as the legitimate visitors.

Wolf, rat, mouse, sow, cow, cat, snake, dragon, dog, toad, are among the many animal prefixes to the names of flowers that the English country people have given for various and often most interesting reasons. Just as dog, used as a prefix, expresses an idea of worthlessness to them, so toad suggests a spurious plant; the toadflax being made to bear what is meant to be an odious name because before flowering it resembles the true flax, linum, from which the generic title is derived.

MARYLAND FIGWORT; BEE PLANT; KNOTTED FIGWORT;
HEAL-ALL; PILEWORT
(Scrophularia Marylandica; S. nodosa of Gray) Figwort family

Flowers - Very small, dull green on outside; vivid, shining brownish purple within; borne in almost leafless terminal clusters on slender stems; Calyx 5-parted.; corolla of 5 rounded lobes, the 2 upper ones erect, side ones ascending, lower one bent downward; 5 staroens, 4 of them twin-like and bearing anthers, the fifth sterile, a mere scale on roof of the globular corolla tube; style with knot-like stigma. Stem: From 3 to 10 ft. high, square, with grooved sides, widely. branching. Leaves: From 3 to 12 in. long, oblong, pointed, coarsely toothed, on slender stems, strong smelling. Preferred Habitat - Moist, shady ground. Flowering Season - July-September. Distribution - New York to the Carolinas, westward to Tennessee and Kansas; possibly beyond.

An insignificant little flower by itself, conspicuous only because it rears itself in clusters on a level with one's eyes, lacking beauty, perfume, and all that makes a blossom charming to the human mind - why has it been elevated by the botanists to the dignity of lending its name to a large and important family, and why is it mentioned at all in a popular flower book beside the more showy ornaments of nature's garden? Both questions have the same answer: Because it is the typical flower of the family, and therefore serves as an illustration of the manner in which many others are fertilized. Beautiful blossoms are by no means always the most important ones.

It well repays one to observe the relative times of maturing anthers and stigmas in the flowers, as thereby hangs a tale in which some insect plays an interesting role. The figwort matures its stigma at the lip of the style before its anthers have ripened their pollen. Why? By having the stigma of a newly opened flower thrust forward to the mouth of the corolla, an insect alighting on the lip, which forms his only convenient landing place, must brush against it and leave upon it some pollen brought from an older flower, whose anthers are already matured. At this early stage of the flower's development its stamens lie curved over in the tube of the corolla; but presently, as the already fertilized style begins to wither, and its stigma is dry and no longer receptive to pollen, then, since there can be no longer any fear of self-pollination - the horror of so many flowers - the figwort uncurls and elevates its stamens. The insect visitor in search of nectar must get dusted with pollen from the late maturing anthers now ready for him. By this ingenious method the flower becomes cross-fertilized and wastes the least pollen.

Bees and wasps evidently pursue opposite routes in going to work, the former beginning at the bottom of a spike or raceme, where the older, more mature flowers are, and working upward; the wasps commencing at the top, among the newly opened ones. In spite of the fact that we usually see hive bees about this plant, pilfering the generous supply of nectar in each tiny cup, it is undoubtedly the wasp that is the flower's truest benefactor, since he carries pollen from the older blossoms of the last raceme visited to the projecting stigmas of the newly opened flowers at the top of the next cluster. Manifestly no flower, even though it were especially adapted to wasps, as this one is, could exclude bees. About one-third of all its visitors are wasps.

HAIRY BEARD-TONGUE
(Pentstemon hirsutus; P. pubescens of Gray) Figwort family

Flowers - Dull violet or lilac and white, about 1 in. long, borne in a loose spike. Calyx 5-parted, the sharply pointed sepals overlapping; corolla, a gradually inflated tube widening where the mouth divides into a 2-lobed upper lip and a 3-lobed lower lip; the throat nearly closed by hairy palate at base of lower lip; sterile fifth stamen densely bearded for half its length; 4 anther-bearing stamens, the anthers divergent. Stem: 1 to 3 ft. high, erect, downy above. Leaves: Oblong to lance shape, upper ones seated on stem; lower ones narrowed into petioles. Preferred Habitat - Dry or rocky fields, thickets, and open woods. Flowering Season - May-July. Distribution - Ontario to Florida, Manitoba to Texas.

It is the densely bearded, yellow, fifth stamen (pente =five, stemon = a stamen) which gives this flower its scientific name and its chief interest to the structural botanist. From the fact that a blossom has a lip in the center of the lower half of its corolla, that an insect must use as its landing place, comes the necessity for the pistil to occupy a central position. Naturally, a fifth stamen would be only in its way, an encumbrance to be banished in time. In the figwort, for example, we have seen the fifth stamen reduced, from long sterility, to a mere scale on the roof of the corolla tube in other lipped flowers, the useless organ has disappeared; but in the beard-tongue, it goes through a series of curious curves from the upper to the under side of the flower to get out of the way of the pistil. Yet it serves an admirable purpose in helping close the mouth of the flower, which the hairy lip alone could not adequately guard against pilferers. A long-tongued bee, thrusting in his head up to his eyes only, receives the pollen in his face. The blossom is male (staminate) in its first stage and female (pistillate) in its second.

While this is the beard-tongue commonly found in the Eastern United States, particularly southward, and one of the most beautiful of its clan, the western species have been selected by the gardeners for hybridizing into those more showy, but often less charming, flowers now quite extensively cultivated. Several varieties of these, having escaped from gardens in the East, are locally common wild.

The LARGE-FLOWERED BEARD-TONGUE (P. grandiflorus), one of the finest prairie species, whose lavender-blue, bell-shaped corolla is abruptly dilated above the calyx, measures nearly two inches long. Its sterile filament, curved over at the summit, is bearded there only.

Handsomest of all is the COBEA BEARD-TONGUE, a native of the Southwest, with a broadly rounded, bell-shaped corolla, hairy without, like the leaves, but smooth within. The pale purple blossom, delicately suffused with yellow, and pencilled with red lines - pathfinders for the bees - has the base of its tube creamy white. Few flowers hang from each stout clammy spike.

The more densely crowded spikes of the large SMOOTH BEARD-TONGUE (P. glaber), a smaller blue or purple flowered, narrower-leaved species, that shows an unusual preference for moist soil throughout its range, is, like the other beard-tongues mentioned, better known to the British gardener, perhaps, than to Americans, who have yet to learn the value of many of their wild flowers under cultivation.

The tall FOXGLOVE BEARD-TONGUE (P. digitalis), with large, showy white blossoms tinged with purple, the one most commonly grown in gardens here, escapes on the slightest encouragement to run wild again from Maine to Virginia, west to Illinois and Arkansas. Small bees crawl into the broad tube, and butterflies drain the nectar evidently secreted for long-tongued bees, but without certainly transferring pollen. To insure cross-fertilization, the flower first develops its anthers, whose saw-edges grating against the visitors thorax, aid in sifting out the dry pollen; and later the style, which when immature clung to the top of the corolla, lowers its receptive stigma to oppose the bee's entrance. Professor Robertson has frequently detected the common wasp nipping holes with her sharp jaws in the base of the tube. With remarkable intelligence she invariably chose to insert her tongue at the precise spots where the nectar is stored on either side of the sterile filament.

BLUE-EYED MARY; INNOCENCE; BROAD-LEAVED COLLINSIA
(Collinsia verna) Figwort family

Flowers - On slender, weak stalks; whorled in axils of upper leaves. Blue on lower lip of corolla, its middle lobe folded lengthwise to enclose 4 adhering stamens and pistil; upper lip white, with scalloped margins; corolla from 1/2 to 3/4 in. long, its throat about equaling the deeply 5-cleft calyx. Stem: Hoary, slender, simple or branched, from 6 in. to 2 ft. high. Leaves: Thin, opposite; upper and more acute ones clasping the stem; lower, ovate ones on short petioles. Fruit: A round capsule to which the enlarged calyx adheres. Preferred Habitat - Moist meadows, woods, and thickets. Flowering Season - April-June. Distribution - Western New York and Pennsylvania to Wisconsin, Kentucky, and Indian Territory.

Next of kin to the great Paulonia tree, whose deliciously sweet, vanilla-scented, trumpet-shaped violet flowers are happily fast becoming as common here as in their native Japan, what has this fragile, odorless blossom of the meadows in common with it? Apparently nothing; but superficial appearances count for little or nothing among scientists, to whom the structure of floral organs is of prime importance; and analysis instantly shows the close relationship between these dissimilar-looking cousins. Even without analysis one can readily see that the monkey flower is not far removed.

Because few writers have arisen as yet in the newly settled regions of the middle West and Southwest, where blue-eyed Mary dyes acres of meadow land with her heavenly color, her praises are little sung in the books, but are loudly buzzed by myriads of bees that are her most devoted lovers. "I regard the flower as especially adapted to the early flying bees with abdominal collecting brushes for pollen - i.e., species of Osmia - and these bees," says Professor Robertson of Illinois, "although not the exclusive visitors, are far more abundant and important than all the other visitors together." For them are the brownish marks on the palate provided as pathfinders. At the pressure of their strong heads the palate yields to give them entrance, and at their removal it springs back to protect the pollen against the inroads of flies, mining bees, and beetles. As the longer stamens shed their pollen before the shorter ones mature theirs, bees must visit the flower several times to collect it all.

MONKEY-FLOWER
 (Minulus ringens) Figwort family

Flowers - Purple, violet, or lilac, rarely whitish; about 1 in. long, solitary, borne on slender footstems from axils of upper leaves. Calyx prismatic, 5-angled, 5-toothed; corolla irregular, tubular, narrow in throat, 2-lipped; upper lip 2-lobed, erect; under lip 3-lobed, spreading; 4 stamens, a long and a short pair, inserted on corolla tube; pistil with 2-lobed, plate-like stigma. Stem: Square, erect, usually branched, 1 to 3 ft. high. Leaves: Opposite, oblong to lance-shaped, saw-edged, mostly seated on stem. Preferred Habitat - Swamps, beside streams and ponds. Flowering Season - June-September. Distribution - Manitoba, Nebraska, and Texas, eastward to Atlantic Ocean.

No wader is the square-stemmed Monkey-flower whose grinning corolla peers at one from grassy tuffets in swamps, from the brookside, the springy soil of low meadows, and damp hollows beside the road; but moisture it must have to fill its nectary and to soften the ground for the easier transit of its creeping rootstock. Imaginative eyes see what appears to them the gaping (ringens) face of a little ape or buffoon (mimulus) in this common flower whose drolleries, such as they are, call forth the only applause desired - the buzz of insects that become pollen-laden during the entertainment.

Now the advanced stigma of this flower is peculiarly irritable, and closes up on contact with an incoming visitor's body, thus exposing the pollen-laden anthers behind it, and, except in rare cases, preventing self-fertilization. Delpino was the first to guess what advantage so sensitive a stigma might mean. Probably the smaller bees find the tube too long for their short tongues. The yellow palate, which partially guards the entrance to the nectary from pilferers, of course serves also as a pathfinder to the long-tongued bees.

AMERICAN BROOKLIME
(Veronica Americana) Figwort family

Flowers - Light blue to white, usually striped with deep blue or purple structure of flower similar to that of V. officinalis, but borne in long, loose racemes branching outward on stems that spring from axils of most of the leaves. Stem: Without hairs, usually branched, 6 in. to 3 ft. long, lying partly on ground and rooting from lower joints. Leaves: Oblong, lance-shaped, saw-edged, opposite, petioled, and lacking hairs; 1 to 3 in. long, 1/4 to 1 in. wide. Fruit: A nearly round, compressed, but not flat, capsule with flat seeds in 2 cells. Preferred Habitat - In brooks, ponds, ditches, swamps. Flowering Season - April-September. Distribution - From Atlantic to Pacific, Alaska to California and New Mexico, Quebec to Pennsylvania.

This, the perhaps most beautiful native speedwell, whose sheets of blue along the brookside are so frequently mistaken for masses of forget-me-nots by the hasty observer, of course shows marked differences on closer investigation; its tiny blue flowers are marked with purple pathfinders, and the plant is not hairy, to mention only two. But the poets of England are responsible for most of whatever confusion stills lurks in the popular mind concerning these two flowers. Speedwell, a common medieval benediction from a friend, equivalent to our farewell or adieu, and forget - me-not of similar intent, have been used interchangeably by some writers in connection with parting gifts of small blue flowers. It was the germander speedwell that in literature and botanies alike was most commonly known as the forget-me-not for over two hundred years, or until only fifty years ago. When the "Mayflower" and her sister ships were launched; "Speedwell" was considered a happier name for a vessel than it proved to be.

The WATER SPEEDWELL, or PIMPERNEL (V. Anagallis-aquatica), differs from the preceding chiefly in having most of its leaves seated on the stalk, only the lower ones possessing stems, and those short ones. In autumn the increased growth of sterile shoots from runners produce almost circular leaves, often two inches broad, a certain aid to identification.

Another close relation, the MARSH or SKULLCAP SPEEDWELL (V. scutellata), on the other hand, has long, very slender, acute leaves, their teeth far apart; and as these three species are the only members of their clan likely to be found in watery places within our limits, a close examination of the leaves of any water-loving plant bearing small four-lobed blue flowers, usually marked with lines of a deeper blue or purple, should enable one to correctly name the species. None of these blossoms can be carried far after being picked; they have a tantalizing habit of dropping off, leaving a bouquet of tiny green calices chiefly.

Many kinds of bees, wasps, flies, and butterflies fertilize all these little flowers, which are first staminate, then pistillate, simply by crawling over them in search of nectar.

COMMON SPEEDWELL; FLUELLIN; PAUL'S BETONY;
GROUND-HELE
(Veronica officinalis) Figwort family

Flowers - Pale blue, very small, crowded on spike-like racemes from axils of leaves, often from alternate axils. Calyx 4-parted; corolla of 4 lobes, lower lobe commonly narrowest ; 2 divergent stamens inserted at base and on either side of upper corolla lobe ; a knob-like stigma on solitary pistil. Stem: From 3 to 10 in. long, hairy, often prostrate, and rooting at joints. Leaves: Opposite, oblong, obtuse, saw-edged, narrowed at base. Fruit: Compressed heart-shaped capsule, containing numerous flat seeds. Preferred Habitat - Dry fields, uplands, open woods. Flowering Season - May-August. Distribution - From Michigan and Tennessee eastward, also from Ontario to Nova Scotia. Probably an immigrant from Europe and Asia.

An ancient tradition of the Roman Church relates that when Jesus was on His way to Calvary, He passed the home of a certain Jewish maiden, who, when she saw the drops of agony on His brow, ran after Him along the road to wipe His face with her kerchief. This linen, the monks declared, ever after bore the impress of the sacred features - vera iconica, the true likeness. When the Church wished to canonize the pitying maiden, an abbreviated form of the Latin words was given her, St. Veronica, and her kerchief became one of the most precious relics at St. Peter's, where it is said to be still preserved. Medieval flower lovers, whose piety seems to have been eclipsed only by their imaginations, named this little flower from a fancied resemblance to the relic. Of course, special healing virtue was attributed to the square of pictured linen, and since all could not go to Rome to be cured by it, naturally the next step was to employ the common, wayside plant that bore the saint's name. Mental healers will not be surprised to learn that because of the strong popular belief in its efficacy to cure all fleshly ills, it actually seemed to possess miraculous powers. For scrofula it was said to be the infallible remedy, and presently we find Linnaeus grouping this flower, and all its relatives under the family name of Scrofulariaceae. "What's in a name?" Religion, theology, medicine, folk-lore, metaphysics, what not?

One of the most common wild flowers in England is this same familiar little blossom of that lovely shade of blue known by Chinese artists as "the sky after rain." "The prettiest of all humble roadside flowers I saw," says Burroughs, in "A Glance at British Wild Flowers." "It is prettier than the violet, and larger and deeper colored than our houstonia. It is a small and delicate edition of our hepatica, done in indigo blue, and wonted to the grass in the fields and by the waysides.

'The little speedwell's darling blue'

sings Tennyson. I saw it blooming with the daisy and buttercup upon the grave of Carlyle. The tender human and poetic element of his stern, rocky nature was well expressed by it."

Only as it grows in masses is the speedwell conspicuous - a sufficient reason for its habit of forming colonies and of gathering its insignificant blossoms together into dense spikes, since by these methods it issues a flaunting advertisement of its nectar. The flower that simplifies dining for insects has its certain reward in rapidly increased and vigorous descendants. To save repetition, the reader interested in the process of fertilization is referred to the account of the Maryland figwort, since many members of the large family to which both belong employ the same method of economizing pollen and insuring fertile seed. In this case visitors have only to crawl over the tiny blossoms.

>From Labrador to Alaska, throughout almost every section of the United States, in South America, Europe, and Asia, roams the THYME-LEAVED SPEEDWELL (V. serpyllifolia), by the help of its numerous flat seeds, that are easily transported on the wind, and by its branching stem, that lies partly on the ground, rooting where the joints touch earth. The small oval leaves, barely half an inch long, grow in pairs. The tiny blue, or sometimes white, flowers, with dark pathfinders to the nectary, are borne on spike-like racemes at the ends of the stem and branches that rear themselves upward in fields and thickets to display their bloom before the passing bee.

PALE, or NAKED, or ONE-FLOWERED BROOM-RAPE (Thalesia uniflora; Aphyllon uniflorum of Gray) Broom-rape family

Flowers - Violet, rarely white, delicately fragrant, solitary at end of erect, glandular peduncles. Calyx hairy, bell-shaped, 5-toothed, not half the length of corolla, which is 1 in. or less long, with curved tube spreading into 2 lips, 5-lobed, yellow-bearded within; 4 stamens, in pairs, inserted on tube of corolla ; 1 pistil. Stem: About 1 in. long, scaly, often entirely underground; the 1 to 4 brownish scape-like peduncles, on which flowers are borne, from 3 to 8 in. high. Leaves: None. Fruit: An elongated, egg-shaped, 1-celled capsule containing numerous seeds. Preferred Habitat - Damp woods and thickets. Flowering Season - April-June. Distribution - British Possessions and United States from coast to coast, southward to Virginia, and Texas.

A curious, beautiful parasite, fastened on the roots of honest plants from which it draws its nourishment. The ancestors of this species, having deserted the path of rectitude ages ago to live by piracy, gradually lost the use of their leaves, upon which virtuous plants depend as upon a part of their digestive apparatus; they grew smaller and smaller, shriveled and dried, until now that the one-flowered broom-rape sucks its food, rendered already digestible through another's assimilation, no leaves remain on its brownish scapes. Disuse of any talent in the vegetable kingdom, as in the spiritual, leads to inevitable loss: "Unto every one which hath shall be given; and from him that hath not, even that he hath shall be taken away."

HAIRY RUELLIA
 (Ruellia ciliosa) Acanthus family

Flowers - Pale violet blue, showy, about 2 in. long, solitary or clustered in the axils or at the end of stem. Calyx of 5 bristle-shaped hairy segments; corolla with very slender tube expanding above in 5 nearly equal obtuse lobes; stamens 4; pistil with recurved style. Stem: Hairy, especially above, erect, 1 to 2 1/2 ft. high. Leaves: Opposite, oblong, narrowed at apex, entire, covered with soft white hairs. Preferred Habitat - Dry soil. Flowering Season - June-September. Distribution - New Jersey southward to the Gulf and westward to Michigan and Nebraska.

Many charming ruellias from the tropics adorn hothouses and window gardens in winter; but so far north as the New Jersey pine barrens, and westward where killing frosts occur, this perennial proves to be perfectly hardy. In addition to its showy blossoms, which so successfully invite insects to transfer their pollen, thereby counteracting the bad effects of close inbreeding, the plant bears inconspicuous cleistogamous or blind ones also. These look like arrested buds that never open; but, being fertilized with their own pollen, ripen abundant seed nevertheless.

One frequently finds holes bitten in these flowers, as in so many others long of tube or spur. Bumblebees, among the most intelligent and mischievous of insects, are apt to be the chief offenders; but wasps are guilty too, and the female carpenter bee, which ordinarily slits holes to extract nectar, has been detected in the act of removing circular pieces of the corolla from this ruellia with which to plug up a thimble-shaped tube in some decayed tree. Here she deposits an egg on top of a layer of baby food, consisting of a paste of pollen and nectar, and seals up the nursery with another bit of leaf or flower, repeating the process until the long tunnel is filled with eggs and food for larvae. Then she dies, leaving her entire race apparently extinct, and living only in embryo for months. This is the bee which commonly cuts her round plugs from rose leaves.

The SMOOTH RUELLIA (R. strepens), an earlier bloomer than the preceding, and with a more southerly range, has a shorter, thicker tube to its handsome blue flower, and lacks the hairs which guard its relative from crawling pilferers.

BLUETS; INNOCENCE; HOUSTONIA; QUAKER LADIES; QUAK-ER BONNETS;
VENUS' PRIDE
(Houstonia caerulea) Madder family

Flowers - Very small, light to purplish blue or white, with yellow center, and borne at end of each erect slender stem that rises from 3 to 7 in. high. Corolla funnel-shaped, with 4 oval, pointed, spreading lobes that equal the slender tube in length; rarely the corolla has more divisions; 4 stamens inserted on tube of corolla; 2 stigmas; calyx 4-lobed. Leaves: Opposite, seated on stem, oblong, tiny; the lower ones spatulate. Fruit: A 2-lobed pod, broader than long, its upper half free from calyx; seeds deeply concave. Root stock: Slender, spreading, forming dense tufts. Preferred Habitat - Moist meadows, wet rocks and banks. Flowering Season - April-July, or sparsely through summer. Distribution - Eastern Canada and United States west to Michigan, south to Georgia and Alabama.

Millions of these dainty wee flowers, scattered through the grass of moist meadows and by the wayside, reflect the blue and the serenity of heaven in their pure, upturned faces. Where the white variety grows, one might think a light snowfall had powdered the grass, or a milky way of tiny floral stars had streaked a terrestrial path. Linnaeus named the flower for Dr. Houston, a young English physician, botanist, and collector, who died in South America in 1733, after an exhausting tramp about the Gulf of Mexico.

To secure cross-fertilization, the object toward which so much marvelous floral organism is directed, this little plant puts forth two forms of blossoms - one with the stamens in the lower portion of the corolla tube, and the stigmas exserted; the other form with the stigmas below, and the stamens elevated to the mouth of the corolla. But the two kinds do not grow in the same patch, seed from either producing after its kind. Many insects visit these blossoms, but chiefly small bees and butterflies. Conspicuous among the latter is the common little meadow fritillary (Brenthis bellona), whose tawny, dark-speckled wings expand and close in apparent ecstasy as he tastes the tiny drop of nectar in each dainty enameled cup. Coming to feast with his tongue dusted from anthers nearest the nectary, he pollenizes the large stigmas of a short-styled blossom without touching its tall anthers. But it is evident that he could not be depended on to fertilize the long-styled form, with its smaller stigma, because of this ability to insert his slender tongue from the side where it avoids contact. Flies and beetles enter the blossoms, but small bees are best adapted as all-round benefactors. This simple-looking blossom, that measures barely half an inch across, is clever enough to multiply its lovely species a thousand fold, while many a larger, and therefore one might suppose a wiser, flower dwindles toward extinction.

John Burroughs found a single bluet in blossom one January, near Washington, when the clump of earth on which it grew was frozen solid. A pot of roots gathered in autumn and placed in a sunny window has sent up a little colony of star-like flowers throughout a winter.

WILD, COMMON, or CARD TEASEL; GYPSY COMBS
(Dipsacus sylvestris) Teasel family

Flowers - Purple or lilac, small, packed in dense, cylindric heads, 3 to 4 in. long; growing singly on ends of footstalks, the flowers set among stiffly pointed, slender scales. Calyx cup-shaped, 4-toothed. Corolla 4-lobed; stamens 4; leaves of involucre, slender, bristled, curved upward as high as flower-head or beyond. Stems: 3 to 6 ft. high, stout, branched, leafy, with numerous short prickles. Leaves:

Opposite, lance-shaped, seated on stem, with bristles along the stout midrib. Preferred Habitat - Roadsides and waste places. Flowering Season - July-September. Distribution - Maine to Virginia, westward to Ontario and the Mississippi. Europe and Asia.

Manufacturers find that no invention can equal the natural teasel head for raising a nap on woolen cloth, because it breaks at any serious obstruction, whereas a metal substitute, in such a case, tears the material. Accordingly, the plant is largely cultivated in the west of England, and quantities that have been imported from France and Germany may be seen in wagons on the way to the factories in any of the woolen-trade towns. After the flower-heads wither, the stems are cut about eight inches long, stripped of prickles, to provide a handle, and after drying, the natural tool is ready for use.

Bristling with armor, the teasel is not often attacked by browsing cattle. Occasionally even the upper leaf surfaces are dotted over with prickles enough to tear a tender tongue. This is a curious feature, for prickles usually grow out of veins. In the receptacle formed where the bases of the upper leaves grow together, rain and dew are found collected - a certain cure for warts, country people say. Venus' Cup, Bath, or Basin, and Water Thistle, are a few of the teasel's folk names earned by its curious little tank. In it many small insects are drowned, and these are supposed to contribute nourishment to the plant; for Mr. Francis Darwin has noted that protoplasmic filaments reach out into the liquid.

Owing to the stiff spines which radiate from the flower cluster, the bumblebees, which principally fertilize it, can reach the florets only with their heads, and not pollenize them by merely crawling over them as in the true compositae. But by first maturing its anthers, then when they have shed their pollen, elevating its stigmas, the teasel prevents self-fertilization.

HAREBELL or HAIRBELL; BLUE BELLS of SCOTLAND; LADY'S THIMBLE
 (Campanula rotundifolia) Bellflower family

Flowers - Bright blue or violet blue, bell-shaped, 1/2 in. long or over, drooping from hair-like stalks. Calyx of 5-pointed, narrow, spreading lobes; slender stamens alternate with lobes of corolla, and borne on summit of calyx tube, which is adherent to ovary; pistil with 3 stigmas in maturity only. Stem: Very slender, 6 in. to 3 ft. high, often several from same root; simple or branching. Leaves: Lower ones nearly round, usually withered and gone by flowering season; stem leaves narrow, pointed, seated on stem. Fruit: An egg-shaped, pendent, 3-celled capsule with short openings near base; seeds very numerous, tiny. Preferred Habitat - Moist rocks, uplands. Flowering Season - June-September. Distribution - Arctic regions of Europe, Asia, and America; southward on this continent, through Canada to New Jersey and Pennsylvania; westward to Nebraska, to Arizona in the Rockies, and to California in the Sierra Nevadas.

The inaccessible crevice of a precipice, moist rocks sprayed with the dashing waters of a lake or some tumbling mountain stream, wind-swept upland meadows, and shady places by the roadside may hold bright bunches of these hardy bells, swaying with exquisite grace on tremulous, hair-like stems that are fitted to withstand the fiercest mountain blasts, however frail they appear. How dainty, slender, tempting these little flowers are! One gladly risks a watery grave or broken bones to bring down a bunch from its aerial cranny.

It was a long stride forward in the evolutionary scale when the harebell welded its five once separate petals together; first at the base, then farther and farther up the sides, until a solid bell-shaped structure resulted. This arrangement which makes insect fertilization a more certain process because none of the pollen is lost through apertures, and because the visitor must enter the flower only at the vital point where the stigmas come in contact with his pollen-laden body, has given to all the flowers that have attained to it, marked ascendency.

Like most inverted blossoms, the harebell hangs its head to protect its nectar and pollen, not only from rain, but from the intrusion of undesirable crawling insects which would simply brush off its pollen in the grass before reaching the pistil of another flower, and so defeat cross-fertilization, the end and aim of so many blossoms.

Advertising for winged insects by its bright color, the harebell attracts bees, butterflies, and many others. These visitors cannot well walk on the upright petals, and sooner or later must clasp the pistil if they would secure the nectar secreted at the base. In doing so, they will dust themselves and the immature pistil with the pollen from the surrounding anthers; but a newly opened flower is incapable of fertilization. The pollen, although partially discharged in the unopened bud, is prevented from falling out by a coat of hairs on the upper part of the style. By the time all the pollen has been removed by visitors, however, and the stamens which matured early have withered, the pistil has grown longer, until it looks like the clapper in a bell; the stigma at its top has separated into three horizontal lobes which, being sticky on the under side, a pollen-laden insect on entering the bell must certainly brush against them and render them fertile. But bumblebees, its chief benefactors, and others may not have done their duty by the flower; what then? Why, the stigmas in that case finally bend backward to reach the left over pollen, and fertilize themselves, obviously the next best thing for them to do. How one's reverence increases when one begins to understand, be it ever so little of, the divine plan!

"Probably the most striking blue and purple wild flowers we have," says John Burroughs, "are of European origin. These colors, except with the fall asters and gentians, seem rather unstable in our flora." This theory is certainly borne out in the case of the RAMPION, EUROPEAN, or CREEPING BELLFLOWER (C. rapunculoides), now detected in the act of escaping from gardens from New Brunswick to Ontario, Southern New York, Pennsylvania, and Ohio, and making itself very much at home in our fields and along the waysides. Compared with the delicate little harebell, it is a plant of rank, rigid habit. Its erect, rather stout stem, set with elongated oval, hairy, alternate leaves, and crowned with a one-sided raceme of widely expanded, purple-blue bells rising about two feet above the ground, has little of the exquisite grace of its cousin. It blooms from July to September. This is the species whose roots are eaten by the omnivorous European peasant.

One of the few native campanulas, the TALL BELLFLOWER (C. Americana), waves long, slender wands studded with blue or sometimes whitish flowers high above the ground of moist thickets and woods throughout the eastern half of this country, but rarely near the sea. Doubtless the salt air, which intensifies the color of so many flowers, would brighten its rather slatey blue. The corolla, which is flat, round, about an inch across, and deeply cleft into five pointed petals, has the effect of a miniature pinwheel in motion. Mature flowers have the style elongated, bent downward, then curved upward, that the stigmas may certainly be in the way of the visiting insect pollen-laden from an earlier bloomer, and be cross-fertilized. The larger bees, its benefactors, which visit it for nectar, touch only the upper side of the style, on which they must alight; but the anthers waste pollen by shedding it on all sides. No insect can take shelter from rain or pass the night in this flower, as he frequently does in its more hospitable relative, the harebell. English gardeners, more appreciative than our own of our native flora, frequently utilize this charming plant in their rockwork, increasing their stock by a division of the dense, leafy rosettes.

VENUS' LOOKING-GLASS; CLASPING BELLFLOWER
  (Legouzia perfoliata; Specularia perfoliata of Gray)
Bellflower family

Flowers - Violet blue, from 1/2 to 3/4 in. across; solitary or 2 or 3 together, seated, in axils of upper leaves. Calyx lobes varying from 3 to 5 in earlier and later flowers, acute, rigid; corolla a 5-spoked wheel; 5 stamens; pistil with 3 stigmas. Stem: 6 in. to 2 ft. long, hairy, densely leafy, slender, weak. Leaves: Round, clasped about stem by heart-shaped base. Preferred Habitat - Sterile waste places, dry woods. Flowering Season - May-September. Distribution - From British Columbia, Oregon, and Mexico, east to Atlantic Ocean.

At the top of a gradually lengthened and apparently overburdened leafy stalk, weakly leaning upon surrounding vegetation, a few perfect blossoms spread their violet wheels, while below them insignificant earlier flowers, which, although they have never opened, nor reared their heads above the hollows of the little shell-like leaves where they lie secluded, have, nevertheless, been producing seed without imported pollen while their showy sisters slept. But the later blooms, by attracting insects, set cross-fertilized seed to counteract any evil tendencies that might weaken the species if it depended upon self-fertilization only. When the European Venus' looking-glass used to be cultivated in gardens here, our grandmothers tell us it was altogether too prolific, crowding out of existence its less fruitful, but more lovely, neighbors.

The SMALL VENUS' LOOKING-GLASS (L. biflora), of similar habit to the preceding, but with egg-shaped or oblong leaves seated on, not clasping, its smooth and very slender stem, grows in the South and westward to California.

GREAT LOBELIA; BLUE CARDINAL-FLOWER
(Lobelia syphilitica) Bellflower family

Flowers - Bright blue, touched with white, fading to pale blue, about 1 in. long, borne on tall, erect, leafy spike. Calyx 5-parted, the lobes sharply cut, hairy. Corolla tubular, open to base on one side, 2-lipped, irregularly 5-lobed, the petals pronounced at maturity only. Stamens 5, united by their hairy anthers into a tube around the style; larger anthers smooth. Stem: 1 to 3 ft. high, stout, simple, leafy, slightly hairy. Leaves: Alternate, oblong, tapering, pointed, irregularly toothed, 2 to 6 in. long, 1/2 to 2 in. wide. Preferred Habitat - Moist or wet soil; beside streams. Flowering Season - July-October Distribution - Ontario and northern United States west to Dakota, south to Kansas and Georgia.

To the evolutionist, ever on the lookout for connecting links, the lobelias form an interesting group, because their corolla, slit down the upper side and somewhat flattened, shows the beginning of the tendency toward the strap or ray flowers that are nearly confined to the composites of much later development, of course, than tubular single blossoms. Next to massing their flowers in showy heads, as the composites do, the lobelias have the almost equally advantageous plan of crowding theirs along a stem so as to make a conspicuous advertisement to attract the passing bee and to offer him the special inducement of numerous feeding places close together.

The handsome GREAT LOBELIA, constantly and invidiously compared with its gorgeous sister the cardinal flower, suffers unfairly. When asked what his favorite color was, Eugene Field replied: "Why, I like any color at all so long as it's red!" Most men, at least, agree with him, and certainly hummingbirds do; our scarcity of red flowers being due, we must believe, to the scarcity of hummingbirds, which chiefly fertilize them. But how bees love the blue blossoms!

There are many cases where the pistil of a flower necessarily comes in contact with its own pollen, yet fertilization does not take place, however improbable this may appear. Most orchids, for example, are not susceptible to their own pollen. It would seem as if our lobelia, in elevating its stigma through the ring formed by the united anthers, must come in contact with some of the pollen they have previously discharged from their tips, not only on the bumblebee that shakes it out of them when he jars the flower, but also within the tube. But when the anthers are mature, the two lobes of the still immature stigma are pressed together, and cannot be fertilized. Nevertheless, the hairy tips of some of the anthers brush off the pollen grains that may have lodged on the stigma as it passes through the ring in its ascent, thus making surety doubly sure. Only after the stigma projects beyond the ring of anthers does it expand its lobes, which are now ready to receive pollen brought from another later flower by the incoming bumblebee to which it is adapted.

Linnaeus named this group of plants for Matthias de l'Obel, a Flemish botanist, or herbalist more likely, who became physician to James I. of England.

Preferably in dry, sandy soil or in meadows, and over a wide range, the slender, straight shoots of PALE SPIKED LOBELIA (L. spicata) bloom early and throughout the summer months, the inflorescence itself sometimes reaching a height of two feet. At the base of the plant there is usually a tuft of broadly oblong leaves; those higher up narrow first into spoon-shaped, then into pointed, bracts, along the thick and gradually lengthened spike of scattered bloom. The flowers are oft en pale enough to be called white. Like their relatives, they first ripen their anthers to prevent self-fertilization.

The lithe, graceful little BROOK LOBELIA (L. Kalmii), whose light-blue flowers, at the end of thread-like footstems, form a loose raceme, sways with a company of its fellows among the grass on wet banks, beside meadow runnels and brooks, particularly in limestone soil, from Nova Scotia to the Northwest Territory and southward to New Jersey. It bears an insignificant capsule, not inflated like the Indian tobacco's; and long, narrow, spoon-shaped leaves. Twenty inches is the greatest height this little plant may hope to attain.

Not only beside water, and in it, but often totally immersed, grows the WATER LOBELIA or GLADIOLE (L. Dortmanna). The slender, hollow, smooth stem rises from a submerged tuft of round, hollow, fleshy leaves longitudinally divided by a partition, and bears at the top a scattered array of pale-blue flowers from August to September.

INDIAN or WILD TOBACCO; GAG-ROOT; ASTHMA-WEED; BLADDER-POD
LOBELIA
(Lobelia inflata) Bellflower family

Flowers - Pale blue or violet, small, borne at short intervals in spike-like leafy racemes. Calyx 5-parted, its awl-shaped lobes 1/4 in. long, or as long as the tubular, 2-lipped, 5-cleft, corolla that opens to base of tube on upper side. Stamens, 5 united by their hairy anthers into a ring around the 2-lobed style. Stem: From 1 to 3 feet high, hairy, very acrid, much branched, leafy. Leaves: Alternate, oblong or ovate, toothed, the upper ones acute, seated on stem; lower ones obtuse, petioled, to 2 1/2 in. long. Fruit: A much inflated, rounded, ribbed, many seeded capsule. Preferred Habitat - Dry fields and thickets; poor soil. Flowering Season - July-November. Distribution - Labrador westward to the Missouri River, south to Arkansas and Georgia.

The most stupid of the lower animals knows enough to let this poisonous, acrid plant alone; but not so man, who formerly made a quack medicine from it in the days when a drug that set one's internal organism on fire was supposed to be especially beneficial. One taste of the plant gives a realizing sense of its value as an emetic.

How the red man enjoyed smoking and chewing the bitter leaves, except for the drowsiness that followed, is a mystery.

On account of the smallness of its flowers and their scantiness, the Indian tobacco is perhaps the least attractive of the lobelias, none of which has so inflated a seed vessel, the distinguishing characteristic of this common plant.

CHICORY; SUCCORY; BLUE SAILORS; BUNK
(Cichorium Intybus) Chicory family

Flower-head - Bright, deep azure to gray blue, rarely pinkish or white, 1 to 1 1/2 in. broad, set close to stem, often in small clusters for nearly the entire length; each head a composite of ray flowers only, 5-toothed at upper edge, and set in a flat green receptacle. Stem: Rigid, branching, to 3 ft. high. Leaves: Lower ones spreading on ground, 3 to 6 in. long, spatulate, with deeply cut or irregular edges, narrowed into petioles, from a deep tap-root; upper leaves of stem and branches minute, bract-like. Preferred Habitat - Roadsides, waste places, fields. Flowering Season - July-October. Distribuition - Common in Eastern United States and Canada, south to the Carolinas; also sparingly westward to Nebraska.

At least the dried and ground root of this European invader is known to hosts of people who buy it undisguised or not, according as they count it an improvement to their coffee or a disagreeable adulterant. So great is the demand for chicory that, notwithstanding its cheapness, it is often in its turn adulterated with roasted wheat, rye, acorns, and carrots. Forced and blanched in a warm, dark place, the bitter leaves find a ready market as a salad known as "barbe de Capucin" by the fanciful French. Endive and dandelion, the chicory's relatives, appear on the table too, in spring, where people have learned the possibilities of salads, as they certainly have in Europe.

>From the depth to which the tap-root penetrates, it is not unlikely the succory derived its name from the Latin succurrere = to run under. The Arabic name chicourey testifies to the almost universal influence of Arabian physicians and writers in Europe after the Conquest. As chicoree, achicoria, chicoria, cicorea, chicorie, cichorei,

cikorie, tsikorei, and cicorie the plant is known respectively to the French, Spanish, Portuguese, Italians, Germans, Dutch, Swedes, Russians, and Danes.

On cloudy days or in the morning only throughout midsummer the "peasant posy" opens its "dear blue eyes"

"Where tired feet
Toil to and fro;
Where flaunting Sin
May see thy heavenly hue,
Or weary Sorrow look from thee
Toward a tenderer blue!"
- Margaret Deland.

In his "Humble Bee" Emerson, too, sees only beauty in the "Succory to match the sky;" but, mirabile dictu, Vergil, rarely caught in a prosaic, practical mood, wrote, "And spreading succ'ry chokes the rising field."

IRON-WEED; FLAT TOP
(Vernonia Noveboracensis) Thistle family.

Flower-head - Composite of tubular florets only, intense reddish-purple thistle-like heads, borne on short, branched peduncles and forming broad, flat clusters; bracts of involucre, brownish purple, tipped with awl-shaped bristles. Stem: 3 to 9 ft. high, rough or hairy, branched. Leaves: Alternate, narrowly oblong or lanceolate, saw-edged, 3 to 10 in. long, rough. Preferred Habitat - Moist soil, meadows, fields. Flowering Season - July-September. Distribution - Massachusetts to Georgia, and westward to the Mississippi.

Emerson says a weed is a plant whose virtues we have not yet discovered; but surely it is no small virtue in the iron-weed to brighten the roadsides and low meadows throughout the summer with bright clusters of bloom. When it is on the wane, the asters, for which it is sometimes mistaken, begin to appear, but an instant's comparison shows the difference between the two flowers. After noting the yellow disk in the center of an aster, it is not likely the iron-weed's thistle-like head of ray florets only will ever again be confused with it. Another rank-growing neighbor with which it has been confounded by the novice is the Joe Pye weed, a far paler, pinkish flower.

To each tiny floret, secreting nectar in its tube, many insects, attracted by the bright color of the iron-weed standing high above surrounding vegetation, come to feast. Long-lipped bees and flies rest awhile for refreshment, but butterflies of many beautiful kinds are by far the most abundant visitors. Pollen carried out by the long, hairy styles as they extend to maturity must attach itself to their tongues. The tiger swallow-tail butterfly appears to have a special preference for this flower. (See Self-Heal.)

COMMON or SCALY BLAZING STAR; COLIC-ROOT; RATTLE-
SNAKE MASTER;
BUTTON SNAKEROOT
  (Lacinaria squarrosa; Liatris squarrosa of Gray) Thistle
family

Flower-heads - Composite, about 1 in. long, bright purple or rose purple, of tubular florets only, from an involucre of overlapping, rigid, pointed bracts; each of the few flower-heads from the leaf axils along a slender stem in a wand-like raceme. Stem: 1/2 to 2 ft. high. Leaves: Alternate, narrow, entire. Preferred habitat - Dry, rich soil. Flowering Season - June-September. Distribution - Ontario to the Gulf of Mexico, westward to Nebraska.

Beginning at the top, the apparently fringed flower-heads open downward along the wand, whose length depends upon the richness of the soil. All of the flowers are perfect and attract long-tongued bees and flies (especially Exoprosopa fasciata) and butterflies, which, as they sip from the corolla tube, receive the pollen carried out and exposed on the long divisions of the style. Some people have pretended to cure rattlesnake bites with applications of the globular tuber of this and the next species.

The LARGE BUTTON SNAKEROOT, BLUE BLAZING STAR, or GAY FEATHER (L. scariosa), may attain six feet, but usually not more than half that height; and its round flower-heads normally stand well away from the stout stem on foot-stems of their own. The bristling scales of the involucre, often tinged with purple at the tips, are a conspicuous feature. With much the same range and choice of habitat as the last species, this Blazing Star is a later bloomer, coming into flower in August, and helping the goldenrods and asters brighten the landscape throughout the early autumn. The name of gay feather, miscellaneously applied to several blazing stars, is especially deserved by this showy beauty of the family.

Unlike others of its class, the DENSE BUTTON SNAKEROOT, DEVIL'S BIT, ROUGH or BACKACHE ROOT, PRAIRIE PINE or THROATWORT (L. spicata), the commonest species we have, chooses moist soil, even salt marshes near the coast, and low meadows throughout a range nearly corresponding with that of the scaly blazing star. Resembling its relatives in general manner of growth, we note that its oblong involucre, rounded at the base, has blunt, not sharply pointed, bracts; that the flower-heads are densely set close to the wand for from four to fifteen inches; that the five to thirteen bright rose-purple florets which compose each head occasionally come white; that its leaves are long and very narrow, and that October is not too late to find the plant in bloom.

BLUE and PURPLE ASTERS or STARWORTS
 Thistle family

Evolution teaches us that thistles, daisies, sunflowers, asters, and all the triumphant horde of composites were once very different flowers from what we see today. Through ages of natural selection of the fittest among their ancestral types, having finally arrived at the most successful adaptation of their various parts to their surroundings in the whole floral kingdom, they are now overrunning the earth. Doubtless the aster's remote ancestors were simple green leaves around the vital organs, and depended upon the wind, as the grasses do - a most extravagant method - to transfer their pollen. Then some rudimentary flower changed its outer row of stamens into petals, which gradually took on color to attract insects and insure a more economical method of transfer. Gardeners today take advantage of a blossom's natural tendency to change stamens into petals when they wish to produce double flowers. As flowers and insects developed side by side, and there came to be a better and better understanding between them of each other's requirements, mutual adaptation followed. The flower that offered the best advertisement, as the composites do, by its showy rays; that secreted nectar in tubular flowers where no useless insect could pilfer it; that fastened its stamens to the inside wall of the tube where they must dust with pollen the underside of every insect, unwittingly cross-fertilizing the blossom as he crawled over it; that massed a great number of these tubular florets together where insects might readily discover them and feast with the least possible loss of time - this flower became the winner in life's race. Small wonder that our June fields are white with daisies and the autumn landscape is glorified with goldenrod and asters!

Since North America boasts the greater part of the two hundred and fifty asters named by scientists, and as variations in many of our common species frequently occur, the tyro need expect no easy task in identifying every one he meets afield. However, the following are possible acquaintances to everyone:

In dry, shady places the LARGE or BROAD-LEAVED ASTER (A. macrophyllus), so called from its three or four conspicuous, heart-shaped leaves on long petioles, in a clump next the ground, may be more easily identified by these than by the pale lavender or violet flower-heads of about sixteen rays each which crown its reddish angular stem in August and September. The disk turns reddish brown.

In prairie soil, especially about the edges of woods in western New York, southward and westward to Texas and Minnesota, the beautiful SKY-BLUE ASTER (A. azureus) blooms from August till after frost. Its slender, stiff, rough stem branches above to display the numerous bright blue flowers, whose ten to twenty rays measure only about a quarter of an inch in length. The upper leaves are reduced to small flat bracts; the next are linear; and the lower ones, which approach a heart shape, are rough on both sides, and may be five or six inches long.

Much more branched and bushy is the COMMON BLUE, BRANCHING, WOOD, or HEART-LEAVED ASTER (A. cordifolius), whose generous masses of small, pale lavender flower-heads look like a mist hanging from one to five feet above the earth in and about the woods and shady roadsides from September even to December in favored places.

The WAVY or VARIOUS-LEAVED ASTER or SMALL FLEABANE (A. undulatus) has a stiff, rough, hairy, widely branching stalk, whose thick, rough lowest leaves are heart-shaped and set on long foot-stems; above these, the leaves have shorter stems, dilating where they clasp the stalk; the upper leaves, lacking stems, are seated on it, while those of the branches are shaped like tiny awls. The flowers, which measure less than an inch across, often grow along one side of an axis as well as in the usual raceme. Eight to fifteen pale blue to violet rays surround the disks which, yellow at first, become reddish brown in maturity. We find the plant in dry soil, blooming in September and October.

By no means tardy, the LATE PURPLE ASTER, so-called, or PURPLE DAISY (A. patens), begins to display its purplish-blue, daisy-like flower-heads early in August, and farther north may be found in dry, exposed places only until October. Rarely the solitary flowers, that are an inch across or more, are a deep, rich violet. The twenty to thirty rays which surround the disk, curling inward to dry, expose the vase-shaped, green, shingled cups that terminate each little branch. The thick, somewhat rigid, oblong leaves, tapering at the tip, broaden at the base to clasp the rough, slender stalk. Range similar to the next species.

Certainly from Massachusetts, northern New York, and Minnesota southward to the Gulf of Mexico one may expect to find the NEW ENGLAND ASTER or STARWORT (A. Novae-Angliae) one of the most striking and widely distributed of the tribe, in spite of its local name. It is not unknown in Canada. The branching clusters of violet or magenta-purple flower-heads, from one to two inches across - composites containing as many as forty to fifty purple ray florets around a multitude of perfect five-lobed, tubular, yellow disk florets in a sticky cup - shine out with royal splendor above the swamps, moist fields, and roadsides from August to October. The stout, bristle-hairy stem bears a quantity of alternate lance-shaped leaves lobed at the base where they clasp it.

In even wetter ground we find the RED-STALKED, PURPLE-STEMMED, or EARLY PURPLE ASTER, COCASH, SWANWEED, or MEADOW SCABISH (A. puniceus) blooming as early as July or as late as November. Its stout, rigid stem, bristling with rigid hairs, may reach a height of eight feet to display the branching clusters of pale violet or lavender flowers. The long, blade-like leaves, usually very rough above and hairy along the midrib beneath, are seated on the stem. The lovely SMOOTH or BLUE ASTER (A. laevis), whose sky-blue or violet flower-heads, about one inch broad, are common through September and October in dry soil and open woods, has strongly clasping, oblong, tapering leaves, rough margined, but rarely with a saw-tooth, toward the top of the stem, while those low down on it gradually narrow into clasping wings.

In dry, sandy soil, mostly near the coast, from Massachusetts to Delaware, grows one of the loveliest of all this beautiful clan, the LOW, SHOWY, or SEASIDE PURPLE ASTER (A. spectabilis). The stiff, usually unbranched stem does its best in attaining a height of two feet. Above, the leaves are blade-like or narrowly oblong, seated on the stem, whereas the tapering, oval basal leaves are furnished with long footstems, as is customary with most asters. The handsome, bright, violet-purple flower-heads, measuring about an inch and a half across, have from fifteen to thirty rays, or only about half as many as the familiar New England aster. Season August to November.

The low-growing BOG ASTER (A. nemoralis), not to be confused with the much taller Red-stalked species often found growing in the same swamp, and having, like it, flower-heads measuring about an inch and a half across, has rays that vary from light violet purple to rose pink. Its oblong to lance-shaped leaves, only two inches long at best, taper to a point at both ends, and are seated on the stem. We look for this aster in sandy bogs from New Jersey northward and westward during August and September.

The STIFF or SAVORY-LEAVED ASTER, SANDPAPER, or PINE STARWORT (Ionactis linariifolius), now separated from the other asters into a genus by itself, is a low, branching little plant with no basal leaves, but some that are very narrow and blade-like, rigid, entire and one-nerved, ascending the stiff stems. The leaves along the branches are minute and awl-shaped, like those on a branch of pine. Only from ten to fifteen violet ray flowers (pistillate) surround the perfect disk florets. From Quebec to the Gulf of Mexico, and westward beyond the Mississippi this prim little shrub grows in tufts on dry or rocky soil, and blooms from July to October.

ROBIN'S, or POOR ROBIN'S, or ROBERT'S PLANTAIN; BLUE
SPRING
DAISY; DAISY-LEAVED FLEABANE
  (Erigeron pulchellus; E. bellifolium of Gray) Thistle family

Flower-heads - Composite, daisy-like, 1 to 1/2 in. across; the outer circle of about 50 pale bluish-violet ray florets; the disk florets greenish yellow. Stem: Simple, erect, hairy, juicy, flexible, from 10 in. to 2 ft. high, producing runners and offsets from base. Leaves: Spatulate, in a flat tuft about the root; stem leaves narrow, more acute, seated, or partly clasping. Preferred Habitat - Moist ground, hills, banks, grassy fields. Flowering Season - April-June. Distribution - United States and Canada, east of the Mississippi.

Like an aster blooming long before its season, Robin's plantain wears a finely cut lavender fringe around a yellow disk of minute florets; but one of the first, not the last, in the long procession of composites has appeared when we see gay companies of these flowers nodding their heads above the grass in the spring breezes as if they were village gossips.

Doubtless it was the necessity for attracting insects which led the Robin's plantain and other composites to group a quantity of minute florets, each one of which was once an independent, detached blossom, into a common head. In union there is strength. Each floret still contains, however, its own tiny drop of nectar, its own stamens, its own pistil connected with embryonic seed below; therefore, when an insect alights where he can get the greatest amount of nectar for the least effort, and turns round and round to exhaust each nectary, he is sure to dust the pistils with pollen, and so fertilize an entire flower-head in a trice. The lavender fringe and the hairy involucre and stem serve the end of discouraging crawling insects, which cannot transfer pollen from plant to plant, from pilfering sweets that cannot be properly paid for. Small wonder that, although the composites have attained to their socialistic practices at a comparatively recent day as evolutionists count time, they have become as individuals and as species the most numerous in the world; the thistle family, dominant everywhere, containing not less than ten thousand members.

COMMON or PHILADELPHIA FLEABANE, or SKEVISH (E. Philadelphicus), a smaller edition of Robin's plantain, with a more findely cut fringe, its reddish-purple ray florets often numbering one hundred and fifty, may be found in low fields and woods throughout North America, except in the circumpolar regions.

THISTLES
(Carduus) Thistle family

Is land fulfilling the primal curse because it brings forth thistles? So thinks the farmer, no doubt, but not the goldfinches which daintily feed among the fluffy seeds, nor the bees, nor the "painted lady," which may be seen in all parts of the world where thistles grow, hovering about the beautiful rose-purple flowers. In the prickly cradle of leaves, the caterpillar of this thistle butterfly weaves a web around its main food store.

When the Danes invaded Scotland, they stole a silent night march upon the Scottish camp by marching barefoot; but a Dane inadvertently stepped on a thistle, and his sudden, sharp cry, arousing the sleeping Scots, saved them and their country: hence the Scotch emblem.

>From July to November blooms the COMMON, BURR, SPEAR, PLUME, BANK, HORSE, BULL, BLUE, BUTTON, BELL, or ROADSIDE THISTLE (C. lanceolatus or Circium lanceolatum of Gray), a native of Europe and Asia, now a most thoroughly naturalized American from Newfoundland to Georgia, westward to Nebraska. Its violet flower-heads, about an inch and a half across, and as high

as wide, are mostly solitary at the ends of formidable branches, up which few crawling creatures venture. But in the deep tube of each floret there is nectar secreted for the flying visitor who can properly transfer pollen from flower to flower. Such a one suffers no inconvenience from the prickles, but, on the contrary, finds a larger feast saved for him because of them. Dense, matted, wool-like hairs, that cover the bristling stems of most thistles, make climbing mighty unpleasant for ants, which ever delight in pilfering sweets. Perhaps one has the temerity to start upward.

"Fain would I climb, yet fear to fall."
"If thy heart fail thee, climb not at all,"

might be the ant's passionate outburst to the thistle, and the thistle's reply, instead of a Sir Walter and Queen Elizabeth couplet. Long, lance-shaped, deeply cleft, sharply pointed, and prickly dark green leaves make the ascent almost unendurable; nevertheless the ant bravely mounts to where the bristle-pointed, overlapping scales of the deep green cup hold the luscious flowers. Now his feet becoming entangled in the cottony fibers wound about the scaly armor, and a bristling bodyguard thrusting spears at him in his struggles to escape, death happily releases him. All this tragedy to insure the thistle's cross-fertilized seed that, seated on the autumn winds, shall be blown far and wide in quest of happy conditions for the offspring!

Sometimes the PASTURE or FRAGRANT THISTLE (C. odoratus or C. pumilum of Gray) still further protects its beautiful, odorous purple or whitish flower-head, that often measures three inches across, with a formidable array of prickly small leaves just below it. In case a would-be pilferer breaks through these lines, however, there is a slight glutinous strip on the outside of the bracts that compose the cup wherein the nectar-filled florets are packed; and here, in sight of Mecca, he meets his death, just as a bird is caught on limed twigs. The pasture thistle, whose range is only from Maine to Delaware, blooms from July to September.

Even gentle Professor Gray hurls anathema at the CANADA THISTLE; "a vile pest" he calls it. As CURSED, CORN, HARD, and CREEPING THISTLE it is variously known here and in Europe, whence it came to overrun our land from Newfoundland to Virginia, westward to Nebraska. By horizontal rootstocks it creeps and forms patches almost impossible to eradicate. The small reddish-purple flower-heads, barely an inch across, usually contain about a hundred florets each. In their tubes the abundant nectar rises high, so that numerous insects, even with the shortest tongues, are able to enjoy it. Not only bees and butterflies, but wasps, flies, and beetles feast diligently. When a floret opens, a quantity of pollen emerges at the upper end of the anther cylinder, pressed up by the growing style. Owing to their slight stickiness and the sharp processes over their entire surface, the pollen grains, which readily cling to the hairs of insects, are transported to the two-branched, hairy stigma of an older floret. But even should insects not visit the flower (and in fine weather they swarm about it), it is marvelously adapted to fertilize itself. Farmers may well despair of exterminating a plant so perfectly equipped in every part; to win life's battles.

"The colour of purple…was, amongst the ancients, typical of royalty. It was a kind of red richly shot with blue, and the dye producing it was attained from a shell found in considerable numbers off the coast of Tyre, and on the shore near the site of that ancient city, great heaps of such shells are still to be found. The production of the true royal purple dye was a very costly affair, and therefore it was often imitated with a mixture of cochineal and indigo…" - J. JAMES TISSOT.

As many so-called purple flowers are more strictly magenta, the reader is referred to the next group if he has not found the flower for which he is in search here. Also to the "White and Greenish" section since many colored flowers show a tendency to revert to the white type from which, doubtless, all were evolved. He should remember that all flowers are more or less variable in shade, according to varying conditions.

MAGENTA TO PINK FLOWERS

"Botany is a sequel of murder and a chronicle of the dead." - JULIAN HAWTHORNE.

"A plant is not to be studied as an absolutely dead thing, but rather as a sentient being…. To measure petals, to count stamens, to describe pistils without reference to their functions, or the why and wherefore of their existence, is to content one's self with husks in the presence of a feast of fatness - to listen to the rattle of dry bones rather than the heavenly harmonies of life. We have reason to be profoundly thankful for the signs to be seen on every side, that the dreary stuff which was called botany in the teaching of the past will soon cease to masquerade in its stolen costume, and that our children and our children's children will study not dried specimens or drier books, but the living things which Nature furnishes in such profusion….

"The reason of this radical change is not far to seek. Since man has learned that the universal brotherhood of life includes himself as the highest link in the chain of organic creation, his interest in all things that live and move and have a being has greatly increased. The movements of the monad now appeal to him in a way that was impossible under the old conceptions. He sees in each of the millions of living forms with which the earth is teeming, the action of many of the laws which are operating in himself; and has learned that to a great extent his welfare is dependent on these seemingly insignificant relations; that in ways undreamed of a century ago they affect human progress." - CLARENCE MOORES WEED.

MAGENTA TO PINK FLOWERS
SESSILE-LEAVED TWISTED-STALK
  (Streptopus roseus) Lily-of-the-Valley family

Flowers - Dull, purplish pink, 1/2 in. long or less, solitary, on threadlike, curved footstalks longer than the small flower itself, nodding from leaf-axils. Perianth bill-shaped, of 6 spreading segments; stamens 6, 2-horned; style spreading into 3 branches, stigmatic on inner side. Stem: 1 to 2 1/2 ft. high, simple or forked. Leaves: Thin, alternate, green on both sides, many nerved, tapering at end, rounded at base, where they are seated on stem. Fruit: A round, red, many-seeded berry. Preferred Habitat - Moist woods. Flowering Season - May-July. Distribution - North America east and west, southward to Georgia and Oregon.

As we look down on this graceful plant, no blossoms are visible; but if we bend the zig-zagged stem backward, we shall discover the little rosy bells swaying from the base of the leaves on curved footstalks (streptos = twisted, pous = a foot or stalk) very much as the plant's relatives the Solomon's seals grow. In the confident expectation of having its seeds dropped far and wide, it bears showy red berries in August for the birds now wandering through the woods with increased, hungry families.

The CLASPING-LEAVED TWISTED-STALK (S. amplexifolius), which has one or two greenish-white bells nodding from its axils, may be distinguished when not in flower by its leaves, which are hoary - not green - on the under side, or by its oval berry. Indeed most plants living in wet soil have a coating of down on the under sides of their leaves to prevent the pores from clogging with rising vapors.

MOCCASIN FLOWER; PINK, VENUS', or STEMLESS LADY'S SLIPPER
(Cypripedium acaule) Orchid family

Flowers - Fragrant, solitary, large, showy, drooping from end of scape, 6 to 12 in. high. Sepals lance-shaped, spreading, greenish purple, 2 in. long or less; petals narrower and longer than sepals. Lip an inflated sac, often over 2 in. long, slit down the middle, and folded inwardly above, pale magenta, veined with darker pink upper part of interior crested with long white hairs. Stamens united with style into unsymmetrical declined column, bearing an anther on either side, and a dilated triangular petal-like sterile stamen above, arching over the broad concave stigma. Leaves: 2, from the base; elliptic, thick, 6 to 8 in. long. Preferred Habitat - Deep, rocky, or sandy woods. Flowering Season - May-June. Distribution - Canada southward to North Carolina, westward to Minnesota and Kentucky.

Because most people cannot forbear picking this exquisite flower that seems too beautiful to be found outside a millionaire's hothouse, it is becoming rarer every year, until the finding of one in the deep forest, where it must now hide, has become the event of a day's walk. Once it was the commonest of the orchids.

"Cross-fertilization," says Darwin, "results in offspring which vanquish the offspring of self-fertilization in the struggle for existence." This has been the motto of the orchid family for ages. No group of plants has taken more elaborate precautions against self-pollination or developed more elaborate and ingenious mechanism to compel insects to transfer their pollen than this.

The fissure down the front of the pink lady's slipper is not so wide but that a bee must use some force to push against its elastic sloping sides and enter the large banquet chamber where he finds generous entertainment secreted among the fine white hairs in the upper part. Presently he has feasted enough. Now one can hear him buzzing about inside, trying to find a way out of the trap. Toward the two little gleams of light through apertures at the end of a passage beyond the nectary hairs, he at length finds his way. Narrower and narrower grows the passage until it would seem as if he could never struggle through; nor can he until his back has rubbed along the sticky, overhanging stigma, which is furnished with minute, rigid, sharply pointed papillae, all directed forward, and placed there for the express purpose of combing out the pollen he has brought from another flower on his back or head. The imported pollen having been safely removed, he still has to struggle on toward freedom through one of the narrow openings, where an anther almost blocks his way.

As he works outward, this anther, drawn downward on its hinge, plasters his back with yellow granular pollen as a parting gift, and away he flies to another lady's slipper to have it combed out by the sticky stigma as described above. The smallest bees can squeeze through the passage without paying toll. To those of the Andrena and Halictus tribe the flower is evidently best adapted. Sometimes the largest bumblebees, either unable or unwilling to get out by the legitimate route, bite their way to liberty. Mutilated sacs are not uncommon. But when unable to get out by fair means, and too bewildered to escape by foul, the large bee must sometimes perish miserably in his gorgeous prison.

SHOWY, GAY, or SPRING ORCHIS
(Orchis spectabilis) Orchid family

Flowers - Purplish pink, of deeper and lighter shade, the lower lip white, and thick of texture; from 3 to 6 on a spike; fragrant. Sepals pointed, united, arching above the converging petals, and resembling a hood; lip large, spreading, prolonged into a spur, which is largest at the tip and as long as the twisted footstem. Sterm: 4 to 12 in. high, thick, fleshy, 5-sided. Leaves: 2 large, broadly ovate, glossy green, silvery on under side, rising from a few scales from root. Fruit: A sharply angled capsule, 1 in. long. Preferred Habitat - Rich, moist woods, especially under hemlocks. Flowering Season - April-June. Distribution - From New Brunswick and Ontario southward to our Southern States, westward to Nebraska.

Of the six floral leaves which every orchid, terrestrial or aerial, possesses, one is always peculiar in form, pouch-shaped, or a cornucopia filled with nectar, or a flaunted, fringed banner, or a broad platform for the insect visitors to alight on. Some orchids look to imaginative eyes as if they were masquerading in the disguise of bees, moths, frogs, birds, butterflies. A number of these queer freaks are to be found in Europe. Spring traps, adhesive plasters, and hair-triggers attached to explosive shells of pollen are among the many devices by which orchids compel insects to cross-fertilize them, these flowers as a family showing the most marvelous mechanism adapted to their requirements from insects in the whole floral kingdom. No other blossoms can so well afford to wear magenta, the ugliest shade nature produces, the "lovely rosy purple" of Dutch bulb growers, a color that has an unpleasant effect on not a few American stomachs outside of Hoboken.

But an orchid, from the amazing cleverness of its operations, is attractive under any circumstances to whomever understands it. This earliest member of the family to appear charms the female bumble-bee, to whose anatomy it is especially adapted. The males, whose faces are hairy where the females' are bare, and therefore not calculated to retain the sticky pollen masses, are not yet flying when the showy orchis blooms. Bombus Americanorum, which can drain the longest spurs, B. separatus, B. terricola, and, rarely, butterflies as well, have been caught with its pollen masses attached. The bee alights on the projecting lip, pushes her head into the mouth of the corolla, and, as she sips the nectar from the horn of plenty, ruptures by the slight pressure a membrane of the pouch where two sticky buttons, to which two pollen masses are attached, lie imbedded. Instantly after contact these adhere to the round bare spots on her face, the viscid cement hardening before her head is fairly withdrawn. Now the diverging pollen masses, that look like antennae, fall from the perpendicular, by remarkable power of contraction, to a horizontal attitude, that they may be in the precise position to fertilize the stigma of the next flower visited - just as if they possessed a reasoning intelligence! Even after all the pollen has been deposited on the sticky stigmas of various blossoms, stump-like caudicles to which the two little sacs were attached have been found still plastered on a long-suffering bee. But so rich in nectar are the moisture-loving orchids that, to obtain a draught, the sticky plasters which she must carry do not seem too dear a price to pay. In this showy orchis the nectar often rises an eighth of an inch in the tube, and sufficient pressure to cause a rupture will eject it a foot.

ROSE or SWEET POGONIA; SNAKE-MOUTH
  (Pogonia ophioglossoides) Orchid family

Flowers - Pale rose pink, fragrant, about 1 in. long, usually solitary at end of stem 8 to 15 in. high, and subtended by a leaf-like bract. Sepals and petals equal, oval, about 1/2 in. long, the lip spoon-shaped, crested, and fringed. Column shorter than petals, thick, club-shaped. Anther terminal, attached to back of column, pollen mass in each of its 2 sacs. Stigma a flattened disk below anther. Leaves: 1 to 3, erect, lance-oblong, sometimes one with long footstem from fibrous root. Preferred Habitat - Swamps and low meadows. Flowering Season - June-July. Distribution - Canada to Florida, westward to Kansas.

Rearing its head above the low sedges, often brightened with colonies of the grass pink at the same time, this shy recluse of the swamps woos the passing bee with lovely color, a fragrance like fresh red raspberries, an alluring alighting place all fringed and crested, and with the prospect of hospitable entertainment in the nectary beyond. So in she goes, between the platform and the column overhead, pushing first her head, then brushing her back against the stigma just below the end of the thick column that almost closes the passage. Any powdery pollen she brought on her back from another pogonia must now be brushed off against the sticky stigma. Her feast ended, out she backs. And now a wonderful thing happens. The lid of the anther which is at the end of the column, catching in her shoulders, swings outward on its elastic hinge, releasing a little shower of golden dust, which she must carry on the hairs of her head or back until the sticky stigma of the next pogonia entered kindly wipes it off! This is one of the few orchids whose pollen, usually found in masses, is not united by threads. Without the bee's aid in releasing it from its little box, the lovely species would quickly perish from the face of the earth.

ARETHUSA; INDIAN PINK
(Arethusa bulbosa) Orchid family

Flowers - 1 to 2 in. long, bright purple pink, solitary, violet scented, rising from between a pair of small scales at end of smooth scape from 5 to 10 in. high. Lip dropping beneath sepals and petals, broad, rounded, toothed, or fringed, blotched with purple, and with three hairy ridges down its surface. Leaf: Solitary, hidden at first, coming after the flower, but attaining length of 6 in. Root: Bulbous. Fruit: A 6-ribbed capsule, 1 in. long, rarely maturing. Preferred Habitat - Northern bogs and swamps. Flowering Season - May-June.

Distribution - From North Carolina and Indiana northward to the Fur Countries.

One flower to a plant, and that one rarely maturing seed; a temptingly beautiful prize which few refrain from carrying home, to have it wither on the way pursued by that more persistent lover than Alpheus, the orchid-hunter who exports the bulbs to European collectors - little wonder this exquisite orchid is rare, and that from certain of those cranberry bogs of Eastern New England, which it formerly brightened with its vivid pink, it has now gone forever. Like Arethusa, the nymph whom Diana changed into a fountain that she might escape from the infatuated river god, Linnaeus fancied this flower a maiden in the midst of a spring bubbling from wet places where presumably none may follow her.

But the bee, our Arethusa's devoted lover, although no villain, still pursues her. He knows that moisture-loving plants secrete the most nectar. When the head of the bee enters the flower to sip, nothing happens; but as he raises his head to depart, it cannot help lifting the lid of the helmet-shaped anther and so letting fall a few soft pellets of pollen on it. Now, after he has drained the next arethusa, his pollen-laden head must rub against the long sticky stigma before it touches the helmet-like anther lid and precipitates another volley of pollen. In some such manner most of our orchids compel insects to work for them in preventing self-fertilization.

Another charming, but much smaller, orchid, that we must don our rubber boots to find where it hides in cool, peaty bogs from Canada and the Northern United States to California, and southward in the Rockies to Arizona, is the CALYPSO (Calypso bulbosa). It is a solitary little flower, standing out from the top of a jointed scape that never rises more than six inches from the solid bulb, hidden in the moss, nor boasts more than one nearly round leaf near its base. The blossom itself suggests one of the lady's slipper orchids, with its rosy purple, narrow, pointed sepals and petals clustered at the top above a large, sac-shaped, whitish lip. The latter is divided into two parts, heavily blotched with cinnamon brown, and woolly with a patch of yellow hairs near the point of the division. May - June.

CALOPOGON; GRASS PINK (Limodorum tuberosum; Calopogon pulchellus of Gray) Orchid family

Flowers - Purplish pink, 1 in. long, 3 to 15 around a long, loose spike. Sepals and petals similar, oval, acute; the lip on upper side of flower is broad at the summit, tapering into a claw, flexible as if hinged, densely bearded on its face with white, yellow, and magenta hairs (Calopogon = beautiful beard). Column below lip (ovary not twisted in this exceptional case); sticky stigma at summit of column, and just below it a 2-celled anther, each cell containing 2 pollen masses, the grain lightly connected by threads. Scape: 1 to 1 1/2 ft. high, slender, naked. Leaf: Solitary, long, grass-like, from a round bulb arising from bulb of previous year. Preferred Habitat - Swamps, cranberry bogs, and low meadows. Flowering Season - June-July. Distribution - Newfoundland to Florida, and westward to the Mississippi.

Fortunately this lovely orchid, one of the most interesting of its highly organized family, is far from rare, and where we find the rose pogonia and other bog-loving relatives growing, the calopogon usually outnumbers them all. Limodorum translated reads meadow-gift; but we find the flower less frequently in grassy places than those who have waded into its favorite haunts could wish.

Owing to the crested lip being oddly situated on the upper part of the flower, which appears to be growing upside down in consequence, one might suppose a visiting insect would not choose to alight on it. The pretty club-shaped, vari-colored hairs, which he may mistake for stamens, and which keep his feet from slipping, irresistibly invite him there, however, when, presto! down drops the fringed lip with startling suddenness. Of course, the bee strikes his back against the column when he falls. Now, there are two slightly upturned little wings on either side of the column, which keep his body from slipping off at either side and necessitate its exit from the end where the stigma smears it with viscid matter. The pressure of the insect on this part starts the pollen masses from their pocket just below; and as the bee slides off the end of the column, the exposed, cobwebby threads to which the pollen grains are attached cling to his sticky body. The sticky substance instantly hardening, the pollen masses, which are drawn out from their pocket as he escapes, are cemented to his abdomen in the precise spot where they must strike against the stigma of the next calopogon he tumbles in; hence cross-fertilization results. What recompense does the bee get for such rough handling? None at all, so far as is known. The flower, which secretes no nectar, is doubtless one of those gay deceivers that Sprengel named "Scheinsaftblumen," only it leads its visitors to look for pollen instead of nectar, on the supposition that the club-shaped hairs on the crests are stamens. The wonder is that the intelligent little bees (a species of Andrenidae), which chiefly are its Victims, have not yet learned to boycott it.

"Calopogon," says Professor Robertson, who knows more about the fertilization of American wild flowers by insects than most writers, "is one of a few flowers which move the insect toward the stigma…. There is no expenditure in keeping up a supply of nectar, and the flower, although requiring a smooth insect of a certain size and weight, suffers nothing from the visits of those it cannot utilize. Then, there is no delay caused by the insect waiting to suck; but as soon as it alights it is thrown down against the stigma. This occurs so quickly that, while standing net in hand, I have seen insects effect pollination and escape before I could catch them. So many orchids fasten their pollinia upon the faces and tongues of insects that it is interesting to find one which applies them regularly to the first abdominal segment. Mr. Darwin has observed that absence of hair on the tongues of Lepidoptera (butterflies and moths) and on the faces of Hymenoptera (bees; wasps, etc.) has led to the more usual adaptations, and sparseness of hair has its influence in this case. Species of Augochlora are the only insects on which I found pollinia. These bees are very smooth, depending for ornament on the metallic sheen of their bodies. An Halictus repeatedly pulled down the labella (lips) of flowers from which pollinia had not been removed; and the only reason I can assign for its failure to extract pollinia is that it is more hairy than the Augochlora.

COMMON PERSICARIA, PINK KNOTWEED, or JOINT-WEED; SMARTWEED
  (Polygonum Pennsylvanicum) Buckwheat family

Flowers - Very small, pink, collected in terminal, dense, narrow, obtuse spikes, 1 to 2 in. long. Calyx pink or greenish, 5-parted, like petals; no corolla; stamens 8 or less; style 2-parted. Stem: 1 to 3 ft. high, simple or branched, often partly red, the joints swollen and sheathed; the branches above, and peduncles glandular. Leaves: Oblong, lance-shaped, entire edged, 2 to 11 in. long, with stout midrib, sharply tapering at tip, rounded into short petioles below. Preferred Habitat - Waste places, roadsides, moist soil. Flowering Season - July-October. Distribution - Nova Scotia to the Gulf of Mexico; westward to Texas and Minnesota.

Everywhere we meet this commonest of plants or some of its similar kin, the erect pink spikes brightening roadsides, rubbish heaps, fields, and waste places, from midsummer to frost. The little flowers, which open without method anywhere on the spike they choose, attract many insects, the smaller bees (Andrena) conspicuous among the host. As the spreading divisions of the perianth make nectar-stealing all too easy for ants and other crawlers that would not come in contact with anthers and stigma where they enter a flower near its base, most buckwheat plants whose blossoms secrete sweets protect themselves from theft by coating the upper stems with glandular hairs that effectually discourage the pilferers. Shortly after fertilization, the little rounded, flat-sided fruit begins to form inside the persistent pink calyx. At any time the spike-like racemes contain more bright pink buds and shining seeds than flowers. Familiarity alone breeds contempt for this plant, that certainly possesses much beauty.

The LADY'S THUMB (P. Persicaria), often a troublesome weed, roams over the whole of North America, except at the extreme north - another illustration of the riotous profusion of European floral immigrants rejoicing in the easier struggle for existence here. Its pink spikes are shorter and less slender than those of the preceding taller, but similar species, and its leaves, which are nearly seated on the stem, have dark triangular or lunar marks near the center in the majority of cases.

An insignificant little plant, found all over our continent, Europe, and Asia, is the familiar KNOT-GRASS or DOORWEED (P. aviculare), often trailing its leafy, jointed stems over the ground, but at times weakly erect, to display its tiny greenish or white pink-edged flowers, clustered in the axils of oblong, bluish-green leaves that are considerably less than an inch long. Although in bloom from June to October, insects seldom visit it, for it secretes very little, if any, nectar. As might be expected in such a case, its stem is smooth.

When the amphibious WATER PERSICARIA (P. amphibium) lifts its short, dense, rose-colored ovoid or oblong club of bloom above ponds and lakes, it is sufficiently protected from crawling pilferers, of course, by the water in which it grows. But suppose the pond dries up and the plant is left on dry ground, what then? Now, a remarkable thing happens: protective glandular, sticky hairs appear on the epidermis of the leaves and stems, which were perfectly smooth when the flowers grew in water. Such small wingless insects as might pilfer nectar without bringing to their hostess any pollen from other blossoms are held as fast as on bird-lime. The stem, which sometimes floats, sometimes is immersed, may attain a length of twenty feet; the rounded, elliptic, petioled leaves may be four inches long or only half that size. From Quebec to New Jersey, and westward to the Pacific, the solitary, showy inflorescence, which does well to attain a height of an inch, may be found during July and August.

Throughout the summer, narrow, terminal, erect, spike-like racemes of small, pale pink, flesh-colored, or greenish flowers are sent upward by the MILD WATER PEPPER (P. hydropiperoides). It is like a slender, pale variety of the common pink persicaria. One finds its inconspicuous, but very common, flowers from June to September. The plant, which grows in shallow water, swamps, and moist places throughout the Union and considerably north and south of it, rises three feet or less. The cylindric sheaths around the swollen joints of the stem are fringed with long bristles - a clue to identification. Another similar WATER PEPPER or SMARTWEED (P. hydropiper) is so called because of its acrid, biting juice.

The CLIMBING FALSE BUCKWHEAT (P. scandens) straggles over bushes in woods, thickets, and by the waysides throughout a very wide range; yet its small, dull, greenish-yellow and pinkish flowers, loosely clustered in long pedicelled racemes, are so inconspicuous during August and September, when the showy composites are in their glory, that we give them scarcely a glance. The alternate leaves, which are heart-shaped at the base and pointed at the lip, suggesting those of the morning glory, are on petioles arising from sheaths over the enlarged joints which, in this family, are always a most prominent characteristic - (Poly = many, gonum = a knee). The three outer sepals, keeled when in flower, are irregularly winged when the three-angled, smooth achene hangs from the matured blossom in autumn, the season at which the vine assumes its greatest attractiveness.

The ARROW-LEAVED TEAR THUMB (P. sagittatum), found in ditches and swampy wet soil, weakly leans on other plants, or climbs over them with the help of the many sharp, recurved prickles which arm its four-angled stem. Even the petioles and underside of the leaf's midrib are set with prickles. The light green leaves, that combine the lance and the arrow shapes, take on a beautiful russet-red tint in autumn. The little, five-parted rose-colored or greenish-white flowers grow in small, close terminal heads from July to September from Nova Scotia to the Gulf and far westward.

SEASIDE or COAST JOINTWEED or KNOT-GRASS (Polygonella articulata; Polygonum articulatum of Gray) a low, slender, wiry, diffusely spreading little plant, with thread-like leaves seated on its much-jointed stem, rises cleanly from out the sand of the coast from Maine to Florida, and the shores of the Great Lakes. Very slender racemes of tiny, nodding, rose-tinted white flowers, with a dark midrib to each of the five calyx segments, are insignificant of themselves; but when seen in masses, from July to October, they tinge the upper beaches and sandy meadows with a pink blush that not a few artists have transferred to the foreground of their marine pictures.

CORN COCKLE; CORN ROSE; CORN or RED CAMPION;
CROWN-OF-THE-FIELD
  (Agrostemma Githago; Lychnis Githago of Gray) Pink family

  Flowers - Magenta or bright purplish crimson, to 3 in. broad, solitary at end of long, stout footstem; 5 lobes of calyx leaf-like, very long and narrow, exceeding petals. Corolla of 5 broad, rounded petals; 10 stamens; 5 styles alternating with calyx lobes, opposite petals. Stem: 1 to 3 ft. high, erect, with few or no branches, leafy, the plant covered with fine white hairs. Leaves: Opposite, seated on stem, long, narrow, pointed, erect. Fruit: a 1-celled, many-seeded capsule. Preferred Habitat - Wheat and other grain fields; dry, waste places. Flowering Season - July-September. Distribution - United States at large; most common in Central and Western States. Also in Europe and Asia.

"Allons! allons! sow'd cockle, reap'd no corn," exclaims Biron in "Love's Labor Lost." Evidently the farmers even in Shakespeare's day counted this brilliant blossom the pest it has become in many of our own grain fields just as it was in ancient times, when Job, after solemnly protesting his righteousness, called on his own land to bear record against him if his words were false. "Let thistles grow instead of wheat, and cockle instead of barley," he cried, according to James the First's translators; but the "noisome weeds" of the original text seem to indicate that these good men were more anxious to give the English people an adequate conception of Job's willingness to suffer for his honor's sake than to translate literally. Possibly the cockle grew in Southern Asia in Job's time : today its range is north.

Like many another immigrant to our hospitable shores, this vigorous invader shows a tendency to outstrip native blossoms in life's race. Having won in the struggle for survival in the old country, where the contest has been most fiercely waged for centuries, it finds life here easy, enjoyable. What are its methods for insuring an abundance of fertile seed? We see that the tube of the flower is so nearly closed by the stamens and five-styled pistil as to be adapted only to the long, slender tongues of moths and butterflies, for which benefactors it became narrow and deep to reserve the nectar. "A certain night-flying moth (one of the Dianthaecia) fertilizes flowers of this genus exclusively, and its larvae feed on their unripe seeds as a staple. Bees and some long-tongued flies seen about the corn cockle doubtless get pollen only; but there are few flowers so deep that the longest-tongued bees cannot sip them. Butterflies, attracted by the bright color of the flower - and to them color is the most catchy of advertisements - are guided by a few dark lines on the petals to the nectary.

Soon after the blossom opens, five of the stamens emerge from the tube and shed their pollen on the early visitor. Later, the five other stamens empty the contents of their anthers on more tardy comers. Finally, when all danger of self-fertilization is past, the styles stretch upward, and the butterfly, whose head is dusted with pollen brought from earlier flowers, necessarily leaves some on their sticky surfaces as he takes the leavings in the nectary.

So much cross-fertilized seed as the plant now produces and scatters through the grain fields may well fill the farmer's prosaic mind with despair. To him there is no glory in the scarlet of the poppy comparable with the glitter of a silver dollar; no charm in the heavenly blue of the corn-flower, that likewise preys upon the fertility of his soil; the vivid flecks of color with which the cockle lights up his fields mean only loss of productiveness in the earth that would yield him greater profit without them. Moreover, seeds of this so-called weed not only darken his wheat when they are threshed out together, but are positively injurious if swallowed in any quantity. Emerson said every plant is called a weed until its usefulness is discovered. Linnaeus called this flower Agrostemma = the crown-of-the-field. Agriculturalists never realize that beauty is in itself a sufficient plea for respected existence. Not a few of the cockle's relatives adorn men's gardens.

WILD PINK or CATCHFLY
(Silene Caroliniana; S. Pennsylvanica of Gray) Pink family

Flowers - Rose pink, deep or very pale; about inch broad, on slender footstalks, in terminal clusters. Calyx tubular, 5-toothed, much enlarged in fruit, sticky; 5 petals with claws enclosed in calyx, wedged-shaped above, slightly notched. Stamens 10; pistil with 3 styles. Stem: 4 to 10 in. high, hairy, sticky above, growing in tufts. Leaves: Basal ones spatulate; 2 or 3 pairs of lance-shaped, smaller leaves seated on stem. Preferred Habitat - Dry, gravelly, sandy, or rocky soil. Flowering Season - April-June. Distribution - New England, south to Georgia, westward to Kentucky.

Fresh, dainty, and innocent-looking as Spring herself are these bright flowers. Alas, for the tiny creatures that try to climb up the rosy tufts to pilfer nectar, they and their relatives are not so innocent as they appear! While the little crawlers are almost within reach of the cup of sweets, their feet are gummed to the viscid matter that coats it, and here their struggles end as flies' do on sticky fly-paper, or birds' on limed twigs. A naturalist counted sixty-two little corpses on the sticky stem of a single pink. All this tragedy to protect a little nectar for the butterflies which, in sipping it, transfer the pollen from one flower to another, and so help them to produce the most beautiful and robust offspring.

The pink, which has two sets of stamens of five each, elevates first one set, then the other, for economy's sake and to run less risk of failure to get its pollen transferred in case of rain when its friends are not flying. After all the golden dust has been shed, however, up come the three recurved styles from the depth of the tube to receive pollen brought by butterflies from younger flowers. There are few cups so deep that the largest bumblebees cannot suck them. Flies which feed on the pink's pollen only, sometimes come by mistake to older blossoms in the stigmatic stage, and doubtless cross-fertilize them once in a while.

In waste places and woods farther southward and westward, and throughout the range of the Wild Pink as well, clusters of the SLEEPY CATCHFLY (S. antirrhina) open their tiny pink flowers for a short time only in the sunshine. At any stage they are mostly calyx, but in fruit this part is much expanded. Swollen, sticky joints are the plant's means of defense from crawlers. Season: Summer.

When moths begin their rounds at dusk, the NIGHT-FLOWERING CATCHFLY (S. noctiflora) opens its pinkish or white flowers to emit a fragrance that guides them to a feast prepared for them alone. Day-blooming catchflies have no perfume, nor do they need it; their color and markings are a sufficient guide to the butterflies. Sticky hairs along the stems of this plant ruthlessly destroy, not flies, but ants chiefly, that would pilfer nectar without being able to render the flower any service. Yet the calyx is beautifully veined, as if to tantalize the crawlers by indicating the path to a banquet hail they may never reach. Only a very few flowers, an inch across or less, are clustered at the top of the plant, which blooms from July to September in waste places east of the Mississippi and in Canada.

SOAPWORT; BOUNCING BET; HEDGE PINK; BRUISEWORT; OLD MAID'S PINK; FULLER'S HERB
(Saponaria officinalis) Pink family

Flowers - Pink or whitish, fragrant, about 1 inch broad, loosely clustered at end of stem, also sparingly from axils of upper leaves. Calyx tubular, 5-toothed, about 3/4 in. long; 5 petals, the claws inserted in deep tube. Stamens 10, in 2 sets; 1 pistil with 2 styles. Flowers frequently double. Stem: to 2 ft. high, erect, stout, sparingly branched, leafy. Leaves: Opposite, acutely oval, 2 to 3 in. long, about 1 in. wide, 3 to 5 ribbed. Fruit: An oblong capsule, shorter than calyx, opening at top by 4 short teeth or valves. Preferred Habitat - Roadsides, banks, and waste places. Flowering Season - June-September. Distribution - Generally common. Naturalized from Europe.

A stout, buxom, exuberantly healthy lassie among flowers is bouncing Bet, who long ago escaped from gardens whither she was brought from Europe, and ran wild beyond colonial farms to roadsides, along which she has traveled over nearly our entire area. Underground runners and abundant seed soon form thrifty colonies. This plant, to which our grandmothers ascribed healing virtues, makes a cleansing, soap-like lather when its bruised leaves are agitated in water.

Butterflies, which delight in bright colors and distinct markings, find little to charm them here; but the pale shade of pink or white, easily distinguished in the dark, and the fragrance, strongest after sunset, effectively advertise the flower at dusk when its benefactors begin to fly. The sphinx moth, a frequent visitor, works as rapidly in extracting nectar from the deep tube as any hawk moth, so frequently mistaken for a hummingbird. The little cliff-dwelling bees (Halictus), among others, visit the flowers by day for pollen only. At first five outer stamens protrude slightly from the flower and shed their pollen on the visitor, immediately over the entrance. Afterward, having spread apart to leave the entrance free, the path is clear for the five inner stamens to follow the same course. Now the styles are still enclosed in the tube but when there is no longer fear of self-fertilization - that is to say, when the pollen has all been carried off, and the stamens have withered - up they come and spread apart to expose their rough upper surfaces to pollen brought from younger flowers by the moths.

DEPTFORD PINK
 (Dianthus Armeria) Pink family

Flowers - Pink, with whitish dots, small, borne in small clusters at end of stem. Calyx tubular, 5-toothed, with several bract-like leaves at base; 5 petals with toothed edges, clawed at base within deep calyx; 10 stamens; 1 pistil with 2 styles. Stem: 6 to 18 in. high, stiff, erect, finely hairy, few branches. Leaves: Opposite, blade-shaped, or lower ones rounded at end. Preferred Habitat - Fields, roadsides. Flowering Season - June-September. Distribution - Southern Ontario, New England, south to Maryland, west to Michigan.

The true pinks of Europe, among which are the SWEET WILLIAM or BUNCH PINK (D. barbatus) of our gardens, occasionally wild here, and the deliciously spicy CLOVE PINK (D. Carophyllus), ancestor of the superb carnations of the present day, that have reached a climax in the Lawson pink of newspaper fame, were once held sacred to Jupiter, hence Dianthus = Jove's own flower. The Deptford pink, a rather insignificant little European immigrant, without fragrance, has a decided charm, nevertheless, when seen in bright patches among the dry grass of early autumn, with small butterflies, that are its devoted admirers, hovering above.

PINK OR PALE CORYDALIS (Capnoides sempervirens; Corydalis glauca of Gray) Poppy family

Flowers - Pink, with yellow tip, about 1/2 in. long, a few borne in a loose, terminal raceme. Calyx of 2 small sepals; corolla irregular, of 4 erect, closed, and flattened petals joined, 1 of outer pair with short rounded spur at base, the interior ones narrow and keeled on back. Stamens 6, in 2 sets, Opposite outer petals; 1 pistil. Stem: Smooth, curved, branched, 1 to 2 feet high. Leaves: Pale grayish green, delicate, divided into variously and finely cut leaflets. Fruit: Very narrow, erect pod, 1 to 2 in. long. Preferred Habitat - Rocky, rich, cool woods. Flowering Season - April-September. Distribution - Nova Scotia westward to Alaska, south to Minnesota and North Carolina.

Dainty little pink sacs, yellow at the mouth, hang upside down along a graceful stem, and instantly suggest the Dutchman's breeches, squirrel corn, bleeding heart, and climbing fumitory, to which the plant is next of kin. Because the lark (Korydalos) has a spur, the flower, which boasts a small one also, borrows its Greek name.

Hildebrand proved by patient experiments that some flowers of this genus have not only lost the power of self-fertilization, but that they produce fertile seed only when pollen from another plant is carried to them. Yet how difficult they make dining for their bene-factors! The bumblebee, which can reach the nectar, but not lap it conveniently, often "gets square" with the secretive blossom by nipping holes through its spur, to which the hive bees and others hasten for refreshment. We frequently find these punctured flowers. But hive and other bees visiting the blossom for pollen, some rubs off against their breast when they depress the two middle petals, a sort of sheath that contains pistil and stamens.

HARDHACK; STEEPLE BUSH
 (Spiraea tomentosa) Rose family

Flowers - Pink or magenta, rarely white, very small, in dense, py-ramidal clusters. Calyx of 5 sepals; corolla of 5 rounded petals; sta-mens, 20 to 60; usually 5 pistils, downy. Stem: 2 to 3 ft. high, erect, shrubby, simple, downy. Leaves: Dark green above, covered with whitish woolly hairs beneath; oval, saw-edged, 1 to 2 in. long. Preferred Habitat - Low moist ground, roadside ditches, swamps. Flowering Season - July-September. Distribution - Nova Scotia westward, and southward to Georgia and Kansas.

These bright spires of pink bloom attract our attention no less than the countless eyes of flies, beetles, and bees, ever on the lookout for food to be eaten on the spot or stored up for future progeny. Pollen-feeding insects such as these, delight in the spireas, most of which secrete little or no nectar, but yield an abundance of pollen, which they can gather from the crowded panicles with little loss of time, transferring some of it to the pistils, of course, as they move over the tiny blossoms. But most spireas are also able to fertilize themselves, insects failing them.

An instant's comparison shows the steeple bush to be closely related to the fleecy, white meadow-sweet, often found growing near. The pink spires, which bloom from the top downward, have pale brown tips where the withered flowers are, toward the end of summer.

Why is the under side of the leaves so woolly? Not as a protection against wingless insects crawling upward, that is certain; for such could only benefit these tiny clustered flowers. Not against the sun's rays, for it is only the under surface that is coated. When the upper leaf surface is hairy, we know that the plant is protected in this way from perspiring too freely. Doubtless these leaves of the steeple bush, like those of other plants that choose a similar habitat, have woolly hairs beneath as an absorbent to protect their pores from clogging with the vapors that must rise from the damp ground where the plant grows. If these pores were filled with moisture from without, how could they possibly throw off the waste of the plant? All plants are largely dependent upon free perspiration for health, but especially those whose roots, struck in wet ground, are constantly sending up moisture through the stem and leaves.

PURPLE-FLOWERING OR VIRGINIA RASPBERRY
 (Rubus odoratus) Rose family

Flowers - Royal purple or bluish pink, showy, fragrant, 1 to 2 in. broad, loosely clustered at top of stem. Calyx sticky-hairy, deeply 5-parted, with long pointed tips; corolla of 5 rounded petals; stamens and pistils very numerous. Stem: 3 to 5 ft. high, erect, branched, shrubby, bristly, not prickly. Leaves: Alternate, petioled, 3 to 5 lobed, middle lobe largest, and all pointed; saw-edged lower leaves immense. Fruit: A depressed red berry, scarcely edible. Preferred Habitat - Rocky woods, dells, shady roadsides. Flowering Season - June-August. Distribution - Northern Canada south to Georgia, westward to Michigan and Tennessee.

To be an unappreciated, unloved relative of the exquisite wild rose, with which this flower is so often likened, must be a similar misfortune to being the untalented son of a great man, or the un-happy author of a successful first book never equaled in later attempts. But where the bright blossoms of the Virginia raspberry burst forth above the roadside tangle and shady woodland dells, even those who despise magenta see beauty in them where abundant green tones all discordant notes into harmony. Purple, as we of today understand the color, the flower is not; but rather the purple of ancient Orientals. On cool, cloudy days the petals are a deep, clear purplish rose, that soon fades and dulls with age, or changes into pale, bluish pink when the sun is hot.

Many yellow stamens help conceal the nectar secreted in a narrow ring between the filaments and the base of the receptacle. Bumblebees, the principal and most efficient visitors, which can reach sweets more readily than most insects, although numerous others help to self-fertilize the flower, bring to the mature stigmas of a newly opened blossom pollen carried on their undersides from the anthers of a flower a day or two older. When the inner row of anthers shed their pollen, some doubtless falls on the stigmas below them, and so spontaneous self-fertilization may occur. Fruit sets quickly; nevertheless the shrub keeps on flowering nearly all summer. Children often fold the lower leaves, which sometimes measure a foot across, to make drinking-cups.

QUEEN-OF-THE-PRAIRIE
 (Ulmaria rubra; Spirea lobata of Gray) Rose family

 Flowers - Deep pink, like the peach blossom, fragrant, about 1/3
in. across, clustered in large cymose panicles on a long footstalk.
Calyx 5-lobed; 5-clawed, rose-like petals; stamens numerous; pistils
5 to 15, usually 10. Stem: 2 to 8 ft. tall, smooth, grooved, branched.
Leaves: Mostly near the ground, large, rarely measuring 3 ft. long,
compounded of from 3 to 7 leaflets; end leaflet, of 7 to 9 divisions,
much the largest; side leaflets opposite, seated on stem, 3 to 5 lobed
or parted; all lobes acute, and edges unequally incised. Prominent
kidney-shaped stipules. Preferred Habitat - Moist meadows and
prairies. Flowering Season - June-July. Distribution - Western Pennsylvania to Michigan and Iowa, and southward.

 A stately, beautiful native plant, seen to perfection where it rears
bright panicles of bloom above the ranker growth in the low moist
meadows of the Ohio Valley. When we find it in the East, it has only
recently escaped from man's gardens into Nature's. Butterflies and
bees pay grateful homage to this queen. Indeed, butterflies appear
to have a special fondness for pink, as bees have for blue flowers.
Cattle delight to chew the leaves, which, when crushed, give out a
fragrance like sweet birch.

WILD ROSES
 (Rosa) Rose family

 Just as many members of the lily tribe show a preference for the
rule of three in the arrangements of their floral parts, so the wild
roses cling to the quinary method of some primitive ancestor, a
favorite one also with the buttercup and many of its kin, the geraniums, mallows, and various others. Most of our fruit trees and bushes are near relatives of the rose. Five petals and five sepals, then, we
always find on roses in a state of nature; and although the progres-
sive gardener of today has nowhere shown his skill more than in
the development of a multitude of petals from stamens in the magnificent roses of fashionable society, the most highly cultivated dar-

ling of the greenhouses quickly reverts to the original wild type, setting his work of years at naught, if once it regain its natural liberties through neglect.

To protect its foliage from being eaten by hungry cattle, the rose goes armed into the battle of life with curved, sharp prickles, not true thorns or modified branches, but merely surface appliances which peel off with the bark. To destroy crawling pilferers of pollen, several species coat their calices, at least, with fine hairs or sticky gum; and to insure wide distribution of offspring, the seeds are packed in the attractive, bright red calyx tube or hip, a favorite food of many birds, which drop them miles away. When shall we ever learn that not even a hair has been added to or taken from a blossom without a lawful cause, and study it accordingly? Fragrance, abundant pollen, and bright-colored petals naturally attract many insects; but roses secrete no nectar. Some species of bees, and a common beetle (Trichius piger) for example, seem to depend upon certain wild roses exclusively for pollen to feed themselves and their larvae. Bumblebees, to which roses are adapted, require a firmer support than the petals would give, and so alight on the center of the flower, where the pistil receives pollen carried by them from other roses. Although the numerous stamens and the pistils mature simultaneously, the former are usually turned outward, that the incoming pollen-laden insect may strike the stigma first. When the large bees cease their visits as they may in long-continued dull or rainy weather, the rose, turning toward the sun, stands more or less obliquely, and some of the pollen must fall on its stigma. Occasional self-fertilization matters little.

If plants have insect benefactors, they have their foes as well and hordes of tiny aphids, commonly known as green flies or plant lice, moored by their sucking tubes to the tender sprays of roses, wild and cultivated, live by extracting their juices. A curious relationship exists between these little creatures and the ants, which "milk" them by stroking and caressing them with their antennae until they emit a tiny drop of sweet, white fluid. The yellow ant, that lives an almost subterranean life, actually domesticates flocks and herds of root-feeding aphids; the brown ant appropriates those that live among the bark of trees; and the common black garden ant (Lasius niger), devoting itself to the aphis of the rose bushes, protects it in extraordinary ways, delightfully described by the author of "Ants, Bees, and Wasps."

In literature, ancient and modern, sacred and profane, no flower figures so conspicuously as the rose. To the Romans it was most significant when placed over the door of a public or private banquet hall. Each who passed beneath it bound himself thereby not to disclose anything said or done within; hence the expression sub rosa, common to this day.

The PRAIRIE, CLIMBING, or MICHIGAN ROSE (R. setigera) lifts clusters of deep, bright pink flowers, that after a while fade almost white, above the thickets and rich prairie soil, from southern Ontario and Wisconsin to the Gulf, as far eastward as Florida. Its distinguishing characteristics are: Stout, widely separated prickles along the stem, that grows several feet long; leaves compounded of three, rarely five, oval leaflets, acute or obtuse at the apex; stalks and calyx often glandular; odorless flowers that, opening in June and July, measure about two and a half inches across, their styles cohering in a smooth column on which bees are tempted to alight; and a round hip, or seed vessel, formed by the fruiting calyx, which is more or less glandular. From this parent stock several valuable double-flowering roses have been derived, among others the Queen and the Gem of the Prairies, but it is our only native rose that has ever passed into cultivation.

The SMOOTH, EARLY, or MEADOW ROSE (R. blanda), found blooming in June and July in moist, rocky places from Newfoundland to New Jersey and a thousand miles westward, has a trifle larger and slightly fragrant flowers, at first pink, later pure white. Their styles are separate, not cohering in a column nor projecting as in the climbing rose. This is a leafy, low bush mostly less than three feet high; it is either entirely unarmed, or else provided with only a few weak prickles; the stipules are rather broad, and the leaf is compounded of from five to seven oval, blunt, and pale green leaflets, often hoary below.

In swamps and low wet ground from Quebec to Florida, and westward to the Mississippi, the SWAMP ROSE (R. Carolina) blooms late in May and on to midsummer. The bush may grow taller than a man, or perhaps only a foot high. It is armed with stout, hooked, rather distant prickles, and few or no bristles. The leaflets, from five to nine, but usually seven, to a leaf, are smooth, pale, or perhaps hairy beneath to protect the pores from filling with moisture arising from the wet ground. Long, sharp calyx lobes, which drop off before the cup swells in fruit into a round, glandular, hairy red hip, are conspicuous among the clustered pink flowers and buds.

Surely no description of our COMMON, LOW, DWARF, or PASTURE ROSE (R. humilis; R. lucida of Gray) is needed. One's acquaintance with flowers must be limited indeed, if it does not include this most abundant of all the wild roses from Ontario to Georgia, and westward to Wisconsin. In light, dry, or rocky soil we find the exquisite, but usually solitary, blossom late in May until July, and, like most roses, it has the pleasant practice of putting forth a stray blossom or two in early autumn. The stamens of this species are turned outward so strongly that self- pollination must very rarely take place.

Among the following charming wild roses, not natives, but naturalized immigrants from foreign lands, that have escaped from gardens, is Shakespeare's CANKER-BLOOM, the lovely DOG ROSE or WILD BRIER (R. canina), that spreads its long, straggling branches along the roadsides and banks, covering the waste lands with its smooth, beautiful foliage, and in June and July with pink or white roses. Because it lacks the fragrance of sweetbrier, which it otherwise closely resembles, it has been branded with the dog prefix as a mark of contempt. Professor Koch says that long before it was customary to surround gardens with walls, men had rose hedges. "Each of the four great peoples of Asia," he continues, "possessed its own variety of rose, and carried it during all wanderings, until finally all four became the common property of the four peoples. The great Indo-Germanic stock chose the 'hundred-leaved' and RED ROSE (R. Gallica); nevertheless, after the Niebelungen the common dog rose played an important part among the ancient Germans. The DAMASCUS ROSE (R. Damascena), which blooms twice a year, as well as the MUSK ROSE (R. moschata), were cherished by the Semitic or Arabic stock; while the Turkish-Mongolian people planted by preference the YELLOW ROSE (R. lutea). Eastern Asia (China and Japan) is the fatherland of the INDIAN and TEA ROSES."

How fragrant are the pages of Chaucer, Spenser, and Shakespeare with the Eglantine! This delicious plant, known here as SWEET-BRIAR (R. rubIginosa), emits its very aromatic odor from russet glands on the under, downy side of the small leaflets, always a certain means of identification. From eastern Canada to Virginia and Tennessee the plant has happily escaped from man's gardens back to Nature's.

In spite of its American Indian name, the lovely white CHERO-KEE ROSE (R. Sinica), that runs wild in the South, climbing, rambling and rioting with a truly Oriental abandon and luxurance, did indeed come from China. Would that our northern thickets and roadsides might be decked with its pure flowers and almost equally beautiful dark, glossy, evergreen leaves!

COMMON RED, PURPLE, MEADOW, or HONEYSUCKLE CLOVER
(Trifolium pratense) Pea family

Flowers - Magenta, pink, or rarely whitish, sweet-scented, the tubular corollas set in dense round, oval, or egg-shaped heads about 1 in. long, and seated in a sparingly hairy calyx. Stem: 6 in. to 2 ft. high, branching, reclining, or erect, more or less hairy. Leaves: On long petioles, commonly compounded of 3, but sometimes of 4 to 11 oval or oblong leaflets, marked with white crescent, often dark-spotted near center; stipules egg-shaped, sharply pointed, strongly veined, over 1/2 in. long. Preferred Habitat - Fields, meadows, roadsides. Flowering Season - April-November. Distribution - Common throughout Canada and United States.

Meadows bright with clover-heads among the grasses, daisies, and buttercups in June resound with the murmur of unwearying industry and rapturous enjoyment. Bumblebees by the tens of thousands buzzing above acres of the farmer's clover blossoms should be happy in a knowledge of their benefactions, which doubtless concern them not at all. They have never heard the story of the Australians who imported quantities of clover for fodder, and had glorious fields of it that season, but not a seed to plant next year's crops, simply because the farmers had failed to import the bumble-bee. After her immigration the clovers multiplied prodigiously. No; the bee's happiness rests on her knowledge that only the butterflies' long tongues can honestly share with her the brimming wells of nectar in each tiny floret. Children who have sucked them too appreciate her rapture. If we examine a little flower under the magnifying glass, we shall see why its structure places it in the pea family. Bumblebees so depress the keel either when they sip, or feed on pollen, that their heads and tongues get well dusted with the yellow powder, which they transfer to the stigmas of other flowers; whereas the butterflies are of doubtful value, if not injurious, since their long, slender tongues easily drain the nectar without depressing the keel. Even if a few grains of pollen should cling to their tongues, it would probably be wiped off as they withdrew them through the narrow slit, where the petals nearly meet, at the mouth of the flower. Bombus terrestris delights in nipping holes at the base of the tube, which other pilferers also profit by. Our country is so much richer in butterflies than Europe, it is scarcely surprising that Professor Robertson found thirteen Lepidoptera out of twenty insect visitors to this clover in Illinois, whereas Muller caught only eight butterflies on it out of a list of thirty-nine visitors in Germany. The fritillaries and the sulphurs are always seen about the clover fields among many others, and the "dusky wings" and the caterpillar of several species feeds almost exclusively on this plant.

"To live in clover," from the insect's point of view at least, may well mean a life of luxury and affluence. Most peasants in Europe will tell you that a dream about the flower foretells not only a happy marriage, but long life and prosperity. For ages the clover has been counted a mystic plant, and all sorts of good and bad luck were said to attend the finding of variations of its leaves which had more than the common number of leaflets. At evening these leaflets fold downward, the side ones like two hands clasped in prayer, the end one bowed over them. In this fashion the leaves of the white and other clovers also go to sleep, to protect their sensitive surfaces from cold by radiation, it is thought.

The ZIG-ZAG CLOVER, COW or MARL-GRASS (T. Medium), a native of Europe and Asia, now naturalized in the eastern half of the United States and Canada, may scarcely be told from the common red clover, except by its crooked, angular stems - often provokingly straight - by its unspotted leaves, and the short peduncle in which its heads are elevated above the calyx.

Farmers here are beginning to learn the value of the beautiful CRIMSON, CARNATION or ITALIAN CLOVER or NAPOLEONS (T. incarnatum), and happily there are many fields and waste places in the East already harboring the brilliant runaways. The narrow heads may be two and a half inches long. A meadow of this fodder plant makes one envious of the very cattle that may spend the summer day wading through acres of its deep bright bloom.

GOAT'S RUE; CAT-GUT; HOARY PEA or WILD SWEET PEA
(Cracca Virginiana; Tephrosia Virginiana of Gray) Pea family

Flowers - In terminal cluster, each 1/2 in. long or over, butterfly-shaped, consisting of greenish, cream-yellow standard, purplish-rose wings, and curved keel of greenish yellow tinged with rose; petals clawed; 10 stamens (9 and 1); calyx 5-toothed. Stem: Hoary, with white, silky hairs, rather woody, 1 to 2 feet high. Leaves: Compounded of 7 to 25 oblong leaflets. Root: Long, fibrous, tough. Fruit: A hoary, narrow pod, to 2 in. long. Preferred Habitat - Dry, sandy soil, edges of pine woods. Flowering Season - June-July. Distribution - Southern New England, westward to Minnesota, south to Florida, Louisiana, and Mexico.

Flowers far less showy and attractive than this denizen of sandy wastelands, a cousin of the wisteria vine and the locust tree, have been introduced to American gardens. Striking its long fibrous root deep into the dry soil, the plant spreads in thrifty clumps through heat and drought - and so tough are its fibers they might almost be used for violin strings. As in the case of the lupine, the partridge pea and certain others akin to it, the leaves of the hoary pea "go to sleep" at night, but after a manner of their own, i.e., by lying along the stem and turning on their own bases.

In similar situations from New York south and southwestward, the MILK PEA (Galactia regularis; G. glabella of Gray) lies prostrate along the ground, the matted, usually branched stems sending up at regular intervals a raceme of rose-purple flowers in July and August from the axil of the trefoliate leaf.

TRAILING BUSH CLOVER
(Lespedeza procumbens) Pea family

Flowers - Purplish pink or violet, veined, the butterfly-shaped ones having standard petal, wings, and keel, clustered at end of peduncles; the minute flowers lacking a corolla, nearly sessile. Calyx of 5 slender, nearly equal lobes. Stems: Prostrate, trailing, or sometimes ascending, woolly or downy, leafy. Leaves: Clover-like, trefoliate. Fruit: A very small, hairy, flat, rounded, acute pod. Preferred Habitat - Dry soil open, sandy places. Flowering Season - August-September. Distribution - Massachusetts to the Gulf, and westward to the Mississippi.

Springing upward from a mass of clover-like leaves, these showy little blossoms elevate themselves to arrest, not our attention, but the notice of the passing bee. As the claw of the standard petal and the calyx are short, he need not have a long tongue to drain the nectary pointed out to him by a triangular white mark at the base of the banner. Now, as his weight depresses the incurved keel, wherein the vital organs are protected, the stigma strikes the visitor in advance of the anthers, so that pollen brought on his underside from another flower must come off on this one before he receives fresh pollen to transfer to a third blossom. At first the keel returns to its original position when depressed; later it loses its elasticity. But besides these showy flowers intended to be cross-fertilized by insects, the bush clovers bear, among the others, insignificant-looking, tightly closed, bud-like ones that produce abundant self-fertilized seed. The petaliferous flowers are simply to counteract the inevitable evils resulting from close inbreeding. One usually finds caterpillars of the "dusky wings" butterfly feeding on the foliage and the similar tick trefoils which are its staple. At night the bush clover leaves turn upward, completely changing the aspect of these plants as we know them by day. Michaux named the group of flowers for his patron, Lespedez, a governor of Florida under the Spanish regime.

128

Perhaps the commonest of the tribe is the VIOLET BUSH CLOVER (L. violacea), a variable, branching, erect, or spreading plant, sometimes only a foot high, or again three times as tall. Its thin leaves are more elliptic than the decidedly clover-like ones of the preceding species; its rose-purple flowers are more loosely clustered, and the stems are only sparingly hairy, never woolly.

On the top of the erect, usually unbranched, but very leafy stem of the WAND-LIKE BUSH CLOVER (L. frutescens), the two kinds of flowers grow in a crowded cluster, and more sparingly from the axils below. The clover-like leaflets, dark green and smooth above, are paler and hairy below. Like the rest of its kin, this bush clover delights in dry soil, particularly in open, sandy places near woods of pine and oak. One readily distinguishes the SLENDER BUSH CLOVER (L. Virginica) by the very narrowly oblong leaves along its wand, which bears two kinds of bright rose flowers, clustered at the top chiefly, and in the axils.

Yellowish-white flowers, about a quarter of an inch long, and with a purplish-rose spot on the standard petal to serve as a pathfinder to the nectary, are crowded in oblong spikes an inch and a half long or less on the HAIRY BUSH CLOVER (L. hirta). The stem, which may attain four feet, or half that height, is usually branched; and the entire plant is often downy to the point of silkiness.

Dense clusters of the yellowish-white flowers of the ROUND-HEADED BUSH CLOVER (L. capitata) are seated in the upper axils of the silvery-hairy, wand-like stem. Pink streaks at the base of the standard petal serve as pathfinders, and its infolded edges guide the bee's tongue straight to the opening in the stamen tube through which he sucks.

WILD or SPOTTED GERANIUM or CRANE'S-BILL; ALUM-ROOT
  (Geranium maculatum) Geranium family

Flowers - Pale magenta, purplish pink, or lavender, regular, 1 to 1 1/2 in. broad, solitary or a pair, borne on elongated peduncles, generally with pair of leaves at their base. Calyx of 5 lapping, pointed sepals; 5 petals, woolly at base; 10 stamens; pistil with 5 styles. Fruit:

A slender capsule pointed like a crane's bill. In maturity it ejects seeds elastically far from the parent plant. Stem: 1 to 2 ft. high, hairy, slender, simple or branching above. Leaves: Older ones sometimes spotted with white; basal ones 3 to 6 in. wide, 3 to 5 parted, variously cleft and toothed; 2 stem leaves opposite. Preferred Habitat - Open woods, thickets, and shady roadsides. Flowering Season - April-July. Distribution - Newfoundland to Georgia, and westward a thousand miles.

Sprengel, who was the first to exalt flowers above the level of mere botanical specimens, had his attention led to the intimate relationship existing between plants and insects by studying out the meaning of the hairy corolla of the common wild geranium of Germany (G. sylvaticum), being convinced, as he wrote in 1787, that "the wise Author of Nature has not made even a single hair without a definite design." A hundred years before, Nehemias Grew had said that it was necessary for pollen to reach the stigma of a flower in order that it might set fertile seed; and Linnaeus had to come to his aid with conclusive evidence to convince a doubting world that this was true. Sprengel made the next step forward, but his writings lay neglected over seventy years because he advanced the then incredible and only partially true statement that a flower is fertilized by insects which carry its pollen from its anthers to its stigma. In spite of his discoveries that the hairs inside the geranium's corolla protect its nectar from rain for the insect's benefit, just as eyebrows keep perspiration from falling into the eye; that most flowers which secrete nectar have what he termed "honey guides" - spots of bright color, heavy veining, or some such pathfinder on the petals - in spite of the most patient and scientific research that shed great light on natural selection a half-century before Darwin advanced the theory, he left it for the author of "The Origin of Species" to show that cross-fertilization - the transfer of pollen from one blossom to another, not from anthers to stigma of the same flower - is the great end to which so much marvelous mechanism is chiefly adapted. Cross-fertilized blossoms defeat self-fertilized flowers in the struggle for existence.

No wonder Sprengel's theory was disproved by his scornful contemporaries in the very case of his wild geranium, which sheds its pollen before it has developed a stigma to receive any; therefore no insect that had not brought pollen from an earlier bloom could possibly fertilize this flower. How amazing that he did not see this! Our common wild crane's-bill, which also has lost the power to fertilize itself, not only ripens first the outer, then the inner, row of anthers, but actually drops them off after their pollen has been removed, to overcome the barest chance of self-fertilization as the stigmas become receptive. This is the geranium's and many other flowers' method to compel cross-fertilization by insects. In cold, stormy, cloudy weather a geranium blossom may remain in the male stage several days before becoming female; while on a warm, sunny day, when plenty of insects are flying, the change sometimes takes place in a few hours. Among others, the common sulphur or puddle butterfly, that sits in swarms on muddy roads and makes the clover fields gay with its bright little wings, pilfers nectar from the geranium without bringing its long tongue in contact with the pollen. Neither do the smaller bees and flies which alight on the petals necessarily come in contact with the anthers and stigmas. Doubtless the larger bees are the flowers' true benefactors.

The so-called geraniums in cultivation are pelargoniums, strictly speaking.

In barren soil, from Canada to the Gulf, and far westward, the CAROLINA CRANE'S-BILL (G. Carolinianum), an erect, much-branched little plant resembling the spotted geranium in general features, bears more compact clusters of pale rose or whitish flowers, barely half an inch across. As their inner row of anthers comes very close to the stigmas, spontaneous self-fertilization may sometimes occur; although in fine weather small bees, especially, visit them constantly. The beak of the seed vessel measures nearly an inch long.

HERB ROBERT; RED ROBIN; RED SHANKS; DRAGON'S BLOOD
  (Geranium Robertianum) Geranium family

Flowers - Purplish rose, about 1/2 in. across, borne chiefly in pairs on slender peduncles. Five sepals and petals; stamens 10; pistil with 5 styles. Stem: Weak, slender, much branched, forked, and spreading, slightly hairy, 6 to i8 in. high. Leaves: Strongly scented, opposite, thin, of 3 divisions, much subdivided and cleft. Fruit: Capsular, elastic, the beak 1 in. long, awn-pointed. Preferreed Habitat - Rocky, moist woods and shady roadsides Flowering Season - May-October Distribution - Nova Scotia to Pennsylvania, and westward to Missouri.

Who was the Robert for whom this his "holy herb" was named? Many suppose that he was St. Robert, a Benedictine monk, to whom the twenty-ninth of April - the day the plant comes into flower in Europe - is dedicated. Others assert that Robert Duke of Normandy, for whom the "Ortus Sanitatis," a standard medical guide for some hundred of years, was written, is the man honored; and since there is now no way of deciding the mooted question, we may take our choice.

Only when the stems are young are they green; later the plant well earns the name of red shanks, and when its leaves show crimson stains, of dragon's blood.

At any time the herb gives forth a disagreeable odor, but especially when its leaves and stem have been crushed until they emit a resinous secretion once an alleged cure for the plague. Flies, that never object to a noxious smell, constantly visit the flower, and have their tongues guided through passages between little ridge-like processes on each petal to the nectar secreted by the base of the filaments at the base of each sepal. To prevent self-fertilization the five stigmas are folded close together when the flower opens, nor do they spread apart and become receptive until after the outer row of anthers, then the inner row, have shed their pollen. When the elastic carpels have ripened their seed, bang! go the little guns, scattering them far and wide.

WHITE OR TRUE WOOD~SORREL; ALLELULA
(Oxalis acetosella) Wood-sorrel family

Flowers - White or delicate pink, veined with deep pink, about 1/2 in. long. Five sepals; 5 spreading petals rounded at tips; 10 stamens, 5 longer, 5 shorter, all anther-bearing; 1 pistil with 5 stigmatic styles. Scape: Slender, leafless, 1-flowered, 2 to 5 in, high. Leaf: Clover-like, of 3 leaflets, on long petioles from scaly, creeping rootstock. Preferred Habitat - Cold, damp woods. Flowering Season - May-July. Distribution - Nova Scotia and Manitoba, southward to North Carolina. Also a native of Europe.

Clumps of these delicate little pinkish blossoms and abundant leaves, cuddled close to the cold earth of northern forests, usually conceal near the dry leaves or moss from which they spring blind flowers that never open - cleistogamous the botanists call them - flowers that lack petals, as if they were immature buds; that lack odor, nectar, and entrance; yet they are perfectly mature, self-fertilized, and abundantly fruitful. Fifty-five genera of plants contain one or more species on which these peculiar products are found, the pea family having more than any other, although violets offer perhaps the most familiar instance to most of us. Many of these species bury their offspring below ground; but the wood-sorrel bears its blind flowers nodding from the top of a curved scape at the base of the plant, where we can readily find them. By having no petals, and other features assumed by an ordinary flower to attract insects, and chiefly in saving pollen, they produce seed with literally the closest economy. It is estimated that the average blind flower of the wood-sorrel does its work with four hundred pollen grains, while the prodigal peony scatters with the help of wind and insect visitors over three and a half millions!

Yet no plant, however economically inclined, can afford to deteriorate its species through self-fertilization; therefore, to overcome the evils of in-breeding, the wood-sorrel, like other plants that bear cleistogamous flowers, takes special pains to produce showy blossoms to attract insects, on which they absolutely depend to transfer their pollen from flower to flower. These have their organs so arranged as to make self-fertilization impossible.

Every child knows how the wood-sorrel "goes to sleep" by drooping its three leaflets until they touch back to back at evening, regaining the horizontal at sunrise - a performance most scientists now agree protects the peculiarly sensitive leaf from cold by radiation. During the day, as well, seedling, scape, and leaves go through some interesting movements, closely followed by Darwin in his

"Power of Movement in Plants," which should be read by all interested.

Oxalis, the Greek for sour, applies to all sorrels because of their acid juice; but acetosella = vinegar salt, the specific name of this plant, indicates that from it druggists obtain salt of lemons. Twenty pounds of leaves yield between two and three ounces of oxalic acid by crystallization. Names locally given the plant in the Old World are wood sour or sower, cuckoo's meat, sour trefoil, and shamrock - for this is St. Patrick's own flower, the true shamrock of the ancient Irish, some claim. Alleluia, another folk-name, refers to the joyousness of the Easter season, when the plant comes into bloom in England.

VIOLET WOOD-SORREL
(Oxalis violacea) Wood-sorrel family

Flowers - Pinkish purple, lavender, or pale magenta; less than 1 in. long; borne on slender stems in umbels or forking clusters, each containing from 3 to 12 flowers. Calyx of 5 obtuse sepals; 5 petals; 10 (5 longer, 5 shorter) stamens; 5 styles persistent above 5-celled ovary. Stem: From brownish, scaly bulb 4 to 9 in. high. Leaves: About 1 in. wide, compounded of 3 rounded, clover-like leaflets with prominent midrib, borne at end of slender petioles, springing from root. Preferred Habitat - Rocky and sandy woods. Flowering Season - May-June. Distribution - Northern United States to Rocky Mountains, south to Florida and New Mexico; more abundant southward.

Beauty of Leaf and blossom is not the only attraction possessed by this charming little plant. As a family the wood-sorrels have great interest for botanists since Darwin devoted such exhaustive study to their power of movement, and many other scientists have described the several forms assumed by perfect flowers of the same species to secure cross-fertilization. Some members of the clan also bear blind flowers, which have been described in the account of the white wood-sorrel given above. Even the rudimentary leaves of the seedlings "go to sleep" at evening, and during the day are in constant movement up and down. The stems, too, are restless; and as for the mature leaves, every child knows how they droop their three leaflets back to back against the stem at evening, elevating them to the perfect horizontal again by day. Extreme sensitiveness to light has been thought to be the true explanation of so much activity, and yet this is not a satisfactory theory in many cases. It is certain that drooping leaves suffer far less from frost than those whose upper surfaces are flatly exposed to the zenith. This view that the sleep of leaves saves them from being chilled at night by radiation is Darwin's own, supported by innumerable experiments; and probably it would have been advanced by Linnaeus, too, since so many of his observations in "Somnus Plantarum" verify the theory, had the principle of radiation been discovered in his day. The violet wood-sorrel produces two sorts of perfect flowers reciprocally adapted to each other, but on different plants in the same neighborhood. The two are essentially alike, except in arrangement of stamens and pistil; one flower having high anthers and low stigmas, the other having lower anthers and higher stigmas; and as the high stigmas are fertile only when pollenized with grains from a flower having high anthers, it is evident insect aid to transfer pollen is indispensable here. Small bees, which visit these blossoms abundantly, are their benefactors; although there is nothing to prevent pollen from falling on the stigmas of the short-styled form. Hildebrand proved that productiveness is greatest, or exists only, after legitimate fertilization. To accomplish cross-pollination, many plants bear flowers of opposite sexes on different individuals; but the violet wood-sorrel's plan, utilized by the bluet and partridge-vine also, has the advantage in that both kinds of its flowers are fruitful.

COMMON, FIELD, or PURPLE MILKWORT; PURPLE POLYGALA
(Polygala viridescens; P. sanguinca of Gray) Milkwort family

Flowers - Numerous, very small, variable; bright magenta, pink, or almost red, or pale to whiteness, or greenish, clustered in a globular clover-like head, gradually lengthening to a cylindric spike. Stem: 6 to 15 in. high, smooth, branched above, leafy. Leaves: Alternate, narrowly oblong, entire. Preferred Habitat - Fields and meadows, moist or sandy. Flowering Season - June-September. Distribution - Southern Canada to North Carolina, westward to the Mississippi.

When these bright clover-like heads and the inconspicuous greenish ones grow together, the difference between them is so striking it is no wonder Linnaeus thought they were borne by two distinct species, sanguinea and viridescens, whereas they are now known to be merely two forms of the same flower. At first glance one might mistake the irregular little blossom for a member of the pea family; two of the five very unequal sepals - not petals - are colored wings. These bright-hued calyx-parts overlap around the flower-head like tiles on a roof. Within each pair of wings are three petals united into a tube, split on the back, to expose the vital organs to contact with the bee, the milkwort's best friend.

Plants of this genus were named polygala, the Greek for much milk, not because they have milky juice - for it is bitter and clear - but because feeding on them is supposed to increase the flow of cattle's milk.

In sandy swamps, especially near the coast from Maine to the Gulf, and westward to the Mississippi, grows the MARSH or CROSS-LEAVED MILKWORT (P. cruciata). Most of its leaves, especially the lower ones, are in whorls of four, and from July to September its dense, bright purple-pink, white, or greenish flowerheads, the wings awn-pointed, are seated on the ends of the square branching stem of this low, mossy little plant.

FRINGED MILKWORT or POLYGALA; FLOWERING WINTERGREEN; GAY WINGS
  (Polygala paucifolia) Milkwort family

Flowers - Purplish rose, rarely white, showy, over 1/2 in. long, from 1 to 4 on short, slender peduncles from among upper leaves. Calyx of 5 unequal sepals, of which 2 are wing-like and highly colored like petals. Corolla irregular, its crest finely fringed; 6 stamens; pistil. Also pale, pouch-like, cleistogamous flowers underground. Stem: Prostrate, 6 to 15 in. long, slender, from creeping rootstock, sending up flowering shoots 4 to 7 in. high. Leaves: Clustered at summit, oblong, or pointed egg-shaped, 1 1/2 in. long or less; those on lower part of shoots scale-like. Preferred habitat - Moist, rich woods, pine lands, light soil. Flowering Season - May-July. Distribution - Northern Canada, southward and westward to Georgia and Illinois.

Gay companies of these charming, bright little blossoms hidden away in the woods suggest a swarm of tiny mauve butterflies that have settled among the wintergreen leaves. Unlike the common milkwort and many of its kin that grow in clover-like heads, each one of the gay wings has beauty enough to stand alone, Its oddity of structure, its lovely color and enticing fringe, lead one to suspect it of extraordinary desire to woo some insect that will carry its pollen from blossom to blossom and so enable the plant to produce cross-fertilized seed to counteract the evil tendencies resulting from the more prolific self-fertilized cleistogamous flowers buried in the ground below. It has been said that the fringed polygala keeps "one flower for beauty and one for use"; "one playful flower for the world, another for serious use and posterity"; but surely the showy flowers, the "giddy sisters," borne by all cleistogamous species to save them from degenerating through close inbreeding, are no idle, irresponsible beauties. Let us watch a bumblebee as she alights on the convenient fringe which edges the lower petal of this milkwort. Now the weight of her body so depresses the keel, or tubular petals, wherein the stamens and pistil lie protected from the rain and useless insects, that as soon as it is pressed downward a spoon-tipped pistil pushes out the pollen through the slit on the top on the bee's abdomen. The stigmatic surface of the pistil is on the opposite side of the spoon, nearest the base of the flower, to guard against self-pollination. After the pollen has been removed, a bumblebee, already dusted from other blossoms, must leave some on the stigma as she sucks the nectar. Indeed, every feature possessed by this pretty flower has been developed for the most serious purpose of life - the salvation of the species.

Only locally common throughout a wide area, embracing the eastern half of the United States and Canada, is the RACEMED MILKWORT (P. polygama), whose small, purple-pink, but showy flowers, clustered along the upper part of numerous leafy stems, are found in dry soil during June and July. Like the fringed milkwort, this one bears many cleistogamous, or blind flowers, on underground branches, flowers that always set an abundance of fertile self-planted seed in case of failure to form any on the part of their showy sisters, which are utterly dependent upon the bee's ministrations. During prolonged stormy weather few insects are abroad.

SWAMP ROSE-MALLOW; MALLOW ROSE
  (Hibiscus Moscheutos) Mallow family

   Flowers - Very large, clear rose pink, sometimes white, often with crimson center, 4 to 7 in. across, solitary, or clustered on peduncles at summit of stems. Calyx 5-cleft, subtended by numerous narrow bractlets; 5 large, veined petals; stamens united into a valvular column bearing anthers on the outside for much of its length; 1 pistil partly enclosed in the column, and with five button-tipped stigmatic branches above. Stem: 4 to 7 ft. tall, stout, from perennial root. Leaves: 3 to 7 in. long, tapering, pointed, egg-shaped, densely white, downy beneath lower leaves, or sometimes all, lobed at middle. Preferred Habitat - Brackish marshes, riversides, lake shores, saline situations. Flowering Season - August-September. Distribution - Massachusetts to the Gulf of Mexico, westward to Louisiana; found locally in the interior, but chiefly along Atlantic seaboard.

   Stately ranks of these magnificent flowers, growing among the tall sedges and "cat-tails" of the marshes, make the most insensate traveler exclaim at their amazing loveliness. To reach them one must don rubber boots and risk sudden seats in the slippery ooze; nevertheless, with spade in hand to give one support, it is well worthwhile to seek them out and dig up some roots to transplant to the garden. Here, strange to say, without salt soil or more water than the average garden receives from showers and hose, this handsomest of our wild flowers soon makes itself delightfully at home under cultivation. Such good, deep earth, well enriched and moistened, as the hollyhock thrives in, suits it perfectly. Now we have a better opportunity to note how the bees suck the five nectaries at the base of the petals and collect the abundant pollen of the newly opened flowers, which they perforce transfer to the five button-shaped stigmas intentionally impeding the entrance to older blossoms. Only its cousin the hollyhock, a native of China, can vie with the rose-mallow's decorative splendor among the shrubbery; and the ROSE OF CHINA (Hibiscus Rosa-Sinensis), cultivated in greenhouses here, eclipse it in the beauty of the individual blossom. This latter flower, whose superb scarlet corolla stains black, is employed by the Chinese married women, it is said, to discolor their teeth; but in the West Indies it sinks to even greater ignominy as a dauber for blacking shoes!

MARSH MALLOW (Althaea officinalis), a name frequently misapplied to the swamp rose-mallow, is properly given to a much smaller pink flower, measuring only an inch and a half across at the most, and a far rarer one, being a naturalized immigrant from Europe found only in the salt marshes from the Massachusetts coast to New York. It is also known as WYMOTE. This is a bushy, leafy plant, two to four feet high, and covered with velvety down as a protection against the clogging of its pores by the moisture arising from its wet retreats. Plants that live in swamps must "perspire" freely and keep their pores open. From the marsh mallow's thick roots the mucilage used in confectionery is obtained, a soothing demulcent long esteemed in medicine. Another relative, the OKRA or GUMBO PLANT of vegetable gardens (Hibiscus esculentus), has mucilage enough in its narrow pods to thicken a potful of soup. Its pale yellow, crimson-centered flowers are quite as beautiful as any hollyhock, but not nearly so conspicuous, because of the plant's bushy habit of growth. In spite of its name, the ALTHAEA of our gardens, or ROSE OF SHARON (Hibiscus Syriacus), is not so closely allied to Althaea officinalis as to the swamp rose-mallow.

Another immigrant from Europe and Asia sparingly naturalized in waste places and roadsides in Canada, the United States, and Mexico is the COMMON HIGH MALLOW, CHEESEFLOWER, or ROUND DOCK (Malva sylvestris). Its purplish-rose flowers, from which the French have derived their word mauve, first applied to this plant, appear in small clusters on slender pedicels from the leaf axils along a leafy, rather weak, but ascending stem, maybe only a foot high, or perhaps a yard, throughout the summer months. The leaf, borne on a petiole two to six inches long, is divided into from five to nine shallow, angular, or rounded saw-edged lobes. Country children eat unlimited quantities of the harmless little circular, flat-tened "cheeses" or seed vessels, a characteristic of the genus Malva. Since the flower invites a great number of insects to feast on its nectar, secreted in five little pits (protected for them from the rain by hairs at the base of the petals), and compels its visitors to wipe off pollen brought from the pyramidal group of anthers in a newly opened blossom to the exserted, radiating stigmas of older ones, the mallow produces more cheeses than all the dairies of the world. So rich is its store of nectar that the hive-bee, shut out from a legitimate entrance to the flower when it closes in the late afternoon, climbs up the outside of the calyx, and inserting his tongue between the five petals, empties the nectaries one after another - intelligent rogue that he is!

The LOW, DWARF, or RUNNING MALLOW (M. rotundifolia), a very common little weed throughout our territory, Europe, and Asia, depends scarcely at all upon insects to transfer its pollen, as might be inferred from its unattractive pale blue to white flowers, that measure only about half an inch across. In default of visitors, its pollen-laden anthers, instead of drooping to get out of the way of the stigmas, as in the showy high mallow, remain extended so as to come in contact with the rough, sticky sides of the long curling stigmas. The leaves of this spreading plant, which are nearly round, with five to nine shallow, saw-edged lobes, are thin, and furnished with long petioles; whereas the flowers which spring from their axils keep close to the main stem. Usually there are about fifteen rounded carpels that go to make up the Dutch, doll, or fairy cheeses, as the seed vessels are called by children. Only once is the mallow mentioned in the Bible, and then as food for the most abject and despised poor (Job 30: 4); but as eighteen species of mallow grow in Palestine, who is the higher critic to name the species eaten?

Occasionally we meet by the roadside in Canada, the Eastern, Middle, and Southern States pink, sometimes white, flowers, about two inches across, growing in small clusters at the top of a stem a foot or two high, the whole plant emitting a faint odor of musk. If the stem leaves are deeply divided into several narrow, much-cleft segments, and the little cheeses are densely hairy, we may safely call the plant MUSK MALLOW (M. moschata), and expect to find it blooming throughout the summer.

MARSH ST.-JOHN'S-WORT
(Triadenum Virginicum; Elodea Virginica of Gray)
St.-John's-wort family

Flowers - Pale magenta, pink, or flesh color, about 1/2 in. across, in terminal clusters, or from leaf axils. Calyx of 5 equal sepals, persistent on fruit; 5 petals; 9 or more stamens united in 3 sets; pistil of 3 distinct styles. Stem: to 1 1/2 ft. high, simple, leafy. Leaves: Opposite, pale, with black, glandular dots, broadly oblong, entire edged, seated on stem or clasping by heart-shaped base. Fruit: An oblong, acute, deep red capsule. Preferred Habitat - Swamps and cranberry bogs. Flowering Season - July-September. Distribution - Labrador to the Gulf, and westward to Nebraska.

Late in the summer, after the rather insignificant pink flowers have withered, this low plant, which almost never lacks some color in its green parts, greatly increases its beauty by tinting stems, leaves, and seed vessels with red. Like other members of the family, the flower arranges its stamens in little bundles of three, and when an insect comes to feast on the abundant pollen - no nectar being secreted - he cannot avoid rubbing some off on the stigmas that are on a level with the anthers. He may sometimes carry pollen from blossom to blossom, it is true, but certainly the St.-John's-wort takes no adequate precautions against self-fertilization at any time. Toward the close of its existence the flower draws its petals together toward the axils, thus bringing anthers and stigmas in contact.

SPIKED WILLOW-HERB; LONG PURPLES; SPIKED or PURPLE LOOSESTRIFE
(Lythrum Salicaria) Loosestrife family

Flowers - Bright magenta (royal purple) or pinkish purple, about 1/2 in. broad, crowded in whorls around long bracted spikes. Calyx tubular, ribbed, 5 to 7 toothed, with small projections between. Corolla of 5 or 6 slightly wrinkled or twisted petals. Stamens, in 2 whorls of 5 or 6 each, and 1 pistil, occurring in three different lengths. Stem: 2 to 3 ft. high, leafy, branched. Leaves: Opposite, or sometimes in whorls of 3; lance-shaped, with heart-shaped base clasping stem. Preferred Habitat - Wet meadows, watery places, ditches, and banks of streams. Flowering Season - June-August. Distribution - Eastern Canada to Delaware, and westward through Middle States; also in Europe.

Through Darwin's patient study of this trimorphic flower, it has assumed so important a place in his theory of the origin of species that its fertilization by insects deserves special attention. On page 5, the method by which the pickerel weed, another flower whose stamens and pistil occur in three different lengths, should be read to avoid much repetition. Now the loosestrife produces six different kinds of yellow and green pollen on its two sets of three stamens; and when this pollen is applied by insects to the stigmatic surface of three different lengths of pistil, it follows that there are eighteen ways in which it may be transferred. But Darwin proved that only pollen brought from the shortest stamens to the shortest pistil, from the middle-length stamens to the middle-length pistil, and from the long stamens to the long pistil effectually fertilizes the flower. And as all the flowers on any one plant are of the same kind, we have here a marvelous mechanism to secure cross-fertilization. His experiments with this loosestrife also demonstrated that "reproductive organs, when of different length, behave to one another like different species of the same genus in regard both to direct productiveness and the character of the offspring; and that consequently mutual barrenness, which was once thought conclusive proof of difference of species, is worthless as such, and the last barrier that was raised between species and varieties is broken down." (Muller.)

Naturally the bright-hued, hospitable flower, which secretes abundant nectar at the base of its tube, attracts many insects, among others, bees of larger and middle size, and the butterflies for which it is especially adapted. They alight on the stamens and pistil on the upper side of the flower. Those with the longest tongues stand on one blossom to sip from the next one: this is the butterfly's customary attitude. But nearly every visitor comes in contact with at least one set of organs. When Darwin first interpreted the trimorphism of the loosestrife, we can realize something of the enthusiasm such a man must have felt in writing to Gray: "I am almost stark, staring mad over lythrum…. For the love of Heaven have a look at some of your species, and if you can get me some seed, do!"

Long ago this beautiful plant reached our shores from Europe, and year by year is extending its triumphal march westward, brightening its course of empire through low meadows and marshes with torches that lengthen even as they glow. It is not a spring flower, even in England; and so when Shakespeare, whose knowledge of floral nature was second only to that of human nature, wrote of Ophelia,

"With fantastic garlands did she come,
 Of crow-flowers, nettles, daisies, and long purples,"

is it probable he so combined flowers having different seasons of bloom? Dr. Prior suggests that the purple orchis (0. mascula) might have been the flower Ophelia wore; but, as long purples has been the folk name of this loosestrife from time immemorial in England, it seems likely that Shakespeare for once may have made a mistake.

BLUE WAX-WEED; CLAMMY CUPHEA; TAR-WEED (Parsonia petiolata; Cuphea viscosissima of Gray) Loosestrife family

Flowers - Purplish pink, about 1/4 in. across, on short peduncles from leaf axils, solitary or clustered. Calyx sticky, tubular, 12-ribbed, with 6 primary teeth, oblique at mouth, extending into a rounded swelling on upper side at base; 6 unequal, wrinkled petals, on short claws; 11 or 12 stamens inserted on calyx throat; pistil with 2-lobed stigma. Stem: 6 to 20 in. high, branched, very sticky-hairy. Leaves: Opposite, on slender petioles, lance-shaped, rounded at base, harsh to the touch. Preferred Habitat - Dry soil, waste places, fields, roadsides. Flowering Season - July-October. Distribution - Rhode Island to Georgia, westward to Louisiana, Kansas, and Illinois.

A first cousin of the familiar Mexican cigar plant, or fire-cracker plant (Cuphea platycentra), whose abundant little vermilion tubes, with black-edged lower lip tipped with white, brighten the borders of so many Northern flower-beds. Kyphos, the Greek for curved, from which cuphea was derived, has reference to the peculiar, swollen little seedpod. From a slit on one side of the clammy cuphea's capsule the placenta, set with tiny flattened seeds, sticks out like a handle. Probably the flower has already fertilized itself in the bud, although, from the fact that the plant has taken such pains to punish crawling insect foes by coating itself with sticky hairs, one might imagine it was wholly dependent upon winged insects to transfer its pollen. What an unworthy relative of the purple loosestrife, whose elaborate scheme to insure cross-fertilization is one of the botanical wonders!

MEADOW-BEAUTY; DEER GRASS
(Rhexia Virginica) Meadow-beauty family

Flowers - Purplish pink, 1 to 1 1/2 in. across, pedicelled, clustered at top of stem. Calyx 4-lobed, tubular or urn-shaped, narrowest at neck; 4 rounded, spreading petals, joined for half their length; 8 equal, prominent stamens in 2 rows; pistil. Stem: 1 to 1 1/2 ft. high, square, more or less hairy, erect, sometimes branching at top. Leaves: Opposite, ascending, seated on stem, oval, acute at tip, mostly 5-nerved, the margins saw-edged. Preferred Habitat - Sandy swamps or near water. Flowering Season - July-September. Distribution - United States, chiefly east of Mississippi.

Suggesting a brilliant magenta evening primrose in form, the meadow-beauty is likewise a rather niggardly bloomer, only a few flowers in each cluster opening at once; but where masses adorn our marshes, we cannot wonder so effective a plant is exported to European peat gardens. Its lovely sister, the MARYLAND MEADOW-BEAUTY (R. Mariana), a smaller, less brilliant flower, found no farther north than the swamps and pine barrens of New Jersey, also goes abroad to be admired; yet neither is of any value for cutting, for the delicate petals quickly discolor and drop off when handled. Blossoms so attractively colored naturally have many winged visitors to transfer their pollen. All too soon after fertilization the now useless petals fall, leaving the pretty urn-shaped calyx, with the large yellow protruding stamens, far more conspicuous than some flowers. "Its seed-vessels are perfect little cream pitchers of graceful form," said Thoreau. Within the smooth capsule the minute seeds are coiled like snail-shells.

GREAT OR SPIKED WILLOW-HERB; FIRE-WEED
(Chamaenerion angustifolium; Epilobium angustifolium of Gray)
Evening Primrose family

Flowers - Magenta or pink, sometimes pale, or rarely white, more or less than 1 in. across, in an elongated, terminal, spike-like raceme. Calyx tubular, narrow, in 4 segments; 4 rounded, spreading petals; 8 stamens; 1 pistil, hairy at base; the stigma 4-lobed. Stem: 2 to 8 ft. high, simple, smooth, leafy. Leaves: Narrow, tapering, willow-like, 2 to 6 in. long. Fruit: A slender, curved, violet-tinted capsule, from 2 to 3 in. long, containing numerous seeds attached to tufts of fluffy, white, silky threads. Preferred Habitat - Dry soil, fields, roadsides, especially in burnt-over districts. Flowering Season - June-September. Distribution - From Atlantic to Pacific, with few inter-ruptions; British Possessions and United States southward to the Carolinas and Arizona. Also Europe and Asia.

Spikes of these beautiful brilliant flowers towering upward above dry soil, particularly where the woodsman's axe and forest fires have devastated the landscape, illustrate Nature's abhorrence of ugliness. Other kindly plants have earned the name of fire-weed, but none so quickly beautifies the blackened clearings of the pio-neer, nor blossoms over the charred trail in the wake of the locomo-tive. Beginning at the bottom of the long spike, the flowers open in slow succession upward throughout the summer, leaving behind the attractive seed-vessels, which, splitting lengthwise in Septem-ber, send adrift white silky tufts attached to seeds that will one day cover far distant wastes with beauty. Almost perfect rosettes, made by the young plants, are met with on one's winter walks.

Epi, upon, and lobos, a pod, combine to make a name applicable to many flowers of this family. In general structure the fire-weed closely resembles its relative the evening primrose. Bees, not moths, however, are its benefactors. Coming to a newly opened flower, the bee finds abundant pollen on the anthers and a sip of nectar in the cup below. At this stage the flower keeps its still immature style curved downward and backward lest it should become self-fertilized - an evil ever to be guarded against by ambitious plants. In a few days, or after the pollen has been removed, up stretches the style, spreading its four receptive stigmas just where an incoming bee, well dusted from a younger flower, must certainly leave some pollen on their sticky surfaces.

The GREAT HAIRY WILLOW-HERB (Epilobium hirsutum), whose white tufted seeds came over from Europe in the ballast to be blown over Ontario and the Eastern States, spreads also by underground shoots, until it seems destined to occupy wide areas. In these showy magenta flowers, about one inch across, the stigmas and anthers mature simultaneously but cross-fertilization is usually insured because the former surpass the latter, and naturally are first touched by the insect visitor. In default of visits, however, the stigmas, at length curling backward, come in contact with the pollen-laden anthers. The fire-weed, on the contrary, is unable to fertilize itself.

A pale magenta-pink or whitish, very small-flowered, branching species, one to two feet high, found in swamps from New Brunswick to the Pacific, and southward to Delaware, is the LINEAR-LEAVED WILLOW-HERB (F. lineare), whose distinguishing features are its very narrow, acute leaves, its hoariness throughout, the dingy threads on its tiny seeds, and the occasional bulblets it bears near the base of the stem. It is scarcely to be distinguished by one not well up in field practice from another bog lover, the DOWNY or SOFT WILLOW-HERB (F. strictum), which, however, is a trifle taller, glandular throughout, and with sessile, not petioled, leaves. The PURPLE-LEAVED WILLOW-HERB (E. coloratum), common in low grounds, may best be named by the reddish-brown coma to which its seeds are attached. Both leaves and stem are often highly colored.

BOG WINTERGREEN
 (Pyrola uliginosa; P. rotundifolia, var. uliginosa of Gray)
Wintergreen family

Flowers - Magenta pink, fragrant, about 1/2 in. across, 7 to 15 on a leafless scape 6 to 15 in. high. Calyx 5-parted; 5 concave petals; 10 stamens; style curved upward, exserted. Leaves: From the root, broadly oval or round, rather thick and dull, on petioles. Preferred Habitat - Swamps and bogs. Flowering Season - June. Distribution - Nova Scotia to British Columbia, southward to New York and Colorado.

Fragrant colonies of this little plant cuddled close to the moss of cool, northern peat bogs draw forth our admiration when we go orchid hunting in early summer. A similar species, the LIVER-LEAF WINTERGREEN (P. asarifolia), with shining, not dull, leaves and rose-colored flowers, not to mention minor differences, is likewise found in swamps and wet woods. These two wintergreens, formerly counted mere varieties of the white-flowered rotundifolia, a lover of dry woods, have now been given specific individuality by later-day systematists. Short-lipped bees and flies may be detected in the act of applying their mouths to the orifices of the anthers through which pollen is shed, and some must be carried to the stigma of another flower.

PIPSISSEWA; PRINCE'S PINE
(Chimaphila umbellata) Wintergreen family

Flowers - Flesh-colored, or pinkish, fragrant, waxy, usually with deep pink ring around center, and the anthers colored; about 1/2 in. across; several flowers in loose, terminal cluster. Calyx 5-cleft; corolla of 5 concave, rounded, spreading petals; 10 stamens, the filaments hairy style short, conical, with a round stigma. Stem: Trailing far along ground, creeping, or partly subterranean, sending up sterile and flowering branches 3 to 10 in. high. Leaves: Opposite or in whorls, evergreen, bright, shining, spatulate to lance-shaped, sharply saw-edged. Preferred Habitat - Dry woods, sandy leaf-mould. Flowering Season - June-August. Distribution - British Possessions and the United States north of Georgia from the Atlantic to the Pacific. Also Mexico, Europe, and Asia.

A lover of winter indeed (cheima = winter and phileo = to love) is the prince's pine, whose beautiful dark leaves keep their color and gloss in spite of snow and intense cold. A few yards of the trailing stem, easily ripped from the light soil of its woodland home, make a charming indoor decoration, especially when the little brown seed-cases remain. Few flowers are more suggestive of the woods than these shy, dainty, deliciously fragrant little blossoms.

The SPOTTED WINTERGREEN, or PIPSISSEWA (C. maculata), closely resembles the prince's pine, except that its slightly larger white or pinkish flowers lack the deep pink ring; and the lance-shaped leaves, with rather distant saw-teeth, are beautifully mottled with white along the veins. When we see short-lipped bees and flies about these flowers, we may be sure their pollen-covered mouths come in contact with the moist stigma on the summit of the little top-shaped style, and so effect cross-fertilization.

WILD HONEYSUCKLE; PINK, PURPLE, or WILD AZALEA;
PINXTER-FLOWER
 (Azalea nudiflora) Heath family

Flowers - Crimson pink, purplish or rose pink, to nearly white, 1 1/2 to 2 in. across, faintly fragrant, clustered, opening before or with the leaves, and developed from cone-like, scaly brown buds. Calyx minute, 5-parted; corolla funnel-shaped, the tube narrow, hairy, with 5 regular, spreading lobes; 5 long red stamens; 1 pistil, declined, protruding. Stem: Shrubby, usually simple below, but branching above, 2 to 6 ft. high. Leaves: Usually clustered, decidu-ous, oblong, acute at both ends, hairy on midrib. Preferred Habitat - Moist, rocky woods, or dry woods and thickets. Flowering Season - April-May. Distribution - Maine to Illinois, and southward to the Gulf.

Woods and hillsides are glowing with fragrant, rosy masses of this lovely azalea, the Pinxter-bloem or Whitsunday flower of the Dutch colonists, long before the seventh Sunday after Easter. Among our earliest exports, this hardy shrub, the swamp azalea, and the superb flame-colored species of the Alleghanies, were sent early in the eighteenth century to the old country, and there crossed with A. Pontica of southern Europe by the Belgian horticulturalists, to whom we owe the Ghent azaleas, the final triumphs of the hybridizer, that glorify the shrubberies on our own lawns to-day. The azalea became the national flower of Flanders. These hardy species lose their leaves in winter, whereas the hothouse varieties of A. Indica, a native of China and Japan, have thickish leaves, almost if not quite evergreen. A few of the latter stand our northern winters, especially the pure white variety now quite commonly planted in cemetery lots. In that delightfully enthusiastic little book, "The Garden's Story," Mr. Ellwanger says of the Ghent azalea "In it I find a charm presented by no other flower. Its soft tints of buff, sulphur, and primrose; its dazzling shades of apricot, salmon, orange, and vermilion are always a fresh revelation of color. They have no parallel among flowers, and exist only in opals, sunset skies, and the flush of autumn woods." Certainly American horticulturists were not clever in allowing the industry of raising these plants from our native stock to thrive on foreign soil.

Naturally the azalea's protruding style forms the most convenient alighting place for the female bee, its chief friend; and there she leaves a few grains of pollen, brought on her hairy underside from another flower, before again dusting herself there as she crawls over the pretty colored anthers on her way to the nectary. Honey produced from azaleas by the hive bee is in bad repute. All too soon after fertilization the now useless corolla slides along to the tip of the pistil, where it swings a while before dropping to earth.

Our beautiful wild honeysuckle, called naked (nudiflora), because very often the flowers appear before the leaves, has a peculiar Japanese grace on that account. Every farmer's boy's mouth waters at sight of the cool, juicy May-apple, the extraordinary pulpy growth on this plant and the swamp pink. This excrescence seems to have no other use than that of a gratuitous, harmless gift to the thirsty child, from whom it exacts no reward of carrying seeds to plant distant colonies, as the mandrake's yellow, tomato-like May-apple does. But let him beware, as he is likely to, of the similar looking, but hollow, stringy apples growing on the bushy Andromeda, which turn black with age.

>From Maine to Florida and westward to Texas, chiefly near the coast, in low, wet places only need we look for the SWAMP PINK or HONEYSUCKLE, WHITE or CLAMMY AZALEA (A. viscosa), a more hairy species than the Pinxter-flower, with a very sticky, glandular corolla tube, and deliciously fragrant blossoms, by no means invariably white. John Burroughs is not the only one who has passed "several patches of swamp honeysuckles, red with blossoms" ("Wake-Robin"). But as this species does not bloom until June and July, when the sun quickly bleaches the delicate flowers, it is true we most frequently find them white, merely tinged with pink. The leaves are well developed before the blossoms appear. Concerning azaleas' poisonous property, see the discussion under mountain laurel that follows.

RHODORA (Rhodora Canadensis; Rhododendron Rhodora of Gray) Heath family

Flowers - Purplish pink, rose, or nearly white, 1 1/2 in. broad or less, in clusters on short, stiff, hairy pedicels, and usually appearing before the leaves, from scaly, terminal buds. Calyx minute; corolla 2-lipped, upper lip unequally 2-3 lobed; lower lip 2-cleft; 10 stamens; pistil, the style slightly protruding. Stem: 1 to 3 ft. high, shrubby, branching. Leaves: Deciduous, oval to oblong, dark green above, pale and hairy beneath. Preferred Habitat - Wet hillsides, damp woods, beside sluggish streams, cool bogs. Flowering Season - May. Distribution - Newfoundland to Pennsylvania mountains.

A superficial glance at this low, little, thin shrub might mistake it for a magenta variety of the leafless Pinxter-flower. It does its best to console the New Englanders for the scarcity of the magnificent rhododendron, with which it was formerly classed. The Sage of Concord, who became so enamored of it that Massachusetts people often speak of it as "Emerson's flower," extols its loveliness in a sonnet:

"Rhodora! If the sages ask thee why
This charm is wasted on the earth and sky,
Tell them, dear, if eyes were made for seeing,
Then Beauty is its own excuse for being."

AMERICAN or GREAT RHODODENDRON; GREAT LAUREL;
ROSE TREE, or BAY
(Rhododendron maximum) Heath family

Flowers - Rose pink, varying to white, greenish in the throat, spotted with yellow or orange, in broad clusters set like a bouquet among leaves, and developed from scaly, cone-like buds; pedicels sticky-hairy. Calyx 5-parted, minute; corolla 5-lobed, broadly bell-shaped, 2 in. broad or less usually 10 stamens, equally spreading; pistil. Stem: Sometimes a tree attaining a height of 40 ft., usually 6 to 20 ft., shrubby, woody. Leaves: Evergreen, drooping in winter, leathery, dark green on both sides, lance-oblong, 4 to 10 in. long, entire edged, narrowing into stout petioles. Preferred Habitat - Mountainous woodland, hillsides near streams. Flowering Season - June-July. Distribution - Uncommon from Ohio and New England to Nova Scotia; abundant through the Alleghanies to Georgia.

When this most magnificent of our native shrubs covers whole mountain sides throughout the Alleghany region with bloom, one stands awed in the presence of such overwhelming beauty. Nowhere else does the rhododendron attain such size or luxuriance. There it produces a tall trunk, and towers among the trees; it spreads its branches far and wide until they interlock and form almost impenetrable thickets locally called "hells;" it glorifies the loneliest mountain road with superb bouquets of its delicate flowers set among dark, glossy foliage scarcely less attractive. The mountain in bloom is worth travelling a thousand miles to see.

Farther south the more purplish-pink or lilac-flowered CAROLINA RHODODENDRON (R. Catawbiense) flourishes. This southern shrub, which is perfectly hardy, unlike its northern sister, has been used by cultivators as a basis for producing the fine hybrids now so extensively grown on lawns in this country and Europe. Crossed with the Nepal species (R. arboreum) the best results follow. Americans, ever too prone to make the eagle scream on their trips abroad, need not monopolize all the glory for the cultivated rhododendron, as they are apt to do when they see it on fine estates in England. The Himalayas, which are covered with rhododendrons of brighter hue than ours, furnish many of the shrubs of commerce. Our rhododendron produces one of the hardest and strongest of woods, weighing thirty-nine pounds per cubic foot.

Rhododendrons, azaleas, and laurels fall under a common ban pronounced by bee-keepers. The bees which transfer pollen from blossom to blossom while gathering nectar, manufacture honey said to be poisonous. Cattle know enough to let all this foliage alone. Apparently the ants fear no more evil results from the nectar than the bees themselves; and were it not for the sticky parts nearest the flowers, on which they crawl to meet their death, the blossom's true benefactors would find little refreshment left.

MOUNTAIN or AMERICAN LAUREL; CALICO BUSH; SPOON-WOOD; CALMOUN;
BROAD-LEAVED KALMIA
(Kalmia latifolia) Heath family

Flowers - Buds and new flowers bright rose pink, afterward fading white, and only lined with pink, 1 in. across, or less, numerous, in terminal clusters. Calyx small, 5-parted, sticky corolla like a 5-pointed saucer, with 10 projections on outside; 10 arching stamens, an anther lodged in each projection; 1 pistil. Stem: Shrubby, woody, stiffly branched, 2 to 20 ft. high. Leaves: Evergreen, entire, oval to elliptic, pointed at both ends, tapering into petioles. Fruit: A round, brown capsule, with the style long remaining on it. Preferred Habitat - Sandy or rocky woods, especially in hilly or mountainous country. Flowering Season - May-June. Distribution - New Brunswick and Ontario, southward to the Gulf of Mexico, and westward to Ohio.

It would be well if Americans, imitating the Japanese in making pilgrimages to scenes of supreme natural beauty, visited the mountains, rocky, woody hillsides, ravines, and tree-girt uplands when the laurel is in its glory; when masses of its pink and white blossoms, set among the dark evergreen leaves, flush the landscape like Aurora, and are reflected from the pools of streams and the serene depths of mountain lakes. Peter Kalm, a Swedish pupil of Linnaeus, who traveled here early in the eighteenth century, was more impressed by its beauty than that of any other flower. He introduced the plant to Europe, where it is known as kalmia, and extensively cultivated on fine estates that are thrown open to the public during the flowering season. Even a flower is not without honor, save in its own country. We have only to prepare a border of leaf-mould, take up the young plant without injuring the roots or allowing them to dry, hurry them into the ground, and prune back the bush a little, to establish it in our gardens, where it will bloom freely after the second year.

All the kalmias resort to a most ingenious device for compelling insect visitors to carry their pollen from blossom to blossom. A newly opened flower has its stigma erected where the incoming bee must leave on its sticky surface the four minute orange-like grains carried from the anther of another flower on the hairy underside of her body. Now, each anther is tucked away in one of the ten little pockets of the saucer-shaped blossom, and the elastic filaments are strained upward like a bow. After hovering above the nectary, the bee has only to descend toward it, when her leg, touching against one of the hair-triggers of the spring trap, pop goes the little anther-gun, discharging pollen from its bores as it flies upward. So delicately is the mechanism adjusted, the slightest jar or rough handling releases the anthers; but, on the other hand, should insects be excluded by a net stretched over the plant, the flowers will fall off and wither without firing off their pollen-charged guns. At least, this is true in the great majority of tests. As in the case of hothouse flowers no fertile seed is set when nets keep away the laurel's bene-factors. One has only to touch the hair-trigger with the end of a pin to see how exquisitely delicate is this provision for cross-fertilization.

However much we may be cautioned by the apiculturalists against honey made from laurel nectar, the bees themselves ignore all warnings and apparently without evil results - happily for flowers dependent upon them and their kin. Mr. Frank R. Cheshire, in "Bees and Bee-keeping," the standard English work on the subject, writes: "During the celebrated Retreat of the Ten Thousand, as recorded by Xenophon in his 'Anabasis,' the soldiers regaled themselves upon some honey found near Trebizonde where were many beehives. Intoxication with vomiting was the result. Some were so overcome, he states, as to be incapable of standing. Not a soldier died, but very many were greatly weakened for several days. Tournefort endeavored to ascertain whether this account was corroborated by anything ascertainable in the locality, and had good reason to be satisfied respecting it. He concluded that the honey had been gathered from a shrub growing in the neighborhood of Trebizonde, which is well known there as producing the before-mentioned effects. It is now agreed that the plants were species of rhododendron and azaleas. Lamberti confirms Xenophon's account by stating that similar effects are produced by honey of Colchis, where the same shrubs are common. In 1790, even, fatal cases occurred in America in consequence of eating wild honey, which was traced to Kahmia latifolia by an inquiry instituted under direction of the American government. Happily, our American cousins are now never likely to thus suffer, thanks to drainage, the plow, and the bee-farm."

One of the beautiful swallow-tail butterflies lays its eggs on laurel leaves, that the larvae may feed on them later; yet the foliage often proves deadly to more highly organized creatures. Most cattle know enough to let it alone; nevertheless some fall victims to it every year. Even the intelligent grouse, hard pressed with hunger when deep snow covers much of their chosen food are sometimes found dead and their crops distended by these leaves. How far more unkind than the bristly armored thistle's is the laurel's method of protecting itself against destruction! Even the ant, intent on pilfering sweets secreted for bees, it ruthlessly glues to death against its sticky stems and calices. According to Dr. Barton the Indians drink a decoction of kalmia leaves when they wish to commit suicide.

As laurel wood is very hard and solid, weighing forty-four pounds to the cubic foot, it is in great demand for various purposes, one of them indicated in the plant's popular name of Spoon-wood.

SHEEP-LAUREL, LAMB-KILL, WICKY, CALF-KILL, SHEEP-POISON NARROW-LEAVED LAUREL (K. angustifolia), and so on through a list of folk names testifying chiefly to the plant's wickedness in the pasture, may be especially deadly food for cattle, but it certainly is a feast to the eyes. However much we may admire the small, deep crimson-pink flowers that we find in June and July in moist fields or swampy ground or on the hillsides, few of us will agree with Thoreau, who claimed that it is "handsomer than the mountain laurel." The low shrub may be only six inches high, or it may attain three feet. The narrow evergreen leaves, pale on the underside, have a tendency to form groups of threes, standing upright when newly put forth, but bent downward with the weight of age. A peculiarity of the plant is that clusters of leaves usually terminate the woody stem, for the flowers grow in whorls or in clusters at the side of it below.

The PALE or SWAMP LAUREL (K. glauca), found in cool bogs from Newfoundland to New Jersey and Michigan, and westward to the Pacific Coast, coats the under side of its mostly upright leaves with a smooth whitish bloom like the cabbage's. It is a straggling little bush, even lower than the lamb-kill, and an earlier bloomer, putting forth its loose, niggardly clusters of deep rose or lilac-colored flowers in June.

TRAILING ARBUTUS; MAYFLOWER; GROUND LAUREL
(Epigaea repens) Heath family

Flowers - Pink, fading to nearly white, very fragrant about 1/2 in. across when expanded, few or many in clusters at ends of branches. Calyx of 5 dry overlapping sepals; corolla salver-shaped, the slender, hairy tube spreading into 5 equal lobes; 10 stamens; 1 pistil with a column-like style and a 5-lobed stigma. Stem: Spreading over the ground (Epigaea = on the earth); woody, the leafy twigs covered with rusty hairs. Leaves: Alternate, oval, rounded at the base, smooth above, more or less hairy below, evergreen, weather-worn, on short, rusty, hairy petioles. Preferred Habitat - Light sandy loam in woods, especially under evergreen trees, or in mossy, rocky places. Flowering Season - March-May. Distribution - Newfoundland to Florida, west to Kentucky, and the Northwest Territory.

Can words describe the fragrance of the very breath of spring - that delicious commingling of the perfume of arbutus, the odor of pines, and the snow-soaked soil just warming into life? Those who know the flower only as it is sold in the city streets, tied with wet, dirty string into tight bunches, withered and forlorn, can have little idea of the joy of finding the pink, pearly blossoms freshly opened among the withered leaves of oak and chestnut, moss, and pine needles in which they nestle close to the cold earth in the leafless, windy northern forest. Even in Florida, where broad patches carpet the woods in February, one misses something of the arbutus's accustomed charm simply because there are no slushy remnants of snow drifts, no reminders of winter hardships in the vicinity. There can be no glad surprise at finding dainty spring flowers in a land of perpetual summer. Little wonder that the Pilgrim Fathers, after the first awful winter on the "stern New England coast," loved this early messenger of hope and gladness above the frozen ground at Plymouth. In an introductory note to his poem "The Mayflowers," Whittier states that the name was familiar in England, as the application of it to the historic vessel shows; but it was applied by the English, and still is, to the hawthorn. Its use in New England in connection with the trailing arbutus dates from a very early day, some claiming that the first Pilgrims so used it in affectionate memory of the vessel and its English flower association.

"Sad Mayflower I watched by winter stars,
   And nursed by winter gales,
 With petals of the sleeted spars,
   And leaves of frozen sails!

"But warmer suns ere long shall bring
   To life the frozen sod,
 And through dead leaves of hope shall spring
   Afresh the flowers of God!"

Some have attempted to show that the Pilgrims did not find the flowers until the last month of spring, and that, therefore, they were named Mayflowers. Certainly the arbutus is not a typical May blossom even in New England. Bryant associates it with the hepatica, our earliest spring flower, in his poem, "The, Twenty-seventh of March":

      "Within the woods
 Tufts of ground laurel, creeping underneath

The leaves of the last summer, send their sweets
Upon the chilly air, and by the oak,
The squirrel cups, a graceful company
Hide in their bells a soft aerial blue."

There is little use trying to coax this shyest of sylvan flowers into our gardens where other members of its family, rhododendrons, laurels, and azaleas make themselves delightfully at home. It is wild as a hawk, an untamable creature that slowly pines to death when brought into contact with civilization. Greedy street venders, who ruthlessly tear up the plant by the yard, and others without even the excuse of eking out a paltry income by its sale, have already exterminated it within a wide radius of our Eastern cities. How curious that the majority of people show their appreciation of a flower's beauty only by selfishly, ignorantly picking every specimen they can find!

In many localities the arbutus sets no fruit, for it is still undergoing evolutionary changes looking toward the perfecting of an elaborate system to insure cross-fertilization. Already it has attained to perfume, nectar, and color to attract quantities of insects, chiefly flies and small female bees but in some flowers the anthers produce no pollen for them to carry, while others are filled with grains, yet all the stigmas in the neighboring clusters may be defective. The styles and the filaments are of several different lengths, showing a tendency toward trimorphism, perhaps, like the wonderful purple loosestrife; but at present the flower pursues a most wasteful method of distributing pollen, and in different sections of the country acts so differently that its phases are impossible to describe except to the advanced student. They may, however, be best summarized in the words of Professor Asa Gray: "The flowers are of two kinds, each with two modifications; the two main kinds characterized by the nature and perfection of the stigma, along with more or less abortion of the stamens; their modifications by the length of the style."

When our English cousins speak of the arbutus, they have in mind a very different species from ours. Theirs is the late flowering strawberry-tree, an evergreen shrub with clustering white blossoms and beautiful rough, red berries. Indeed, the name arbutus is derived from the Celtic word Arboise, meaning rough fruit.

LARGE or AMERICAN CRANBERRY
 (Oxycoccus macrocarpus; Vaccinium macrocarpon of Gray)
Huckleberry family

   Flowers - Light pink, about 1/2 in. across, nodding on slender pedicels from sides and tips of erect branches. Calyx round, 4-or 5-parted; corolla a long cone in bud, its four or five nearly separate, narrow petals turned far backward later; 8 or 10 stamens, the anthers united into a protruding cone, its hollow tubes shedding pollen by a pore at tip. Stem: Creeping or trailing, slender, woody, 1 to 3 ft. long, its leafy branches 8 in. high or less. Leaves: Small, alternate, oblong, evergreen, pale beneath, the edges rolled backward. Fruit: An oblong or ovoid, many seeded, juicy red berry (Oxycoccus = sour berry). Preferred Habitat - Bogs; sandy, swampy meadows. Flowering Season - June-August. Distribution - North Carolina, Michigan, and Minnesota northward and westward.

   A hundred thousand people are interested in the berry of this pretty vine to one who has ever seen its flowers. Yet if the blossom were less attractive, to insects at least, and took less pains to shake out its pollen upon them as they cling to the cone to sip its nectar, few berries would accompany the festive Thanksgiving turkey. Cultivators of the cranberry know how important it is to have the flooded bogs well drained before the flowering season. Water (or ice) may cover the plants to the depth of a foot or more all winter and until the 10th of May; and during the late summer it is often advisable to overflow the bogs to prevent injury of the fine, delicate roots from drought, and to destroy the worm that is the plant's worst enemy; but until the flowers have wooed the bees, flies, and other winged benefactors, and fruit is well formed, every cultivator knows enough not to submerge his bog. With flowers under water there are no insect visitors, consequently no berries. Dense mats of the wiry vines should yield about one hundred and fifty bushels of berries to the acre, under skilful cultivation - a most profitable industry, since the cranberry costs less to cultivate, gather, and market than the strawberry or any of the small perishable fruits. Planted in muck and sand in the garden, the vines yield surprisingly good results. The Cape Cod Bell is the best known market berry. One of the interesting sights to the city loiterer about the New England coast in early autumn is the berry picking that is conducted on an immense scale. Men, women, and children drop all other work; whole villages are nearly depopulated while daylight lasts; temporary buildings set up on the edges of the bogs contain throngs of busy people sorting, measuring, and packing fruit; and lonely railroad stations, piled high with crates, give the branch line its heaviest freight business of the year.

SHOOTING STAR; AMERICAN COWSLIP; PRIDE OF OHIO
(Dodecatheon Meadia) Primrose family

Flowers - Purplish pink or yellowish white, the cone tipped with yellow; few or numerous, hanging on slender, recurved pedicels in an umbel at top of a simple scape 6 in. to 2 ft. high. Calyx deeply 5-parted; corolla of 5 narrow lobes bent backward and upward; the tube very short, thickened at throat, and marked with dark reddish-purple dots; 5 stamens united into a protruding cone; 1 pistil, protruding beyond them. Leaves: Oblong or spatulate 3 to 12 in. long, narrowed into petioles, all from fibrous roots. Fruit: A 5-valved capsule on erect pedicels. Preferred Habitat - Prairies, open woods, moist cliffs. Flowering Season - April-May. Distribution - Pennsylvania southward and westward, and from Texas to Manitoba.

Ages ago Theophrastus called an entirely different plant by this same scientific name, derived from dodeka = twelve, and theos = gods; and although our plant is native of a land unknown to the ancients, the fanciful Linnaeus imagined he saw in the flowers of its umbel a little congress of their divinities seated around a miniature Olympus! Who has said science kills imagination? These handsome, interesting flowers so familiar in the Middle West and Southwest, especially, somewhat resemble the cyclamen in oddity of form, indeed, these prairie wildflowers are not unknown in florists' shops in Eastern cities.

Many flowers like the shooting star, cyclamen, and nightshade, with protruding cones made up of united stamens, are so designed that, as the bees must cling to them while sucking nectar, they receive pollen jarred out from the end of the cone on their undersides. The reflexed petals serve three purposes: First, in making the flower more conspicuous; secondly, in facilitating access to nectar and pollen; and, finally, in discouraging crawling intruders. Where the short tube is thickened, the bee finds her foothold while she forces her tongue between the anther tips. The nectar is well concealed and quite deeply seated, thanks to the rigid cone. Few bee workers are flying at the shooting star's early blooming season. Undoubtedly the female bumblebees, which, by striking the protruding stigma before they jar out any pollen, cross-fertilize it, are the flower's benefactors; but one frequently sees the little yellow puddle butterfly clinging to the pretty blossoms.

Very different from the bright yellow cowslip of Europe is our odd, misnamed blossom.

BITTER-BLOOM; ROSE-PINK; SQUARE-STEMMED SABBATIA; ROSY CENTAURY
(Sabbatia angularis) Gentian family

Flowers - Clear rose pink, with greenish star in center, rarely white, fragrant, 1 1/2 in. broad or less, usually solitary on long peduncles at ends of branches. Calyx lobes very narrow; corolla of 5 rounded segments; stamens 5; style 2-cleft. Stem: Sharply 4-angled, 2 to 3 ft. high, with opposite branches, leafy. Leaves: Opposite, 5-nerved, oval, tapering at tip, and clasping stem by broad base. Preferred Habitat - Rich soil, meadows, thickets. Flowering Season - July-August. Distribution - New York to Florida, westward to Ontario, Michigan, and Indian Territory.

During the drought of midsummer the lovely rose-pink blooms inland with cheerful readiness to adapt itself to harder conditions than most of its moisture-loving kin will tolerate; but it may be noticed that although we may oftentimes find it growing in dry soil, it never spreads in such luxuriant clusters as when the roots are struck beside meadow runnels and ditches. Probably the plant would be commoner than it is about populous Eastern districts were it not so much sought after as a tonic medicine.

It was the Centaurea, represented here by the blue ragged sailor of gardens, and not our Centaury, a distinctly American group of plants, which, Ovid tells us, cured a wound in the foot of the Centaur Chiron, made by an arrow hurled by Hercules.

Three exquisite members of the Sabbatia tribe keep close to the Atlantic coast in salt meadows and marshes, along the borders of brackish rivers, and very rarely in the sand at the edges of fresh-water ponds a little way inland. From Maine to Florida they range, and less frequently are met along the shores of the Gulf of Mexico so far as Louisiana. How bright and dainty and are! Whole meadows are radiant with their blushing lovliness. Probably if they consented to live far away from the sea, they would lose some of the deep, clear pink from out their lovely petals, since all flowers show a tendency to brighten their colors as they approach the coast. In England some of the same wildflowers we have here are far deeper-hued, owing, no doubt to the fact that they live on a sea-girt, mo-isture-laden island, and also that the sun never scorches and blan-ches at the far north as it does in the United States.

As might be expected, blossoms so bright of hue as the marsh pinks attract many insects. Guided by the yellow eye that serves as a pathfinder to the nectary, they feast on the generour supply of sweets; but all unwittingly they must pay for their entertainment by carrying pollen from early to later flowers. Like so many other blossoms, the sabbatias guard themselves against the evils of self-fertilization by shedding their pollen before they mature and spread their two-cleft style, which is now ready to receive the golden, qui-ckening dust on its stigmatic inner surfaces.

The SEA or MARSH PINK, or ROSE OF PLYMOUTH (S. stella-ris), whose graceful alternate branching stem attains a height of two feet only under most favorable conditions, from July to September opens a succession of pink flowers that often fade to white. The yellow eye is bordered with carmine. They measure about one inch across, and are usually solitary at the ends of branches, or else sway on slender peduncles from the axils. The upper leaves are narrow and bract-like; those lower down gradually widen as they approach the root.

Similar to the Rose of Plymouth is the even more graceful SLEN-DER MARSH PINK (S. Campanulata - the S. gracilis of Gray), whose upper leaves are almost thread-like in their narrowness. Its five calyx lobes, too, are exceedingly slender, and often as long as the corolla lobes. One of our soldiers in Cuba, during the Spanish War, sent home to his sister in Massachusetts some of these same little flowers in a letter. "You would just love to see the marshes here," he wrote. "They are filled with beautiful little pink flowers. I wish I knew their names." That soldier had passed by New England marshes aglow with the blossoms all his life, but he had never noticed them until all his perceptions became quickened by the stimulus of travel and the excitement of war. How blind and deaf we all are in some directions; having eyes we see not, and ears we hear not, in the natural as in the spiritual realm.

No danger of confusing the LARGE MARSH PINK (S. dode-candra - S. chloroides of Gray) with its smaller, more branching relatives. It displays few flowers to a plant, but each measures two and a half inches or less across, and has from nine to twelve pink (or rarely white) petals. This sabbatia often chooses the sandy borders of ponds for its habitat.

SPREADING DOGBANE; FLY-TRAP DOGBANE; HONEY-BLOOM; BITTER-ROOT
 (Apocynum androsaemifolium) Dogbane family

Flowers - Delicate pink, veined with a deeper shade, fragrant, bell-shaped, about 1/3 in. across, borne in loose terminal cymes. Calyx 5-parted; corolla of 5 spreading, recurved lobes united into a tube; within the tube 5 tiny, triangular appendages alternate with stamens; the arrow-shaped anthers united around the stigma and slightly adhering to it. Stem: 1 to 4 ft. high, with forking, spreading, leafy branches. Leaves: Opposite, entire-edged, broadly oval, narrow at base, paler, and more or less hairy below. Fruit: Two pods about 4 in. long. Preferred Habitat - Fields, thickets, beside roads, lanes, and walls. Flowering Season - June-July. Distribution - Northern part of British Possessions south to Georgia, westward to Nebraska.

Everywhere at the North we come across this interesting, rather shrubby plant, with its pretty but inconspicuous little rose-veined bells suggesting pink lilies-of-the-valley. Now that we have learned to read the faces of flowers, as it were, we instantly suspect by the color, fragrance, pathfinders, and structure that these are artful wilers, intent on gaining ends of their own through their insect admirers. What are they up to?

Let us watch. Bees, flies, moths, and butterflies, especially the latter, hover near. Alighting, the butterfly visitor unrolls his long tongue and inserts it where the five pink veins tell him to, for five nectar-bearing glands stand in a ring around the base of the pistil. Now, as he withdraws his slender tongue through one of the V-shaped cavities that make a circle of traps, he may count himself lucky to escape with no heavier toll imposed than pollen cemented to it. This granular dust he is required to rub off against the stigma of the next flower entered. Some bees, too, have been taken with the dogbane's pollen cemented to their tongues. But suppose a fly call upon this innocent-looking blossom? His short tongue, as well as the butterfly's, is guided into one of the V-shaped cavities after he has sipped; but, getting wedged between the trap's horny teeth, the poor little victim is held a prisoner there until he slowly dies of starvation in sight of plenty. This is the penalty he must pay for trespassing on the butterfly's preserves! The dogbane, which is perfectly adapted to the butterfly, and dependent upon it for help in producing fertile seed, ruthlessly destroys all poachers that are not big or strong enough to jerk away from its vise-like grasp. One often sees small flies and even moths dead and dangling by the tongue from the wicked little charmers. If the flower assimilated their dead bodies as the pitcher plant, for example, does those of its victims, the fly's fate would seem less cruel. To be killed by slow torture and dangled like a scarecrow simply for pilfering a drop of nectar is surely an execution of justice medieval in its severity.

In July the most splendid of our native beetles, the green dandy (Eumolpus auratus) fastens itself to the dogbane's foliage in numbers until often the leaves appear to be studded with these brilliant little jewels. "It is not easy," says William Hamilton Gibson, "to describe its burnished hue, which is either shimmering green, or peacock blue, or purplish-green, or refulgent ruby, according to the position in which it rests." But it is not golden, as its specific name would imply. It confines itself exclusively to the dogbane. To prevent capture, it has a trick of drawing up its legs and rolling off into the grass its body so cleverly matches.

>From the silky coma on which the small seeds float away from long pods to found new colonies, from the opposite leaves, milky juice, and certain structural resemblances in the flowers, one might guess this plant belonged to the milkweed tribe. Formerly it was so classed; and although the botanists have now removed its family one step away, the milkweed butterflies, especially the Monarch (Anosia plexippus), ignoring the arbitrary dividing line of man, still includes the dogbane on its visiting list. We know that this plant derived its name from the fact that it was considered poisonous to dogs; and we also know that all the tribe of milkweed butterflies are provided with protective secretions which are distasteful to birds and predaceous insects, enjoying their immunity from attack, it is thought, from the acrid, poisonous character of the foliage on which the caterpillars feed.

COMMON MIIKWEED or SILKWEED
(Asclepias Syriaca; A. cornuti of Gray) Milkweed family

Flowers - Dull pale greenish purple pink, or brownish pink, borne on pedicels, in many flowered, broad umbels. Calyx inferior, 5-parted; corolla deeply 5-cleft, the segments turned backward. Above them an erect, 5-parted crown, each part called a hood, containing a nectary, and with a tooth on either side, and an incurved horn projecting from within. Behind the crown the short, stout stamens, united by their filaments in a tube, are inserted on the corolla. Broad anthers united around a thick column of pistils terminating in a large, sticky, 5-angled disk. The anther sacs tipped with a winged membrane; a waxy, pear-shaped pollen-mass in each sac connected with the stigma in pairs or fours by a dark gland, and suspended by a stalk like a pair of saddle-bags. Stem: Stout, leafy, usually unbranched, 3 to 5 ft. high, juice milky. Leaves: Opposite, oblong, entire-edged smooth above, hairy below, 4 to 9 in. long. Fruit: 2 thick, warty pods, usually only one filled with compressed seeds attached to tufts of silky, white, fluffy hairs. Preferred Habitat - Fields and waste places, roadsides. Flowering Season - June-September Distribution - New Brunswick, far westward and southward to North Carolina and Kansas.

After the orchids, no flowers show greater executive ability, none have adopted more ingenious methods of compelling insects to work for them than the milkweeds. Wonderfully have they perfected their mechanism in every part until no member of the family even attempts to fertilize itself; hence their triumphal, vigorous march around the earth, the tribe numbering over nineteen hundred species located chiefly in those tropical and warm, temperate regions that teem with insect life.

Commonest of all with us is this rank weed, which possesses the dignity of a rubber plant. Much more attractive to human eyes, at least, than the dull, pale, brownish-pink umbels of flowers are its exquisite silky seed-tufts. But not so with insects. Knowing that the slightly fragrant blossoms are rich in nectar, bees, wasps, flies, beetles, and butterflies come to feast. Now, the visitor finding his a-lighting place slippery, his feet claw about in all directions to secure a hold, just as it was planned they should for in his struggles some of his feet must get caught in the fine little clefts at the base of the flower. His efforts to extricate his foot only draw it into a slot at the end of which lies a little dark-brown body. In a newly opened flower five of these little bodies may be seen between the horns of the crown, at equal distances around it. This tiny brown excrescence is hard and horny, with a notch in its face. It is continuous with and forms the end of the slot in which the visitor's foot is caught. Into this he must draw his foot or claw, and finding it rather tightly held, must give a vigorous jerk to get it free. Attached to either side of the little horny piece is a flattened yellow pollen-mass, and so away he flies with a pair of these pollinia, that look like tiny saddle-bags, dangling from his feet. One might think that such rough handling as many insects must submit to from flowers would discourage them from making any more visits; but the desire for food is a mighty passion. While the insect is flying off to another blossom, the stalk to which the saddlebags are attached twists until it brings them together, that, when his feet get caught in other slots, they may be in the position to get broken off in his struggles for freedom precisely where they will fertilize the stigmatic chambers. Now the visitor flies away with the stalks alone sticking to his claws. Bumblebees and hive-bees have been caught with a dozen pollen-masses dangling from a single foot. Outrageous imposition!

Does this wonderful mechanism always work to perfection? Alas! no. It is a common thing to find dead hive-bees and flies hanging from the flowers. While still struggling to escape, the unhappy victims will be attacked by ants, beetles, and spiders, or killed by heavy showers. Larger and stronger insects than honeybees are required to regularly effect pollination and free themselves, especially when they are so unfortunate as to catch several feet in the grooves. Doubtless it is the bumblebee that can transfer pollen with impunity; but very many other insects, not perfectly adapted to the flowers, occasionally benefit them. Among the large butterflies the Papilios, which suck with their wings in motion, are the most useful, because in using their legs to offset the motion of their wings they rapidly repeat those movements which are necessary to draw the pollinia from the anther cells and insert them in the stigmatic chambers of other flowers. "Large butterflies like Danais," says Professor Robertson, "hold their wings still in sucking, spending more time on an umbel, but generally carrying pollinia. Small butterflies are worse than useless. They remain long on the umbels sucking, but resting their feet superficially on the flowers.

Since several moths were found entrapped, pollination must often be brought about by night-flying Lepidoptera. As a rule, Diptera (flies) either do not transfer pollinia at all, or become hopelessly entangled when they do. "Occasionally pollen-masses are found on the tongues of insects, especially on those of bees and wasps, which move about with their unruly member sticking out. Probably no one has ever made the exhaustive and absorbingly interesting study of the milkweeds that Professor Robertson has.

Better than any written description of the milkweed blossom's mechanism is a simple experiment. If you have neither time nor patience to sit in the hot sun, magnifying glass in hand, and watch for an unwary insect to get caught, take an ordinary housefly, and hold it by the wings so that it may claw at one of the newly opened flowers from which no pollinia have been removed. It tries frantically to hold on, and with a little direction it may be led to catch its claws in the slots of the flower. Now pull it gently away, and you will find a pair of saddlebags slung over his foot by a slender curved stalk. If you are rarely skilful, you may induce your fly to withdraw the pollinia from all five slots on as many of his feet. And they are not to be thrown or scraped off, let the fly try as hard as he pleases. You may now invite the fly to take a walk on another flower in which he will probably leave one or more pollinia in its stigmatic cavities.

Dr. Kerner thought the milky juice in milkweed plants, especially abundant in the uppermost leaves and stems, serves to protect the flowers from useless crawling pilferers. He once started a number of ants to climb up a milky stalk. When they neared the summit, he noticed that at each movement the terminal hooks of their feet cut through the tender epiderm, and from the little clefts the milky juice began to flow, bedraggling their feet and the hind part of their bodies. "The ants were much impeded in their movements," he writes, "and in order to rid themselves of the annoyance, drew their feet through their mouths. Their movements however, which accompanied these efforts, simply resulted in making fresh fissures and fresh discharges of milky juice, so that the position of the ants became each moment worse and worse. Many escaped by getting to the edge of a leaf and dropping to the ground. Others tried this method of escape too late, for the air soon hardened the milky juice into a tough brown substance, and after this, all the strugglings of the ants to free themselves from the viscid matter were in vain." Nature's methods of preserving a flower's nectar for the insects that are especially adapted to fertilize it, and of punishing all useless intruders, often shock us yet justice is ever stern, ever kind in the largest sense.

If the asclepias really do kill some insects with their juice, others doubtless owe their lives to it. Among the "protected" insects are the milkweed butterflies and their caterpillars, which are provided with secretions that are distasteful to birds and predaceous insects. "These acrid secretions are probably due to the character of the plants upon which the caterpillars feed," says Dr. Holland, in his beautiful and invaluable "Butterfly Book." "Enjoying on this account immunity from attack, they have all, in the process of time, been mimicked by species in other genera which have not the same immunity." "One cannot stay long around a patch of milkweeds without seeing the monarch butterfly. (Anosia plexippus), that splendid, bright, reddish-brown winged fellow, the borders and veins broadly black, with two rows of white spots on the outer borders and two rows of pale spots across the tip of the fore wings. There is a black scent-pouch on the hind wings. The caterpillar, which is bright yellow or greenish yellow, banded with shining black, is furnished with black fleshy 'horns' fore and aft."

Like the dandelion, thistle, and other triumphant strugglers for survival, the milkweed sends its offspring adrift on the winds to found fresh colonies afar. Children delight in making pompons for their hats by removing the silky seed-tufts from pods before they burst, and winding them, one by one, on slender stems with fine thread. Hung in the sunshine, how charmingly fluffy and soft they dry!

Among the comparatively few butterfly flowers - although, of course, other insects not adapted to them are visitors - is the PURPLE MILKWEED (A. purpurasceus), whose deep magenta umbels are so conspicuous through the summer months. Hummingbirds occasionally seek it too. From Eastern Massachusetts to Virginia, and westward to the Mississippi, or beyond, it is to be found in dry fields, woods, and thickets.

The SWAMP MILKWEED (A. incarnata), on the other hand, rears its intense purplish-red or pinkish hoods in wet places. Its leaves are lance-shaped or oblong-lanceolate, whereas the purple milkweed's leaves are oblong or ovate-oblong. This is a smooth plant; and a similar species once reckoned as a mere variety (A. pulchra) is the HAIRY MILKWEED. It differs chiefly in having some hairs on the under side of its leaves, and a great many hairs on its stem. Both plants bear erect, rather slender, tapering pods.

The POKE or TALL MILKWEED (A. exaltata - A. phytolaecoides of Gray) may attain a height of six feet if the moist soil in which it grows be exactly to its liking. Drooping or spreading umbels of flowers whose corolla segments are pale purplish green, and whose crown is clear ivory white or pink, appear from June to August from Maine to Georgia and far westward. Sometimes the tapering oblong leaves may be nine inches long. The erect seedpods are drawn out to an unusually long point.

One may always distinguish the low-growing FOUR-LEAVED MILKWEED (A. quadrifolia) from its relatives of ranker growth by its general air of refinement, as well as by the two pairs of thin, tapering leaves that grow in an upright whorl near the middle of the slender stem. Usually there are no leaves on the lower part. Small terminal umbels of delicate pink and white fragrant flowers, which appear from May till July, give place to very narrow pointed pods in late summer. From Maine to Ontario southward to North Carolina and Arkansas is its range, in woods and thickets chiefly.

HEDGE or GREAT BINDWEED; WILD MORNING-GLORY; RUTLAND BEAUTY; BELL-BIND; LADY'S NIGHTCAP

(Convolvulus sepium; Calystegia sepium of Gray) Morning-glory family

Flowers - Light pink, with white stripes or all white, bell-shaped, about 2 in. long, twisted in the bud, solitary, on long peduncles from leaf axils. Calyx of 5 sepals, concealed by 2 large bracts at base. Corolla 5-lobed, the 5 included stamens inserted on its tube; style with 2 oblong stigmas. Stem: Smooth or hairy, 3 to 10 ft. long, twining or trailing over ground. Leaves: Triangular or arrow-shaped, 2 to 5 in. long, on slender petioles. Preferred Habitat - Wayside hedges, thickets, fields, walls. Flowering Season - June-September. Distribution - Nova Scotia to North Carolina, westward to Nebraska. Europe and Asia.

No one need be told that the pretty, bell-shaped pink and white flower on the vigorous vine clambering over stone walls and winding about the shrubbery of wayside thickets in a suffocating embrace is akin to the morning-glory of the garden trellis (C. major). An exceedingly rapid climber, the twining stem often describes a complete circle in two hours, turning against the sun, or just contrary to the hands of a watch. Late in the season, when an abundance of seed has been set, the flower can well afford to keep open longer hours, also in rainy weather; but early in the summer, at least, it must attend to business only while the sun shines and its benefactors are flying. Usually it closes at sundown. On moonlight nights, however, the hospitable blossom keeps open for the benefit of certain moths. In Europe the plant's range is supposed to be limited to that of a crepuscular moth (Sphinx convolvuli), and where that benefactor is rare, as in England, the bindweed sets few seeds where it does not occur, as in Scotland, this convolvulus is seldom found wild; whereas in Italy Delpino tells of catching numbers of the moths in hedges overgrown with the common plant, by standing with thumb and forefinger over a flower, ready to close it when the insect has entered. We know that every floral clock is regulated by the hours of flight of its insect friends. When they have retired, the flowers close to protect nectar and pollen from useless pilferers. In this country various species of bees chiefly fertilize the bindweed blossoms. Guided by the white streaks, or pathfinders, they crawl into the deep tube and sip through one of the five narrow passages leading to the nectary. A transverse section of the flower cut to show these five passages standing in a circle around the central ovary looks like the end of a five-barreled revolver. Insects without a suitably long proboscis are, of course, excluded by this arrangement.

>From July until hard frost look for that exquisite little beetle, Cassida aurichalcea, like a drop of molten gold, clinging beneath the bindweed's leaves. The small perforations reveal his hiding places. "But you must be quick if you would capture him," says William Hamilton Gibson, "for he is off in a spangling streak of glitter. Nor is this golden sheen all the resource of the little insect; for in the space of a few seconds, as you hold him in your hand, he has become a milky, iridescent opal, and now mother-of-pearl, and finally crawls before you in a coat of dull orange." A dead beetle loses all this wonderful luster. Even on the morning-glory in our gardens we may sometimes find these jeweled mites, or their fork-tailed, black

larvae, or the tiny chrysalids suspended by their tails, although it is the wild bindweed that is ever their favorite abiding place.

The small FIELD BINDWEED (C. arvensis), a common immigrant from Europe, which has taken up its abode from Nova Scotia and Ontario southward to New Jersey, and westward to Kansas, trails over the ground with a deathless persistency which fills farmers with dismay. It is like a small edition of the hedge bind weed, only its calyx lacks the leaf-like bracts at its base, its slender stem rarely exceeds two feet in length, and the little pink and white flowers often grow in pairs. Their habit of closing both in the evening and in rainy weather indicates that they are adapted for diurnal insects only; but if the bell hang down, or if the corolla drop off, the pollen must fall on the stigma and effect self-fertilization. Many more insects visit this flower than the large bindweed, attracted by the peculiar fragrance, and led by the white streaks to the orange-colored under surface of the ovary, where the nectar lies concealed. Stigmas and anthers mature at the same time; but as the former are slightly the longer, they receive pollen brought from another flower before the visitor gets freshly dusted.

GROUND OR MOSS PINK
(Phlox subulata) Phlox family

Flowers - Very numerous, small, deep purplish pink, lavender or rose, varying to white, with a darker eye, growing in simple cymes, or solitary in a Western variety. Calyx with 5 slender teeth; corolla salver-form with 5 spreading lobes; 5 stamens inserted on corolla tube; style 3-lobed. Stems: Rarely exceeding 6 in. in height, tufted like mats, much branched, plentifully set with awl-shaped, ever-green leaves barely 1/2 in. long, growing in tufts at joints of stem. Preferred Habitat - Rocky ground, hillsides. Flowering Season - April-June Distribution - Southern New York to Florida, westward to Michigan and Kentucky.

A charming little plant, growing in dense evergreen mats with which Nature carpets dry, sandy, and rocky hillsides, is often completely hidden beneath its wealth of flowers. Far beyond its natural range, as well as within it, the moss pink glows in gardens, cemeteries and parks, wherever there are rocks to conceal or sterile wastes to beautify. Very slight encouragement induces it to run wild. There are great rocks in Central Park, New York, worth travelling miles to see in early May, when their stern faces are flushed and smiling with these blossoms.

Another low ground species is the CRAWLING PHLOX (P. reptans). It rarely exceeds six inches in height; nevertheless its larger pink, purple, or white flowers, clustered after the manner of the tall garden phloxes, are among the most showy to be found in the spring woods. A number of sterile shoots with obovate leaves, tapering toward the base, rise from the runners and set off the brilliant blossoms among their neat foliage. From Pennsylvania southward and westward is its range, especially in mountainous regions; but this plant, too, was long ago transplanted from Nature's gardens into man's.

Large patches of the DOWNY PHLOX (P. pilosa) brighten dry prairie land with its pinkish blossoms in late spring. Britton and Brown's botany gives its range as "Ontario to Manitoba, New Jersey, Florida, Arkansas, and Texas." The plant does its best to attain a height of two feet; usually its flowers are much nearer the ground. Butterflies, the principal visitors of most phloxes, although long-tongued bees and even flies can sip their nectar, are ever seen hovering above them and transferring pollen, although in this species the style is so short pollen must often fall into the tube and self-fertilize the stigma. To protect the flowers from useless crawling visitors, the calices are coated with sticky matter, and the stems are downy.

OBEDIENT PLANT; FALSE DRAGONHEAD; LION'S HEART
(Physostegia Virginiana) Mint family

Flowers - Pale magenta, purplish rose, or flesh-colored, often variegated with white, 1 in. long or over, in dense spikes from 4 to 8 in. long. Calyx a 5-toothed oblong bell, swollen and remaining open in fruit, held up by lance-shaped bracts. Corolla tubular and much enlarged where it divides into 2 lips, the upper lip concave, rounded, entire, the lower lip 3 lobed. Stamens 4, in two pairs under roof of upper lip, the filaments hairy; 1 pistil. Stem: 1 to 4 ft. high, simple or branched above, leafy. Leaves: Opposite, firm, oblong to oblong-lanceolate, narrowing at base, deeply saw-edged. Preferred Habitat - Moist soil. Flowering Season - July-September. Distribution - Quebec to the Northwest Territory, southward to the Gulf of Mexico as far west as Texas.

Bright patches of this curious flower enliven railroad ditches, gutters, moist meadows and brooksides - curious, for it has the peculiarity of remaining in any position in which it is placed. With one puff a child can easily blow the blossoms to the opposite side of the spike, there to stay in meek obedience to his will. "The flowers are made to assume their definite position," says Professor W. W. Bailey in the "Botanical Gazette," "by friction of the pedicels against the subtending bracts. Remove the bracts, and they at once fall limp."

Qf course the plant has some better reason for this peculiar obedience to every breath that blows than to amuse windy-cheeked boys and girls. Is not the ready movement useful during stormy weather in turning the mouth of the flower away from driving rain, and in fair days, when insects are abroad, in presenting its gaping lips where they can best alight? We all know that insects, like birds, make long flights most easily with the wind, but in rising and a-lighting it is their practice to turn against it. When bees, for example, are out for food on windy days, and must make frequent stops for refreshment among the flowers, they will be found going against the wind, possibly to catch the whiffs of fragrance borne on it that guide them to feast, but more likely that they may rise and alight readily. One always sees bumblebees conspicuous among the obedient plant's visitors. After the anthers have shed their pollen - and tiny teeth at the edges of the outer pair aid its complete removal by insects - the stigma comes up to occupy their place under the roof. Certainly this flower; which is so ill-adapted to fertilize itself, has every reason to court insect messengers in fair and stormy 'weather.

MOTHERWORT
(Leonurus Cardiaca) Mint family.

Flowers - Dull purple pink, pale purple, or white, small, clustered in axils of upper leaves. Calyx tubular, bell-shaped, with 5 rigid awl-like teeth; corolla 2-lipped, upper lip arched, woolly without; lower lip 3-lobed, spreading, mottled; the tube with oblique ring of hairs inside. Four twin-like stamens, anterior pair longer, reaching under upper lip; style 2-cleft at summit. Stem: 2 to 5 ft. tall, straight, branched, leafy, purplish. Leaves: Opposite, on slender petioles; lower ones rounded, 2 to 4 in. broad, palmately cut into 2 to 5 lobes; upper leaves narrower, 3-cleft or 3- toothed. Preferred Habitat - Waste places near dwellings. Flowering Season - June-September. Distribution - Nova Scotia southward to North Carolina, west to Minnesota and Nebraska. Naturalized from Europe and Asia.

"One is tempted to say that the most human plants, after all, are the weeds," says John Burroughs. "How they cling to man and follow him around the world, and spring up wherever he sets foot How they crowd around his barns and dwellings, and throng his garden, and jostle and override each other in their strife to be near him! Some of them are so domestic and familiar, and so harmless withal, that one comes to regard them with positive affection. Motherwort, catnip, plantain, tansy, wild mustard - what a homely, human look they have! They are an integral part of every old homestead. Your smart, new place will wait long before they draw near it."

How the bees love this generous, old-fashioned entertainer! One nearly always sees them clinging to the close whorls of flowers that are strung along the stem, and of course transferring pollen, in recompense, as they journey on. A more credulous generation imported the plant for its alleged healing virtues. What is the significance of its Greek name, meaning a lion's tail? Let no one suggest, by a far-stretched metaphor, that our grandmothers, in Revolutionary days, enjoyed pulling it to vent their animosity against the British.

WILD BERGAMOT
(Monarda fisiulosa) Mint family

Flowers - Extremely variable, purplish, lavender, magenta, rose, pink, yellowish pink, or whitish, dotted; clustered in a solitary, nearly flat terminal head. Calyx tubular, narrow, 5-toothed, very hairy within. Corolla 1 to 1 1/2 in. long, tubular, 2-lipped, upper lip erect, toothed; lower lip spreading, 3-lobed, middle lobe longest; 2 anther-bearing stamens protruding; 1 pistil; the style 2-lobed. Stem: 2 to 3 ft. high, rough, branched. Leaves: Opposite, lance-shaped, saw-edged, on slender petioles, aromatic, bracts and upper leaves whitish or the color of flower. Preferred Habitat - Open woods, thickets, dry rocky hills. Flowering Season - June-September. Distribution - Eastern Canada and Maine, westward to Minnesota, south to Gulf of Mexico.

Half a dozen different shades of bloom worn by this handsome, robust perennial afford an excellent illustration of the trials that beset one who would arbitrarily group flowers according to color. If the capricious blossom shows a decided preference for any shade, it is for magenta, the royal purple of the ancients, scarcely tolerated now except by Hoboken Dutch and the belles of the kitchen, whose Sunday hats are resplendent with intense effects.

Only a few bergamot flowers open at a time; the rest of the slightly rounded head, thickly set with hairy calices, looks as if it might be placed in a glass cup and make an excellent pen wiper. If the cultivated human eye (and stomach) revolt at magenta, It is ever a favorite shade with butterflies. They flutter in ecstasy over the gay flowers; indeed, they are the principal visitors and benefactors, for the erect corollas, exposed organs, and level-topped heads are well adapted to their requirements. That exquisite little feathered jewel, the ruby-throated hummingbird, flashes about the bright patches an instant, and is gone; but he too has paid for his feast in transferring pollen. Insects which land anywhere they please on the flowers, receive pollen on various places, just as in the case of the scarlet Oswego tea, of similar formation. Small bees, which if unable to drain the brimming tubes of nectar, at least sip from them and help themselves to pollen also, without paying the flower's price; and certain mischievous wasps, forever bent on nipping holes in tubes they cannot honestly drain, give a score of other pilferers an opportunity to steal sweets.

SNAKE-HEAD; TURTLE-HEAD; BALMONY; SHELL-FLOWER; COD-HEAD
(Chelone glabra) Figwort family

Flowers - White tinged with pink, or all white, about 1 in. long, growing in a dense terminal cluster. Calyx 5-parted, bracted at base; corolla irregular, broadly tubular, 2-lipped; upper lip arched, swollen, slightly notched; lower lip 3-lobed, spreading, woolly within; 5 stamens, sterile, 4 in pairs, anther-bearing, woolly; 1 pistil. Stem: 1 to 3 ft. high, erect, smooth, simple, leafy. Leaves: Opposite, lance-shaped, saw-edged. Preferred Habitat - Ditches, beside streams, swamps. Flowering Season - July-September. Distribution - Newfoundland to Florida, and half way across the continent.

It requires something of a struggle for even so strong and vigorous an insect as the bumblebee to gain admission to this inhospitable-looking flower before maturity; and even he abandons the attempt over and over again in its earliest stage before the little heart-shaped anthers are prepared to dust him over. As they mature, it opens slightly, but his weight alone is insufficient to bend down the stiff, yet elastic, lower lip. Energetic prying admits first his head, then he squeezes his body through, brushing past the stamens as he finally disappears inside. At the moment when he is forcing his way in, causing the lower lip to spring up and down, the eyeless turtle seems to chew and chew until the most sedate beholder must smile at the paradoxical show. Of course it is the bee that is feeding, though the flower would seem to be masticating the bee with the keenest relish The counterfeit tortoise soon disgorges its lively mouthful, however, and away flies the bee, carrying pollen on his velvety back to rub on the stigma of an older flower. After the anthers have shed their pollen and become effete, the stigma matures, and occupies their place. By this time the flower presents a wider entrance, and as the moisture-loving plant keeps the nectaries abundantly filled, what is to prevent insects too small to come in contact with anthers and stigma in the roof from pilfering to their heart's content? The woolly throat discourages many, to be sure; but the turtle-head, like its cousins the beard-tongues, has a sterile fifth stamen, whose greatest use is to act as a drop-bar across the base of the flower. The long-tongued bumblebee can get his drink over the bar, but smaller, unwelcome visitors are literally barred out.

If bees are the preferred visitors of the turtle-head, why do we find the Baltimore butterfly, that very beautiful, but freaky, creature (Melitaea phaeton) hovering near? - that is, when we find it at all; for where it is present, it swarms, and keeps away from other localities altogether. On the under side of the leaves we shall often see patches of its crimson eggs. Later the caterpillars use the plant as their main, if not exclusive, food store. They are the innocent culprits which nine times out of ten mutilate the foliage.

LARGE PURPLE GERARDIA
(Gerardia purpurea) Figwort family

Flowers - Bright purplish pink, deep magenta, or pale to whitish, about 1 in. long and broad, growing along the rigid, spreading branches. Calyx 5-toothed; corolla funnel-form, the tube much inflated above and spreading into 5 unequal, rounded lobes, spotted within, or sometimes downy; 4 stamens in pairs, the filaments hairy; 1 pistil. Stem: 1 to 2 1/2 ft. high, slender, branches erect or spreading. Leaves: Opposite, very narrow, 1 to 1 1/2 in. long. Preferred Habitat - Low fields and meadows; moist, sandy soil. Flowering Season - August-October. Distribution - Northern United States to Florida, chiefly along Atlantic coast.

Low-lying meadows gay with gerardias were never seen by that quaint old botanist and surgeon, John Gerarde, author of the famous "Herball or General Historie of Plants," a folio of nearly fourteen hundred pages, published in London toward the close of Queen Elizabeth's reign. He died without knowing how much he was to be honored by Linnaeus in giving his name to this charming American genus.

Large patches of the lavender-pink gerardia, peeping above the grass, make the wayfarer pause to feast his eyes, while the practical bee, meanwhile, takes a more substantial meal within the spreading funnels. It is his practice to hang upside down while sucking, using the hairs on the filaments as footholds. Naturally he receives the pollen on his underside - just where it will be rubbed off against the stigma impeding his entrance to the next funnel visited. Any of the very dry pollen that may have fallen on the hairy filaments drops upon him.

"And 'tis my faith that every flower
  Enjoys the air it breathes,"

chanted Wordsworth. It is a special pity to gather the gerardias, which, as they grow, seem to enjoy life to the full, and when picked, to be so miserable they turn black as they dry. Like their relatives the foxgloves, they are difficult to transplant, because it is said they are more or less parasitic, fastening their roots on those of other plants. When robbery becomes flagrant, Nature brands sinners in the vegetable kingdom by taking away their color, and perhaps their leaves, as in the case of the broom-rape and Indian pipe; but the fair faces of the gerardias and foxgloves give no hint of the petty thefts committed under cover of darkness in the soil below.

The SMALL-FLOWERED GERARDIA (G. Paupercula) so like the preceding species it was once thought to be a mere variety, ranges westward as far as Wisconsin, especially about the Great Lakes. But it is a lower plant, with more erect branches, smaller flowers, quite woolly within, and with a decided preference for bogs as well as low meadows.

In salt marshes along the Atlantic Coast and the Gulf of Mexico, from Maine to Louisiana, the SEA-SIDE GERARDIA (G. maritima) flowers in midsummer, or a few weeks ahead of the autumnal, up-land species. The plant, which rarely exceeds a foot in height, is sometimes only four inches above ground; and although at the North the paler magenta blossoms are only about half the length of the purple gerardias, in the South they are sometimes quite as long.

In dry woods and thickets, on banks and hills from Quebec to Georgia, and westward to the Mississippi we find the SLENDER

GERARDIA (G. tenuifolia), its pale magenta, spotted, compressed corolla about half an inch long; its very slender, low stem set with exceedingly narrow leaves.

TWIN-FLOWER; GROUND VINE
(Linnaea borealis) Honeysuckle family

Flowers - Delicate pink or white tinged with rose, bell-shaped, about 1/2 in. long, fragrant, nodding in pairs on slender, curved pedicels from an erect peduncle, 2-bracted where they join. Calyx 5-toothed, sticky; corolla 5-lobed, bell-shaped, hairy within; 4 stamens in pairs inserted near base of tube; 1 pistil. Stem: Trailing, 6 in. to 2 ft. long; the branches erect. Leaves: Opposite, rounded, petioled, evergreen. Preferred Habitat - Deep, cool, mossy woods. Flowering Season - May-July. Distribution - Northern parts of America, Europe, and Asia. In the United States southward as far as the mountains of Maryland, and the Sierra Nevadas in California.

With the consent of modest Linnaeus himself, Dr. Gronovius selected this typical woodland blossom to transmit the great master's flame to posterity -

"Monument of the man of flowers."

But small and shy as it is, does Nature's garden contain a lovelier sight than scores of these deliciously fragrant pink bells swaying above a carpet of the little evergreen leaves in the dim aisle of some deep, cool, lonely forest? Trailing over prostrate logs and mossy rocks, racing with the partridge vine among the ferns and dwarf cornels, the plant sends up "twin-born heads" that seem more fair and sweet than the most showy pampered darlings of the millionaire's conservatory. Little wonder that Linnaeus loved these little twin sisters, or that Emerson enshrined them in his verse.

Contrary to popular impression, this vine, that suggests the dim old forest and exhales the very breath of the spring woods, will consent to run about our rock gardens, although it seems almost a sacrilege to move it from natural surroundings so impressively beautiful. Unlike the arbutus, which remains ever a wildling, pining slowly to death on close contact with civilization, the twin-flower thrives in light, moist garden soil where the sun peeps for a little while only in the morning. By nodding its head the flower protects its precious contents from rain, the hairs inside exclude small pilferers; but bees, attracted by the fragrance and color, are guided to the nectary by five dark lines and a patch of orange color near it.

JOE-PYE WEED; TRUMPET WEED; PURPLE THOROUGHWORT; GRAVEL or
KIDNEY-ROOT; TALL or PURPLE BONESET
(Eupatorium purpureum) Thistle family

Flower-heads - Pale or dull magenta or lavender pink, slightly fragrant, of tubular florets only, very numerous, in large, terminal, loose, compound clusters, generally elongated. Several series of pink overlapping bracts form the oblong involucre from which the tubular floret and its protruding fringe of style-branches arise. Stem: 3 to 10 ft. high, green or purplish, leafy, usually branching toward top. Leaves: In whorls of 3 to 6 (usually 4), oval to lance-shaped, saw-edged, petioled, thin, rough. Preferred Habitat - Moist soil, meadows, woods, low ground. Flowering Season - August-September. Distribution - New Brunswick to the Gulf of Mexico, westward to Manitoba and Texas.

Towering above the surrounding vegetation of low-lying meadows, this vigorous composite spreads clusters of soft, fringy bloom that, however deep or pale of tint, are ever conspicuous advertisements, even when the goldenrods, sunflowers, and asters enter into close competition for insect trade. Slight fragrance, which to the delicate perception of butterflies is doubtless heavy enough, the florets' color and slender tubular form indicate an adaptation to them, and they are by far the most abundant visitors, which is not to say that long-tongued bees and flies never reach the nectar and transfer pollen, for they do. But an excellent place for the butterfly collector to carry his net is to a patch of Joe-Pye weed in September. As the spreading style-branches that fringe each tiny floret are furnished with hairs for three-quarters of their length, the pollen caught in them comes in contact with the alighting visitor. Later, the lower portion of the style-branches, that is covered with stigmatic papillae along the edge, emerges from the tube to receive pollen carried from younger flowers when the visitor sips his reward. If the hairs still contain pollen when the stigmatic part of the style is exposed, insects self-fertilize the flower; and if in stormy, weather no insects are flying, the flower is nevertheless able to fertilize itself, because the hairy fringe must often come in contact with the stigmas of neighboring florets. It is only when we study flowers with reference to their motives and methods that we understand why one is abundant and another rare. Composites long ago utilized many principles of success in life that the triumphant Anglo-Saxon carries into larger affairs today.

Joe-Pye, an Indian medicine-man of New England, earned fame and fortune by curing typhus fever and other horrors with decoctions made from this plant.

COMMON BURDOCK; COCKLE-BUR; BEGGARS BUTTONS; CLOT-BUR; CUCKOO BUTTON
(Arctium minus; Lappa officinalis: var. minor of Gray) Thistle family

Flower-heads - Composite of tubular florets only, about 1/2 in. broad; magenta varying to purplish or white; the prominent round involucre of many overlapping leathery bracts, tipped with hooked bristles. Stem: 2 to 5 ft. high, simple or branching, coarse. Leaves: Large, the lower ones often 1 ft. long, broadly ovate, entire edged, pale or loosely cottony beneath, on hollow petioles. Preferred Habitat - Waste ground, waysides, fields, barnyards. Flowering Season - July-October. Distribution - Common throughout our area. Naturalized from Europe.

A larger burdock than this (A. Lappa) may be more common in a few localities of the East, but wherever one wanders, this plebeian boldly asserts itself. In close-cropped pastures it still flourishes with the well-armed thistles and mulleins, for the great leaves contain an exceedingly bitter, sour juice, distasteful to grazers. Nevertheless the unpaid cattle, like every other beast and man, must nolens volens transplant the burs far away from the parent plant to found new colonies. Literally by hook or by crook they steal a ride on every switching tail, every hairy dog and woolly sheep, every trouser-leg or petticoat. Even the children, who make dolls and baskets of burdock burs, aid them in their insatiate love of travel. Wherever man goes, they follow, until, having crossed Europe - with the Romans? - they are now at home throughout this continent. Their vitality is amazing; persecution with scythe and plow may retard, but never check their victorious march. Opportunity for a seed to germinate may not come until late in the summer; but at once the plant sets to work putting forth flowers and maturing seed, losing no time in developing superfluous stalk and branches. Butterflies, which, like the Hoboken Dutch, ever delight in magenta, and bees of various kinds, find these flowers, with a slight fragrance as an additional attraction, generous entertainers.

Pink, of all colors, is the most unstable in our flora, and the most likely to fade. Magentas incline to purple, on the one hand, or to pure pink on the other, and delicate shades quickly blanch when long exposed to the sun's rays. Thus we frequently find white blossoms of the once pink rhododendron, laurel, azalea, bouncing Bet, and turtle-head. Albinos, too, regularly occur in numerous species. Many colored flowers show a tendency among individuals to revert to the white type of their ancestors. The reader should bear these facts in mind, and search for his unidentified flower in the previous section or in the following one if this group does not contain it.

WHITE AND GREENISH FLOWERS

"The transition from wind-fertilization to insect-fertilization and the first traces of adaptation to insects, could only be due to the influence of quite short-lipped insects with feebly developed color sense. The most primitive flowers are therefore for the most part simple, widely open, regular, devoid of nectar or with their nectar unconcealed and easily accessible, and greenish, white, or yellow in color.... Lepidoptera, by the thinness, sometimes by the length, of their tongues, were able to produce special modifications. Through their agency were developed flowers with long and narrow tubes, whose colors and time of opening were in relation to the tastes and habits of their visitors." - Hermann Muller.

"Of all colors, white is the prevailing one; and of white flowers a considerably larger proportion smell sweetly than of any other color, namely, 14.6 per cent; of red only 8.2 per cent are odoriferous. The fact of a large proportion of white flowers smelling sweetly may depend in part on those which are fertilized by moths requiring the double aid of conspicuousness in the dusk and of odor. So great is the economy of Nature, that most flowers which are fertilized by crepuscular or nocturnal insects emit their odor chiefly or exclusively in the evening." - Charles Darwin.

WATER-PLANTAIN
(Alisma Plantago-aquatica) Water-plantain family

Flowers - Very small and numerous, white, or pale pink, whorled in bracted clusters forming a large, loose panicle 6 to 15 in. long on a usually solitary scape 1/2 to 3 ft. high. Calyx of 3 sepals corolla of 3 deciduous petals; 6 or more stamens; many carpels in a ring on a small flat receptacle. Leaves: Erect or floating, oblong or ovate, with several ribs, or lance-shaped or grass-like, petioled, all from root. Perferred Habitat - Shallow water, mud, marshes. Flowering Season - June-September. Distribution - North America, Europe, Asia.

Unlike its far more showy, decorative cousin the arrow-head, this wee-blossomed plant, whose misty white panicles rise with compensating generosity the world around, bears only perfect, regular flowers. Twelve infinitesimal drops of nectar, secreted in a fleshy ring around the center, are eagerly sought by flies. As the anthers point obliquely outward and away from the stigmas, an incoming fly, bearing pollen on his under side, usually alights in the center, and leaves some of the vitalizing dust just where it is most needed. But a "fly starting from a petal," says Muller, "usually applies its tongue to the nectar-drops one by one, and after each it strokes an anther with its labellae; in so doing it may bring various parts of its body in contact with the anthers. As a rule, however, the parts which come in contact with the anthers are not those which come in contact with the stigmas in the same flower." Any plant that lives in shallow water, which may dry up as summer advances, is under special necessity to produce an extra quantity of cross-fertilized seed to guard against extinction during drought. For the same reason it bears several kinds of leaves adapted to its environment: broad ones that spread their surfaces to the sunshine, and long grass-like ones to glide through currents of water that would tear those of any other shape. What diversity of leaf-form and structure we meet daily, and yet how very little does the wisest man of science understand of the reasons underlying such marvellous adaptability!

BROAD-LEAVED ARROWHEAD (Sagittaria latifolia; S. variabilis of Gray) Water-plantain family

Flowers - White, 1 to 1 1/2 in. wide, in 3-bracted whorls of 3, borne near the summit of a leafless scape 4 in. to 4 ft. tall. Calyx of 3 sepals corolla of 3 rounded, spreading petals. Stamens and pistils numerous, the former yellow in upper flowers usually absent or imperfect in lower pistillate flowers. Leaves: Exceedingly variable; those under water usually long and grasslike; upper ones sharply arrow-shaped or blunt and broad, spongy or leathery, on long petioles. Preferred Habitat - Shallow water and mud. Flowering Season - July-September. Distribution - From Mexico northward throughout our area to the circumpolar regions.

Wading into shallow water or standing on some muddy shore, like a heron, this striking plant, so often found in that bird's haunts, is quite as decorative in a picture, and, happily, far more approachable in life. Indeed, one of the comforts of botany as compared with bird study is that we may get close enough to the flowers to observe their last detail, whereas the bird we have followed laboriously over hill and dale, through briers and swamps, darts away beyond the range of field-glasses with tantalizing swiftness.

While no single plant is yet thoroughly known to scientists, in spite of the years of study devoted by specialists to separate groups, no plant remains wholly meaningless. When Keppler discovered the majestic order of movement of the heavenly bodies, he exclaimed, "Oh God, I think Thy thoughts after Thee!" - the expression of a discipleship every reverent soul must be conscious of in penetrating, be it ever so little a way, into the inner meaning of the humblest wayside weed.

Fragile, delicate, pure white, golden-centered flowers of the arrowhead, usually clustered about the top of the scape, naturally are the first to attract the attention whether of man or insect. Below these, dull green, unattractive collections of pistils, which by courtesy only may be called flowers, also form little groups of three. Like the Quakers at meeting, the male and female arrowhead flowers are separated, often on distinct plants. Of course the insect visitors - bees and flies chiefly - alight on the showy staminate blossoms first, and transfer pollen from them to the dull pistillate ones later, as it was intended they should, to prevent self-fertilization. How endless are the devices of the flowers to guard against this evil and to compel insects to cross-pollinate them! The most minute detail of the mechanism involved, which the microscope reveals, only increases our interest and wonder.

Any plant which elects to grow in shallow water must be amphibious; it must be able to breathe beneath the surface as the fish do, and also be adapted to thrive without those parts that correspond to gills; for ponds and streams have an unpleasant way of drying up in summer, leaving it stranded on the shore. This accounts in part for the variable leaves on the arrowhead, those underneath the water being long and ribbon-like, to bring the greatest possible area into contact with the air with which the water is charged. Broad leaves would be torn to shreds by the current through which grass-like blades glide harmlessly; but when this plant grows on shore, having no longer use for its lower ribbons, it loses them, and expands only broad arrow-shaped surfaces to the sunny air, leaves to be supplied with carbonic acid to assimilate, and sunshine to turn off the oxygen and store up the carbon into their system.

WATER ARUM; MARSH CALLA
(Calla palustris) Arum family

Flowers - Minute, greenish yellow, clustered on a cylinder-like, fleshy spadix about 1 in. long, partly enfolded by a large, white, oval, pointed, erect spathe, the whole resembling a small calla lily open in front. The solitary "flower" on a scape as long as the petioles of leaves, and, like them, sheathed at base. Leaves: Thick, somewhat heart-shaped, their spreading or erect petioles 4 to 8 in. long. Fruit: Red berries clustered in a head. Preferred Habitat - Cool Northern bogs; in or beside sluggish water. Flowering Season - May-June. Distribution - Nova Scotia southward to Virginia, westward to Minnesota and Iowa.

At a glance one knows this beautiful denizen of Northern bogs and ditches to be a poor relation of the stately Ethiopian calla lily of our greenhouses. Where the arum grows in rich, cool retreats, it is apt to be abundant, its slender rootstocks running hither and thither through the yielding soil with thrifty rapidity until the place is carpeted with its handsome dark leaves, from which the pure white "flowers" arise; and yet many flower lovers well up in field practice know it not. Thoreau, for example, was no longer young when he first saw, or, rather, noticed it. "Having found this in one place," he wrote, "I now find it in another. Many an object is not seen, though it falls within the range of our visual ray, because it does not come within the range of our intellectual ray. So, in the largest sense, we find only the world we look for."

Now, the true flowers of the arum and all its spadix-bearing kin are so minute that one scarcely notices them where they are clustered on the club-shaped column in the center of the apparent "flower." The beautiful white banner of the marsh calla, or the green and maroon striped pulpit from which Jack preaches, is no more the flower proper than the papery sheath below the daffodil is the daffodil. In the arum the white advertisement flaunted before flying insects is not even essential to the florets' existence, except as it helps them attract their pollen-carrying friends. Almost all waterside plants, it will be noticed, depend chiefly upon flies and midges, and these lack aesthetic taste. "Such plants have usually acquired small and inconspicuous separate flowers," says Grant Allen; "and then, to make up for their loss in attractiveness, like cheap sweetmeats, they have very largely increased their numbers. Or, to put the matter more simply and physically, in waterside situations those plants succeed best which have a relatively large number of individually small and unnoticeable flowers massed together into large and closely serried bundles. Hence, in such situations, there is a tendency for petals to be suppressed, and for blossoms to grow minute; because the large and bright flowers seldom succeed in attracting big land insects like bees or butterflies, while the small and thick-set ones usually do succeed in attracting a great many little flitting midges." Flies, which are guided far more by their sense of smell than by sight, resort to the petalless, insignificant florets of the ill-scented marsh calla in numbers; and as the uppermost clusters are staminate only, while the lower florets contain stamens and pistil, it follows they must often effect cross-pollination as they crawl over the spadix. But here is no trap to catch the tiny benefactors such as is set by wicked Jack-in-the-pulpit, or the skunk-cabbage, or another cousin, a still more terrible executioner, the cuckoo-pint (Arum maculatum) of Europe.

Few coroner's inquests are held over the dead bodies of our feathered friends; and it is not known whether the innocent-looking marsh calla really poisons the birds on which it depends to carry its bright seeds afar or not. The cuckoo-pint, as is well known, destroys the winged messenger bearing its offspring to plant fresh colonies in a distant bog, because the decayed body of the bird acts as the best possible fertilizer into which the seedling may strike its roots. Most of our noxious weeds, like our vermin, have come to us from Europe; but Heaven deliver us from this cannibalistic pest!

The very common GREEN ARROW-ARUM (Peltandra Virginica), found in shallow water, ditches, swamps, and the muddy shores of ponds throughout the eastern half of the United States, attracts us more by its stately growth and the beauty of its bright, lustrous green arrow-shaped leaves (which have been found thirty inches long), than by the insignificant florets clustered on the spadix within a long pointed green sheath that closely enfolds it. Pistillate florets cover it for only about one-fourth its length. To them flies carry pollen from the staminate florets covering the rest of the spadix. After the club is set with green berries - green, for this plant has no need to attract birds with bright red ones - the flower stalk curves, bends downward, and the pointed leathery sheath acting as an auger, it bores a hole into the soft mud in which the seeds germinate with the help of their surrounding jelly as a fertilizer.

AMERICAN WHITE HELLEBORE; INDIAN POKE; ITCH-WEED
  (Veratrum viride) Bunch-flower family

Flowers - Dingy, pale yellowish or whitish green, growing greener with age, 1 in. or less across, very numerous, in stiff-branching, spike-like, dense-flowered panicles. Perianth of 6 oblong segments; 6 short curved stamens; 3 styles. Stem: Stout, leafy, 2 to 8 ft. tall. Leaves: Plaited, lower ones broadly oval, pointed, 6 to 12 in. long; parallel ribbed, sheathing the stem where they clasp it; upper leaves gradually narrowing; those among flowers small. Preferred Habitat - Swamps, wet woods, low meadows. Flowering Season - May-July. Distribution - British Possessions from ocean to ocean; southward in the United States to Georgia, Tennessee, and Minnesota.

"Borage and hellebore fill two scenes -
Sovereign plants to purge the veins
Of melancholy, and cheer the heart
Of those black fumes which make it smart."

Such are the antidotes for madness prescribed by Burton in his "Anatomie of Melancholy." But like most medicines, so the homeopaths have taught us, the plant that heals may also poison; and the coarse, thick rootstock of this hellebore sometimes does deadly work. The shining plaited leaves, put forth so early in the spring they are especially tempting to grazing cattle on that account, are too well known by most animals, however, to be touched by them - precisely the end desired, of course, by the hellebore, nightshade, aconite, cyclamen, Jamestown weed, and a host of others that resort, for protection, to the low trick of mixing poisonous chemicals with their cellular juices. Pliny told how the horses, oxen, and swine of his day were killed by eating the foliage of the black hellebore. Flies, which visit the dirty, yellowish-green flowers in abundance, must cross-fertilize them, as the anthers mature before the stigmas are ready to receive pollen. Apparently the visitors suffer no ill effects from the nectar. We nave just seen how the green arrow-arum bores a hole in the mud and plants its own seeds in autumn. The hellebore uses its auger in the spring, when we find the stout, shining, solid tool above ground with the early skunk-cabbage.

STAR OF BETHLEHEM; TEN O'CLOCK
(Ornithogalum umbellatum) Lily family

Flowers - Opening in the sunshine, white within, greenish on the outside, veined, borne on slender pedicels in an erect, loose cluster. Perianth of 6 narrowly oblong divisions, 1/2 in. long or over, or about twice as long as the flattened stamens; style short, 3-sided. Scape: Slender, 4 to 12 in. high, with narrow, blade-like bracts above. Leaves: Narrow, grass-like with white midvein, fleshy, all from coated, egg-shaped bulb. Preferred Habitat - Moist, grassy meadows, old lawns. Flowering Season - May-June. Distribution - Escaped from gardens from Massachusetts to Virginia.

The finding of these exquisite little flowers, growing wild among the lush grass of a meadow not far from some old homestead where their ancestors, with crocuses and grape hyacinths, once brightened the lawn in early spring, makes one long to start a Parkinson Society instantly. Some school children not far from New York, receiving their inspiration from Mrs. Ewing's little book, "Mary's Meadow," have spread the gospel of beauty, like the true missionaries they are, by systematically planting in lanes and fields sweet violets, golden coreopsis, hardy poppies, blue corn-flowers, Japanese roses, orange day-lilies, larkspurs, and many other charming garden flowers that need only the slightest encouragement to run wild. Immense quantities of seed, that go to loss in every garden, might so easily be sprinkled at large on our walks. Nearly all the beautiful hardy perennials cultivated here grow in Nature's garden in Europe or Asia, and will do so in America if they are but given the chance. The Star of Bethlehem is a case in point. Several members of the large group of charming spring flowers to which it belongs grow in such abundance in the Old World that for centuries the bulbs have furnished food to the omnivorous Italian and Asiatic peasants. If we cannot spare offsets from the garden, and will wait a few years for seeds to bear, the rich, light loam of our grassy meadows, too, will be streaked with a Milky Way of floral stars, as they are in Italy.

The Greek generic name of the Star of Bethlehem, meaning "bird's milk" (a popular folk expression in Europe for some marvellous thing) was applied by Linnaeus because of the flower's likeness to the wonderful star in the East which guided the Wise Men to the manger where Jesus lay.

STAR-GRASS; COLIC-ROOT
(Aletris farinosa) Lily family

Flowers - Small, oblong-tubular, pure white or yellowish, about 1/4 in. long, set obliquely in a long, wand-like, spiked raceme, at the end of a slender scape 2 to 3 ft. tall. Perianth somewhat bell-shaped, 6-pointed, rough or mealy outside; 6 stamens, inserted below each point; style 3-cleft at tip. (A Southern form or distinct species (?) has yellower, fragrant flowers.) Leaves: >From the base, lance-shaped, 2 to 6 in. long, thin, pale yellowish green, in a spreading cluster. Preferred Habitat - Dry soil; roadsides; open, grassy, sandy woods. Flowering Season - May-July. Distribution - From Ontario and the Mississippi eastward to the Atlantic.

Herb gatherers have searched far and wide for this plant's bitter, fibrous root, because of its supposed medicinal virtues. What decoctions have not men swallowed from babyhood to old age to get relief from griping colic! In partial shade, colonies of the tufted yellow-green leaves send up from the center gradually lengthening spikes of bloom that may finally attain over a foot in length. The plant is not unknown in borders of men's gardens. The Greek word (aletron = meal) from which its generic title is derived, refers to the rough, granular surface of the little oblong white flower.

WILD SPIKENARD; FALSE SOLOMON'S SEAL; SOLOMON'S ZIG-ZAG
(Vagnera racemosa; Smilacina racemosa of Gray)
Lily-of-the-Valley family

Flowers - White or greenish, small, slightly fragrant, in a densely flowered terminal raceme. Perianth of 6 separate, spreading segments; 6 stamens; 1 pistil. Stem: Simple, somewhat angled, 1 to 3 ft. high, scaly below, leafy, and sometimes finely hairy above. Leaves: Alternate and seated along stem, oblong, lance-shaped, 3 to 6 in. long, finely hairy beneath. Rootstock: Thick, fleshy. Fruit: A cluster of aromatic, round, pale red speckled berries. Preferred Habitat - Moist woods, thickets, hillsides. Flowering Season - May-July. Distribution - Nova Scotia to Georgia; westward to Arizona and British Columbia.

As if to offer opportunities for comparison to the confused novice, the true Solomon's seal and the so-called false species - quite as honest a plant - usually grow near each other. Grace of line, rather than beauty of blossom, gives them both their chief charm. But the feathery plume of greenish-white blossoms that crowns the false Solomon's seal's somewhat zig-zagged stem is very different from the small, greenish, bell-shaped flowers, usually nodding in pairs along the stem, under the leaves, from the axils of the true Solomon's seal. Later in summer, when hungry birds wander through the woods with increased families, the wild spikenard offers them branching clusters of pale red speckled berries, whereas the latter plant feasts them with blue-black fruit, in the hope that they will drop the seeds miles away.

By clustering its small, slightly fragrant flowers at the end of its stem, the wild spikenard offers a more taking advertisement to its insect friends than its cousin can show. A few flies and beetles visit them; but apparently the less specialized bees, chiefly those of the Halictus tribe, which predominate in May, are the principal guests. These alight in the center of the widely expanded blossoms set on the upper side of the branching raceme so as to make their nectar and pollen easily accessible; and as the newly opened flower has its stigma already receptive to pollen brought to it while its own anthers are closed, it follows the plant is dependent upon the bees' help, as well as the birds', to perpetuate itself.

The STAR-FLOWERED SOLOMON'S SEAL (V. stellata), found from the Atlantic to the Pacific, from Newfoundland as far south as Kansas, has larger, but fewer, flowers than the wild spikenard, at the end of its erect, low-growing stem. Where the two species grow together - and they often do - it will be noticed that the star-flowered one frequently forms colonies on rich, moist banks, its leaves partly clasp the stem, and its berries, which may be entirely black, are more frequently green, with six black stripes.

The TWO-LEAVED SOLOMON'S SEAL, or FALSE LILY-OF-THE-VALLEY (Unifolium Canadense), very common in moist woods and thickets North and West, is a curious little plant, sometimes with only a solitary, long-petioled leaf; but where many of these sterile plants grow together, forming shining beds. Other individuals lift a white-flowered raceme six inches above the ground; and on the slender, often zig-zagged flowering stem there may be one to three, but usually two, ovate leaves, pointed at the apex, heart-shaped at the base, either seated on it, one above the other, or standing out from it on distinct but short petioles. This flower has only four segments and four stamens. Like the wild spikenard, the little plant bears clusters of pale red speckled berries in autumn.

HAIRY or TRUE or TWIN-FLOWERED SOLOMON'S SEAL
  (Polygonatum biftorum) Lily-of-the-Valley family

Flowers - Whitish or yellowish green, tubular, bell-shaped, 1 to 4, but usually 2, drooping on slender peduncles from leaf axils. Perianth 6-lobed at entrance, but not spreading; 6 stamens, the filaments roughened; 1 pistil. Stem: Simple, slender, arching, leafy, 8 in. to 3 ft. long. Leaves: Oval, pointed, or lance-shaped, alternate, 2 to 4 in. long, seated on stem, pale beneath and softly hairy along veins. Rootstock: Thick, horizontal, jointed, scarred. (Polygonatum = many joints). Fruit: A blue-black berry. Preferred Habitat - Woods, thickets, shady banks. Flowering Season - April-June. Distribution - New Brunswick to Florida, westward to Michigan.

>From a many-jointed, thick rootstock a single graceful curved stem arises each spring, withers after fruiting, and leaves a round scar, whose outlines suggested to the fanciful man who named the genus the seal of Israel's wise king. Thus one may know the age of a root by its seals, as one tells that of a tree by the rings in its trunk.

The dingy little cylindric flowers, hidden beneath the leaves, may be either self-pollenized or cross-pollenized by the bumblebees to which they are adapted. "We may suppose," says Professor Robertson, "that the pendulous position of the flowers owes its origin to the fact that it renders them less convenient to other insects, but equally convenient to the higher bees which are the most efficient pollinators; and that the resulting protection to pollen and nectar is merely an incidental effect." Certain Lepidoptera, and small insects which crawl into the cylinder, visit all the Solomon's seals.

The SMOOTH SOLOMON'S SEAL (P. commutatum; P.giganteum of Gray), with much the same range as its smaller relative, grows in moist woods and along shaded streams. It is a variable, capricious plant, with a stout or slender stem, perhaps only one foot high, or again towering above the tallest man's head; the oval leaves also vary greatly in breadth and length; and a solitary flower may droop from an axil, or perhaps eight dingy greenish cylinders may hang in a cluster. But the plant is always smooth throughout. Even the incurved filaments which obstruct the entrance to this flower are smooth where those of the preceding species are rough-hairy. The style is so short that it may never come in contact with the anthers, although the winged visitors must often leave pollen of the same flower on the stigma.

EARLY or DWARF WAKE-ROBIN
(Trillium nivale) Lily-of-the-Valley family

Flowers - Solitary, pure white, about 1 in. long, on an erect or curved peduncle, from a whorl of 3 leaves at summit of stem. Three spreading, green, narrowly oblong sepals; 3 oval or oblong petals; 6 stamens, the anthers about as long as filaments; 3 slender styles stigmatic along inner side. Stem: 2 to 6 in. high, from a short, tuber-like rootstock. Leaves: 3 in a whorl below the flower, 1 to 2 in. long, broadly oval, rounded at end, on short petioles. Fruit: A 3-lobed reddish berry, about 1/2 in. in diameter, the sepals adhering. Preferred Habitat - Rich, moist woods and thickets. Flowering Season - March-May. Distribution - Pennsylvania, westward to Minnesota and Iowa, south to Kentucky.

Only this delicate little flower, as white as the snow it sometimes must push through to reach the sunshine melting the last drifts in the leafless woods, can be said to wake the robins into song; a full chorus of feathered love-makers greets the appearance of the more widely distributed, and therefore better known, species.

By the rule of three all the trilliums, as their name implies, regulate their affairs. Three sepals, three petals, twice three stamens, three styles, a three-celled ovary, the flower growing out from a whorl of three leaves, make the naming of wake-robins a simple matter to the novice. Rarely do the parts divide into fours, or the petals and sepals revert to primitive green leaves. With the exception of the painted trillium which sometimes grows in bogs, all the clan live in rich, moist woods. It is said the roots are poisonous. In them the next year's leaves lie curled through the winter, as in the iris and Solomon's seal, among others.

One of the most chastely beautiful of our native wild flowers - so lovely that many shady nooks in English rock-gardens and ferneries contain imported clumps of the vigorous plant - is the LARGE-FLOWERED WAKE-ROBIN, or WHITE WOOD LILY (T. grandiflorum). Under favorable conditions the waxy, thin, white, or occasionally pink, strongly veined petals may exceed two inches; and in Michigan a monstrous form has been found. The broadly rhombic leaves, tapering to a point, and lacking petioles, are seated in the usual whorl of three, at the summit of the stem, which may attain a foot and a half in height; from the center the decorative flower arises on a long peduncle. At first the entrance to the blossom is closed by the long anthers which much exceed the filaments; and hive-bees, among other insects, in collecting pollen, transfer it to older and now expanded flowers, in which the low stigmas appear between the tall separated stamens. Nectar stored in septal glands at the base invites the visitor laden with pollen from young flowers to come in contact with the three late maturing stigmas. The berry is black. From Quebec to Florida and far westward we find this tardy wake-robin in May or June.

Certainly the commonest trillium in the East, although it thrives as far westward as Ontario and Missouri, and south to Georgia, is the NODDING WAKE-ROBIN (T. cernuum), whose white or pinkish flower droops from its peduncle until it is all but hidden under the whorl of broadly rhombic, tapering leaves. The wavy margined petals, about as long as the sepals - that is to say, half an inch long or over - curve backward at maturity. According to Miss Carter, who studied the flower in the Botanical Garden at South Hadley, Mass., it is slightly proterandrous, maturing its anthers first, but with a chance of spontaneous self-pollination by the stigmas recurving to meet the shorter stamens. She saw bumblebees visiting it for nectar. In late summer an egg-shaped, pendulous red-purple berry swings from the summit. One finds the plant in bloom from April to June, according to the climate of its long range,

Perhaps the most strikingly beautiful member of the tribe is the PAINTED TRILLIUM (T. undulatum; T. erythrocarpum of Gray). At the summit of the slender stem, rising perhaps only eight inches, or maybe twice as high, this charming flower spreads its long, wavy-edged, waxy-white petals veined and striped with deep pink or wine color. The large ovate leaves, long-tapering to a point, are rounded at the base into short petioles. The rounded, three-angled, bright red, shining berry is seated in the persistent calyx. With the same range as the nodding trillium's, the painted wake-robin comes into bloom nearly a month later - in May and June - when all the birds are not only wide awake, but have finished courting, and are busily engaged in the most serious business of life.

SHOWY LADY'S SLIPPER
(Cypripedium reginae; C. spectabile of Gray) Orchid family

Flowers - Usually solitary, at summit of stem, white, or the inflated white lip painted with purplish pink and white stripes; sepals rounded oval, spreading, white, not longer than the lip; petals narrower, white; the broad sac-shaped pouch open in front, 1 in. long or over. Stem: Stout, leafy, 1 to 2 ft. high. Leaves: 3 to 8 in. long, downy, elliptic, pointed, many ribbed. Preferred Habitat - Peat-bogs; rich, low, wet woods. Flowering Season - June-September. Distribution - Nova Scotia to Georgia, westward to the Mississippi. Chiefly North.

Quite different from the showy orchis, is this far more chaste showy lady's slipper which Dr. Gray has called "the most beautiful of the genus." Because the plants live in inaccessible swampy places, where only the most zealous flower lover penetrates, they have a reputation for rarity at which one who knows a dozen places to find colonies of the stately exquisites during a morning's walk, must smile with superiority. Wine appears to overflow the large white cup and trickle down its sides. Sometimes unstained, pure white chalices are found. C. album is the name by which the plant is known in England. See note after Common Daisy.

LARGE ROUND-LEAVED or GREATER GREEN ORCHIS
(Habenaria orbiculata) Orchid family

Flowers - Greenish white, in a loosely set spike; the upper sepal short, rounded; side ones spreading; petals smaller, arching; the lip long, narrow, drooping, white, prolonged into a spur often 1 1/2 in. long, curved and enlarged at base; anther sacs prominent, converging. Scape: 1 to 2 ft. high. Leaves: 2, spreading flat on ground, glossy above, silvery underneath, parallel-veined, slightly longer than wide, very large, from 4 to 7 in. across. Preferred Habitat - Rich, moist woods in mountainous regions, especially near evergreens. Flowering Season - July-August. Distribution - From British Columbia to the Atlantic; eastern half of the United States southward to the Carolinas.

Wonderfully interesting structure and the comparative rarity of this orchid, rather than superficial beauty, are responsible for the thrill of pleasure one experiences at the sight of the spike of unpretentious flowers. Two great leaves, sometimes as large as dinner plates, attract the eye to where they glisten on the ground. The spur of the blossom, the nectary, "implies a welcome to a tongue two inches long, and will reward none other," says William Hamilton Gibson. "This clearly shuts out the bees, butterflies, and smaller moths. What insect, then, is here implied? The sphinx moth, one of the lesser of the group. A larger individual might sip the nectar, it is true, but its longer tongue would reach the base of the tube without effecting the slightest contact with the pollen, which is, of course, the desideratum." How the moth, in sipping the nectar, thrusts his head against the sticky buttons to which the pollen messes are attached, and, in trying to release himself, loosens them; how he flies off with these little clubs sticking to his eyes; how they automatically adjust themselves to the attitude where they will come in contact with the stigma of the next flower visited, and so cross-fertilize it, has been told in the account of the great purple-fringed orchis of similar construction. To that species the interested reader is, therefore, referred; or, better still, to the luminous description by Dr. Asa Gray.

WHITE-FRINGED ORCHIS
 (Habenaria blephariglottis) Orchid family

Flowers - Pure white, fragrant, borne on a spike from 3 to 6 in. long. Spur long, slender; oval sepals; smaller petals toothed; the oblong lip deeply fringed. Stem: Slender, 1 to 2 ft. high. Leaves: Lance-shaped, parallel-veined, clasping the stem; upper ones smallest.

Preferred Habitat - Peat-bogs and swamps.
Flowering Season - July-August.
Distribution - Northeastern United States and eastern Canada to Newfoundland.

One who selfishly imagines that all the floral beauty of the earth was created for man's sole delight will wonder why a flower so exquisitely beautiful as this dainty little orchid should be hidden in inaccessible peat-bogs, where overshoes and tempers get lost with deplorable frequency, and the water-snake and bittern mock at man's intrusion of their realm by the ease with which they move away from him. Not for man, but for the bee, the moth, and the butterfly, are orchids where they are and what they are. The white-fringed orchis grows in watery places that it may more easily manufacture nectar, and protect itself from crawling pilferers; its flowers are clustered on a spike, their lips are fringed, they have been given fragrance and a snowy-white color that they may effectually advertise their sweets on whose removal by an insect benefactor that will carry pollen from flower to flower as he feeds depends their chance of producing fertile seed. It is probable the flower is white that night-flying moths may see it shine in the gloaming. From the length and slenderness of its spur it is doubtless adapted to the sphinx moth.

At the entrance to the nectary, two sticky disks stand on guard, ready to fasten themselves to the eyes of the first moth that inserts his tongue; and he finds on withdrawing his head that two pollen-masses attached to these disks have been removed with them. This plastering over of insects' eyes by the orchids might be serious business, indeed, were not the lepidoptera gifted with numerous pairs. The fragrance of many orchids, however, would be a sufficient guide even to a blind insect. With the pollen-masses sticking to his forehead, the moth enters another flower and necessarily rubs off some grains from the pollen masses, that have changed their attitude during his flight that they may be in the precise position to fertilize the viscid stigma. In almost the same way the similar Yellow-Fringed Orchis (H. ciliaris) and the great green orchids compel insects to work for them.

A larger-flowered species, the PRAIRIE WHITE-FRINGED ORCHIS (H. lepicophea), found in bloom in June and July, on moist, open ground from western New York to Minnesota and Arkansas, differs from the preceding chiefly in having larger and greenish-white flowers, the lip cleft into wedge-shaped segments deeply fringed. The hawk-moth removes on its tongue one, but not often both, of the pollinia attached to disks on either side of the entrance to the spur.

NODDING LADIES' TRESSES or TRACES
(Gyrostachys cernua; Spiranthes cernua of Gray) Orchid family

Flowers - Small, white or yellowish, without a spur, fragrant, nodding or spreading in 3 rows on a cylindrical, slightly twisted spike 4 or 5 in. long. Side sepals free, the upper ones arching, and united with petals; the oblong, spreading lip crinkle-edged, and bearing minute, hairy callosities at bases Stem: 6 in. to 2 ft. tall, with several pointed, wrapping bracts. Leaves: From or near the base, linear, almost grass-like. Preferred Habitat - Low meadows, ditches, and swamps. Flowering Season - July-October. Distribution - Nova Scotia to the Gulf of Mexico, and westward to the Mississippi.

This last orchid of the season, and perhaps the commonest of its interesting tribe in the eastern United States, at least, bears flowers that, however insignificant in size, are marvelous pieces of mechanism, to which such men as Charles Darwin and Asa Gray have devoted hours of study and, these two men particularly, much correspondence.

Just as a woodpecker begins at the bottom of a tree and taps his way upward, so a bee begins at the lower and older flowers on a spike and works up to the younger ones; a fact on which this little orchid, like many another plant that arranges its blossoms in long racemes, depends. Let us not note for the present what happens in the older flowers, but begin our observations, with the help of a powerful lens, when the bee has alighted on the spreading lip of a newly opened blossom toward the top of the spire. As nectar is already secreted for her in its receptacle, she thrusts her tongue through the channel provided to guide it aright, and by the slight contact with the furrowed rostellum, it splits, and releases a boat-shaped disk standing vertically on its stern in the passage. Within the boat is an extremely sticky cement that hardens almost instantly on exposure to the air. The splitting of the rostellum, curiously enough, never happens without insect aid; but if a bristle or needle be passed over it ever so lightly, a stream of sticky, milky fluid exudes, hardens, and the boat-shaped disk, with pollen masses attached, may be withdrawn on the bristle just as the bee removes them with her tongue. Each pollinium consists of two leaves of pollen united for about half their length in the middle with elastic threads. As the pollinia are attached parallel to the disk, they stick parallel on the bee's tongue, yet she may fold up her proboscis under her head, if she choose, without inconvenience from the pollen masses, or without danger of loosening them. Now, having finished sucking the newly opened flowers at the top of the spike, away she flies to an older flower at the bottom of another one. Here a marvelous thing has happened. The passage which, when the flower first expanded, scarcely permitted a bristle to pass, has now widened through the automatic downward movement of the column in order to expose the stigmatic surfaces to contact with the pollen masses brought by the bee. Without the bee's help this orchid, with a host of other flowers, must disappear from the face of the earth. So very many species which have lost the power to fertilize themselves now depend absolutely on these little pollen carriers, it is safe to say that, should the bees perish, one half our flora would be exterminated with them. On the slight downward movement of the column in the ladies' tresses, then, as well as on the bee's ministrations, the fertilization of the flower absolutely depends. "If the stigma of the lowest flower has already been fully fertilized," says Darwin, "little or no pollen will be left on its dried surface; but on the next succeeding flower, of which the stigma is adhesive, large sheets of pollen will be left. Then as soon as the bee arrives near the summit of the spike she will withdraw fresh pollinia, will fly to the lower flowers on another plant, and fertilize them; and thus, as she goes her rounds and adds to her store of honey, she continually fertilizes fresh flowers and perpetuates the race of autumnal spiranthes, which will yield honey to future generations of bees."

The SLENDER LADIES' TRESSES (G. gracilis; [S. gracilis]), with a range and season of blossom similar to the preceding species, and with even smaller white, fragrant flowers, growing on one side of a twisted spike, chooses dry fields, hillsides, open woods, and sandy places - queer habitats for a member of its moisture-loving tribe. Its leaves have usually fallen by flowering time. The cluster of tuberous, spindle-shaped roots are an aid to identification.

LESSER RATTLESNAKE PLANTAIN [DWARF RATTLESNAKE-PLANTAIN]
(Peramium repens; Goodyera repens of Gray) Orchid family

Flowers - Small, greenish white, the lip pocket-shaped, borne on one side of a bracted spike 5 to 10 in. high, from a fleshy, thick fibrous root. Leaves: From the base, tufted, or ascending the stem on one side for a few inches, 1/2 in. to over 1 in. long, ovate, the silvery-white veins forming a network, or leaf blotched with white. Preferred Habitat - Woods, especially under evergreens. Flowering Season - July-August. Distribution - Colorado eastward to the Atlantic, from Nova Scotia to Florida. Europe and Asia.

Tufts of these beautifully marked little leaves carpeting the ground in the shadow of the hemlocks attract the eye, rather than the spires of insignificantly small flowers. Whoever wishes to know how the bumblebee ruptures the sensitive membrane within the tiny blossom with her tongue, and draws out the pollinia that are instantly cemented to it after much the same plan employed by the ladies' tresses, must use a good lens in studying the operation. To the structural botanist the rattlesnake plantains form an interesting connecting link between orchids of d1stinct forms. In them we see a tendency to lengthen the pollen-masses into caudicles as the showy orchis, for example, has done. "Goodyera probably shows us the state of organs in a group of orchids now mostly extinct," says Darwin; "but the parents of many living descendants."

It has been said that the Indians use this plant to cure bites of the rattlesnake; that they will handle the deadly creature without fear if some of these leaves are near at hand - in fact, a good deal is said about Indians by palefaces that makes even the stolid red man smile when confronted with the white man's tales about him. An intelligent Indian student declares that none of his race will handle a rattlesnake unless its fangs have been removed; that this plant takes its name from the resemblance of its netted-veined leaves to the belly of a serpent, and not to their curative powers; and, finally, that the Southern tribes, especially so reverence the rattlesnake that, far from trying to cure its bite, they count themselves blessed to be bitten to death by one. Indeed, the rattle, a sacred symbol, has been employed in religious ceremonies of most tribes. Snakes may be revered in other lands, but only in America is the rattlesnake worshipped. Among the Moquis there still survives much of the religion of the snake-worshipping Aztecs. Bernal Diaz tells how living rattlesnakes, kept in the great temple at Mexico as sacred and petted objects, were fed with the bodies of the sacrificed. Cortes found a town called by the Spaniards Terraguea, or the city of serpents, whose walls and temples were decorated with figures of the reptiles, which the inhabitants worshiped as gods.

The DOWNY RATTLESNAKE PLANTAIN (P. pubescens), usually a taller plant than the preceding, with larger cream-white, globular-lipped flowers on both sides of its spike, and glandular-hairy throughout, has even more strongly marked leaves. These, the most conspicuous parts, are dark grayish green, heavily netted with greenish or silvery-white veins, silky to the touch, and often wavy edged. This plant scarcely strays westward beyond the Mississippi, but it is common East. It also blooms in midsummer, and shows a preference for dry woods where oak and pine abound.

LIZARD'S TAIL [LIZARD'S-TAIL, WATER-DRAGON]
(Saurus cernuus) Lizard's-tail family

Flowers - Fragrant, very small, white, lacking a perianth, bracted, densely crowded on peduncled, slender spikes 4 to 6 in. long and nodding at the tip. Stamens 6 to 8, the filaments white; carpels 3 or 4, united at base, dangling. Stem: 2 to 5 ft. high, jointed, sparingly branched, leafy. Leaves: Heart-shaped, palmately ribbed, dark green, thin, on stout petioles. Preferred Habitat - Swamps, shallow water. Flowering Season - June-August. Distribution - Southern New England to the Gulf, westward to Minnesota and Texas.

The fragrance arising from these curious, drooping, tail-like spikes of flowers, where they grow in numbers, must lure their insect friends as it does us, since no showy petals or sepals advertise their presence. Nevertheless they are what are known as perfect flowers, each possessing stamens and pistils, the only truly essential parts, however desirable a gaily colored perianth may be to blossoms attempting to woo such large land insects as the bumble-bee and butterfly. Since flies, whose color sense is by no means so acute as their sense of smell, are by far the most abundant fertilizers of waterside plants, we can see a tendency in such to suppress their petals, for the flowers to become minute and massed in series that the little visitors may more readily transfer pollen from one to another, and to become fragrant - just what the lizard's tail has done.

SPRING BEAUTY; CLAYTONIA
  (Claytonia Virginica) Purslane family.

Flowers - White veined with pink, or all pink, the veinings of deeper shade, on curving, slender pedicels, several borne in a terminal loose raceme, the flowers mostly turned one way (secund). Calyx of 2 ovate sepals; corolla of 5 petals slightly united by their bases; 5 stamens, 1 inserted on base of each petal; the style 3-cleft. Stem: Weak, 6 to 12 in. long, from a deep, tuberous root. Leaves: Opposite above, linear to lance-shaped, shorter than basal ones, which are 3 to 7 in. long; breadth variable. Preferred Habitat - Moist woods, open groves, low meadows. Flowering Season - March-May. Distribution - Nova Scotia and far westward, south to Georgia and Texas.

Dainty clusters of these delicate, starry blossoms, mostly turned in one direction, expand in the sunshine only, like their gaudy cousin the portulaca and the insignificant little yellow flowers of another relative, the ubiquitous, invincible "pussley" immortalized in "My Summer in a Garden." At night and during cloudy, stormy weather, when their benefactors are not flying, the claytonias economically close their petals to protect nectar and pollen from rain and pilferers. Pick them, the whole plant droops, and the blossoms close with indignation; nor will any coaxing but a combination of hot water and sunshine induce them to open again. Theirs is a long beauty sleep. They are supersensitive exquisites, however hardy.

Very early in the spring a race is run with the hepatica, arbutus, adder's tongue, blood-root, squirrel corn, and anemone for the honor of being the earliest wild flower; and although John Burroughs and Dr. Abbott have had the exceptional experience of finding the claytonia even before the hepatica - certainly the earliest spring blossom worthy the name in the Middle and New England States - of course the rank skunk-cabbage, whose name is snobbishly excluded from the list of fair competitors, has quietly opened dozens of minute florets in its incurved horn before the others have even started.

Whether the petals of the spring beauty are white or pink, they are always exquisitely marked with pink lines converging near the base and ending in a yellow blotch to serve as pathfinders for the female bumblebees and the little brown bombylius, among other pollen carriers. A newly opened flower, with its stamens surrounding the pistil, must be in peril of self-fertilization one would think who did not notice that when the pollen is in condition for removal by the bees and flies, the stigmatic surfaces of the three-cleft style are tightly pressed together that not a grain may touch them. But when the anthers have shed their pollen, and the filaments have spread outward and away from the pistil, the three stigmatic arms branch out to receive the fertilizing dust carried from younger flowers by their busy friends.

STARRY CAMPION
  (Silene stellata) Pink family

  Flowers - White, about 1/2 in. broad or over, loosely clustered in a showy, pyramidal panicle. Calyx bell-shaped, swollen, 5-toothed, sticky; 5 fringed and clawed petals; 10 long, exserted stamens; 3

styles. Stem: Erect, leafy, 2 to 3 1/2 ft. tall, rough-hairy. Leaves: Oval, tapering to a point, 2 to 4 in. long, seated in whorls of 4 around stem, or loose ones opposite. Preferred Habitat - Woods, shady banks. Flowering Season - June-August. Distribution - Rhode Island westward to Mississippi, south to the Carolinas and Arkansas.

Feathery white panicles of the starry campion, whose protruding stamens and fringed petals give it a certain fleeciness, are dainty enough for spring; by midsummer we expect plants of ranker growth and more gaudy flowers. To save the nectar in each deep tube for the moths and butterflies which cross-fertilize all this tribe of night and day blossoms, most of them - and the campions are notorious examples - spread their calices, and some their pedicels as well, with a sticky substance to entrap little crawling pilferers. Although a popular name for the genus is catchfly, it is usually the ant that is glued to the viscid parts, for the fly that moves through the air alights directly on the flower it is too short-lipped to suck. An ant catching its feet on the miniature lime-twig, at first raises one foot after another and draws it through its mouth, hoping to rid it of the sticky stuff, but only with the result of gluing up its head and other parts of the body. In ten minutes all the pathetic struggles are ended. Let no one guilty of torturing flies to death on sticky paper condemn the Silenes!

The BLADDER CAMPION (S. vulgaris; S. inflata of Gray) to be recognized by its much inflated calyx, especially round in fruit, the two-cleft white petals; and its opposite leaves that are spatulate at the base of the plant, is a European immigrant now naturalized and locally very common from Illinois eastward to New Jersey and north to New Brunswick. Like the night-flowering catchfly this blossom has adapted itself to the night-flying moths; but when either remains open in the morning, bumblebees gladly take the leavings in the deep cup. To insure cross-fertilization, some of the bladder-campion flowers have stamens only, some have a pistil only; some have both organs maturing at different times. In all the night-flowering Silene, each flower, unless unusually disturbed, lasts three days and three nights. Late in the afternoon of the first day, when the petals begin to expand, the five stamens opposite the sepals lengthen in about two hours, and by sunset the anthers, which have matured at the same time, are covered with pollen. So they remain until the forenoon of the second day, and then the emptied anthers hang like shriveled bags, or drop off altogether. Late in the second afternoon, the second set of stamens repeat the actions of their predecessors, bend backward and shed their anthers the following, that is to say the third, morning. But on the third afternoon up rise the S-shaped, twisted stigmas, which until now had been hidden in the center of the flower. Moths, therefore, must transfer pollen from younger to older blossoms.

"With this lengthening and bending of the stamens and stigmas," says Dr. Kerner, "goes hand in hand the opening and shutting of the corolla. With the approach of dusk, the bifid limbs of the petals spread out in a flat surface and fall back against the calyx. In this position they remain through the night, and not till the following morning do they begin (more quickly in sunshine and with a mild temperature, more slowly with a cloudy sky and in cold, wet weather) to curl themselves up in an in-curved spire, while at the same time they form longitudinal creases, and look as though they were gathered in, or wrinkled;…but no sooner does evening return than the wrinkles disappear, the petals become smooth, uncurl themselves, and fall back upon the calyx, and the corolla is again expanded."

Curiously enough, these flowers, which by day we should certainly say were not fragrant, give forth a strong perfume at evening the better to guide moths to their feast. From eight in the evening until three in the morning the fragrance is especially strong. The white blossoms, so conspicuous at night, have little attraction for color-loving butterflies and bees by day; then, as there is no pollen to be carried from the shriveled anther sacs, no visitor is welcome, and the petals close to protect the nectar for the flower's true benefactors. Indeed, few flowers show more thorough adaptation to the night-flying moths than these Silene.

POKEWEED; SCOKE; PIGEON-BERRY; INK-BERRY; GARGET
(Phytolacca decandra) Pokeweed family

Flowers - White, with a green centre, pink-tinted outside, about 1/4 in. across, in bracted racemes 2 to 8 in. long. Calyx of 4 or 5 rounded persistent sepals, simulating petals; no corolla; 10 short stamens; 10-celled ovary, green, conspicuous; styles curved. Stem: Stout, pithy, erect, branching, reddening toward the end of summer, 4 to 10 ft. tall, from a large, perennial, poisonous root. Leaves: Alternate, petioled, oblong to lance-shaped, tapering at both ends, 8 to 12 in. long. Fruit: Very juicy, dark purplish berries, hanging in long clusters from reddened footstalks; ripe, August-October. Preferred Habitat - Roadsides, thickets, field borders, and waste soil, especially in burnt-over districts. Flowering Season - June-October. Distribution - Maine and Ontario to Florida and Texas.

When the pokeweed is "all on fire with ripeness," as Thoreau said; when the stout, vigorous stem (which he coveted for a cane), the large leaves, and even the footstalks, take on splendid tints of crimson lake, and the dark berries hang heavy with juice in the thickets, then the birds, with increased, hungry families, gather in flocks as a preliminary step to traveling southward. Has the brilliant, strong-scented plant no ulterior motive in thus attracting their attention at this particular time? Surely! Robins, flickers, and downy woodpeckers, chewinks and rose-breasted grosbeaks, among other feathered agents, may be detected in the act of gormandizing on the fruit, whose undigested seeds they will disperse far and wide. Their droppings form the best of fertilizers for young seedlings; therefore the plants which depend on birds to distribute seeds, as most berry bearers do, send their children abroad to found new colonies, well equipped for a vigorous start in life. What a hideous mockery to continue to call this fruit the pigeon-berry, when the exquisite bird whose favorite food it once was, has been annihilated from this land of liberty by the fowler's net! And yet flocks of wild pigeons, containing not thousands but millions of birds, nested here even thirty years ago. When the market became glutted with them, they were fed to hogs in the West!

Children, and some grown-ups, find the deep magenta juice of the ink-berry useful. Notwithstanding the poisonous properties of the root, in some sections the young shoots are boiled and eaten like asparagus, evidently with no disastrous consequences. For any service this plant may render to man and bird, they are under special obligation to the little Halictus bees, but to other short-tongued bees and flies as well. These small visitors, flying from such of the flowers as mature their anthers first, carry pollen to those in the female, or pistillate, stage. Exposed nectar rewards their involuntary kindness. In stormy weather, when no benefactors can fly, the flowers are adapted to fertilize themselves through the curving of the styles.

COMMON CHICKWEED
(Aisine media; Stellaria media of Gray) Pink family

Flowers - Small, white, on slender pedicels from leaf axils, also in terminal clusters. Calyx (usually) of 5 sepals, much longer than the 5 (usually) 2-parted petals; 2-10 stamens; 3 or 4 styles. Stem: Weak, branched, tufted, leafy, 4 to 6 in long, a hairy fringe on one side. Leaves: Opposite, acutely oval, lower ones petioled, upper ones seated on stem. Preferred Habitat - Moist, shady soil; woods; meadows. Flowering Season - Throughout the year. Distribution - Almost universal.

The sole use man has discovered for this often pestiferous weed with which nature carpets moist soil the world around is to feed caged song-birds. What is the secret of the insignificant little plant's triumphal progress? Like most immigrants that have undergone ages of selective struggle in the Old World, it successfully competes with our native blossoms by readily adjusting itself to new conditions, filling places unoccupied, and chiefly by prolonging its season of bloom beyond theirs, to get relief from the pressure of competition for insect trade in the busy season. Except during the most cruel frosts, there is scarcely a day in the year when we may not find the little star-like chickweed flowers. Contrast this season with that of a native chickweed, the LONG-LEAVED STITCHWORT [LONG-LEAVED CHICKWEED] (A. longifolia [S. longifolia]), blooming only from May till July, when competition is fiercest! Also, the common chickweed has its parts so arranged that it can fertilize itself when it is too cold for insect pollen-carriers to fly; then, especially, are many of its stamens abortive, not to waste the precious dust. Yet even in winter it produces abundant seed. In sunny, fine spring weather, however, when so much nectar is secreted the fine little drops may be easily seen by the naked eye, small bees, flies, and even thrips visit the blossoms whose anthers shed pollen one by one before the three stigmatic surfaces are ready to receive any from younger flowers.

SWEET-SCENTED WHITE WATER LILY; POND LILY; WATER NYMPH; WATER
CABBAGE [FRAGRANT WATER-LILY]
  (Castalia odorata; Nymphaea odorata of Gray) Water-lily family

Flowers - Pure white or pink tinged, rarely deep pink, solitary, 3 to 8 in. across, deliciously fragrant, floating. Calyx of 4 sepals, green outside; petals of indefinite number, overlapping in many rows, and gradually passing into an indefinite number of stamens; outer row of stamens with petaloid filaments and short anthers, the inner yellow stamens with slender filaments and elongated anthers; carpels of indefinite number, united into a compound pistil, with spreading and projecting stigmas. Leaves: Floating, nearly round, slit at bottom, shining green above, reddish and more or less hairy below, 4 to 12 in. across, attached to petiole at center of lower surface. Petioles and peduncles round and rubber-like, with 4 main air-channels. Rootstock: (Not true stem), thick, simple or with few branches, very long. Preferred Habitat - Still water, ponds, lakes, slow streams. Flowering Season - June-September. Distribution - Nova Scotia to Gulf of Mexico, and westward to the Mississippi.

Sumptuous queen of our native aquatic plants, of the royal family to which the gigantic Victoria regia of Brazil belongs, and all the lovely rose, lavender, blue, and golden exotic water lilies in the fountains of our city parks, to her man, beast, and insect pay grateful homage. In Egypt, India, China, Japan, Persia, and Asiatic Russia, how many millions have bent their heads in adoration of her relative the sacred lotus! From its center Brahma came forth; Buddha, too, whose symbol is the lotus, first appeared floating on the mystic flower (Nelumbo nelumbo, formerly Nelumbium speciosum). Happily the lovely pink or white "sacred bean" or "rose-lily" of the Nile, often cultivated here, has been successfully naturalized in ponds about Bordentown, New Jersey, and maybe elsewhere. If he who planteth a tree is greater than he who taketh a city, that man should be canonized who introduces the magnificent wild flowers of foreign lands to our area of Nature's garden.

Now, cultivation of our native water lilies and all their hardy kin, like charity, begins at home. Their culture in tubs, casks, or fountains on the lawn, is so very simple a matter, and the flowers bloom so freely, every garden should have a corner for aquatic plants. Secure the water-lily roots as early in the spring as possible, and barely cover them with good rich loam or muck spread over the bottom of the sunken tub to a depth of six or eight inches. After it has been filled with water, and replenished from time to time to make good the loss by evaporation, the water garden needs no attention until autumn. Then the tub should be drained, and removed to a cellar, or it may be covered over with a thick mattress of dry leaves to protect from hard freezing. In their natural haunts, water lilies sink to the bottom, where the water is warmest in winter. Possibly the seed is ripened below the surface for the same reason. At no time should the crown of the cultivated plant be lower than two feet below the water. If a number of species are grown, it is best to plant each kind in a separate basket, sunk in the shallow tub, to prevent the roots from growing together, as well as to obtain more effective decoration. Charming results may be obtained with small outlay of either money or time. Nothing brings more birds about the house than one of these water gardens; that serves at once as drinking fountain and bath to our not over-squeamish feathered neighbors. The number of insects these destroy, not to mention the joy of their presence, would alone compensate the householder of economic bent for the cost of a shallow concrete tank.

Opening some time after six o'clock in the morning, the white water lily spreads its many-petalled, deliciously fragrant, golden-centered chalice to welcome the late-flying bees and flower flies, the chief pollinators. Beetles, "skippers," and many other creatures on wings alight too. "I have named two species of bees (Halictus nelumbonis and Prosopis nelumbonis) on account of their close economic relation to these flowers," says Professor Robertson, who has captured over two hundred and fifty species of bees near his home in Carlinville, Illinois, and described nearly a third of them as new.

Linnaeus, no doubt the first to conceive the pretty idea of making a floral clock, drew up a list of blossoms whose times of opening and closing marked the hours on its face; but even Linnaeus failed to understand that the flight of insects is the mainspring on which flowers depend to set the mechanism going. In spite of its whiteness and fragrance, the water lily requires no help from night-flying insects in getting its pollen transferred; therefore, when the bees and flies rest from their labors at sundown, it may close the blinds of its shop, business being ended for the day.

"When doctors disagree, who shall decide?" It is contended by one group of scientists that the water lily, which shows the plainest metamorphosis of some sort, has developed its stamens from petals - just the reverse of Nature's method, other botanists claim. A perfect flower, we know, may consist of only a stamen and a pistil, the essential organs, all other parts being desirable, but of only secondary importance. Gardeners, taking advantage of a wild flower's natural tendency to develop petals from stamens and to become "double," are able to produce the magnificent roses and chrysanthemums of today; and so it would seem that the water lily, which may be either self-fertilized or cross-fertilized by pollen-carriers in its present state of development, is looking to a more ideal condition by increasing its attractiveness to insects as it increases the number of its petals, and by economizing pollen in transforming some of the superfluous stamens into petals.

Scientific speculation, incited by the very fumes of the student lamp, may weary us in winter, but just as surely is it dispelled by the fragrance of the lilies in June. Then, floating about in a birch canoe among the lily-pads, while one envies the very moose and deer that may feed on fare so dainty and spend their lives amid scenes of such exquisite beauty, one lets thought also float as idly as the little clouds high overhead.

LAUREL or SMALL MAGNOLIA; SWEET or WHITE BAY;
SWAMP LAUREL or
SASSAFRAS; BEAVER-TREE [SWEETBAY MAGNOLIA]
  (Magnolia Virginiana; M. glauca of Gray) Magnolia family

Flowers - White, 2 to 3 in. across, globular, depressed, deliciously fragrant, solitary at ends of branches. Calyx of 3 petal-like, spreading sepals. Corolla of 6 to 12 concave rounded petals in rows; stamens very numerous, short, with long anthers; carpels also numerous, and borne on the thick, green, elongated receptacle. Trunk: 4 to 70 ft. high. Leaves: Enfolded in the bud by stipules that fall later and leave rings around gradually lengthening branch; the leaves 3 to 6 in. long in maturity, broadly oblong, thick, almost evergreen, dark above, pale beneath, on short petioles. Fruit: An oblong, reddish pink cone, fleshy, from which the scarlet seeds hang by slender threads. Preferred Habitat - Swampy woods and open swamps. Flowering Season - May-June. Distribution - Atlantic States from Massachusetts southward, and Gulf States from Florida to Texas.

"Every flower its own bo-quet!" shouted by a New York street vender of the lovely magnolia blossoms he had just gathered from the Jersey swamps, emphasized only one of the many claims they have upon popular attention. Far and wide the handsome shrub, which frequently attains a tree's height, is exported from its native hiding-places to adorn men's gardens, and there, where a better opportunity to know it at all seasons is granted, one cannot tell which to admire most, the dark, bluish-green leathery leaves, silvery beneath; the cream-white, deliciously fragrant blossoms that turn pale apricot with age; or the brilliant fruiting cone with the scarlet seeds a-dangling. At all seasons it is a delight. When most members of this lovely tribe confine themselves to warm latitudes, we especially prize the species that naturally endures the rigorous climate of the "stern New England coast."

Beavers (when they used to be common in the East) so often made use of the laurel magnolia, not only of the roots for food, but of the trunk, whose bitter bark, white sapwood, and soft, reddish-brown heartwood were gnawed in constructing their huts, that in some sections it is still known as the beaver-tree. According to Delpino, the conspicuous, pollen-laden magnolia flowers, with their easily accessible nectar, attract beetles chiefly. These winged messengers, entering the heart of a newly opened blossom, find shelter beneath the inner petals that form a vault above their heads, and warmth that may be felt by the finger, and abundant food; consequently they remain long in an asylum so delightful, or until the expanding petals turn them out to carry the pollen, with which they have been thoroughly dusted during their hospitable entertainment, to younger flowers. As the blossoms mature their stigmas in the first stage and the anthers in the second, it follows the beetles must regularly cross-fertilize them as they fly from one shelter to another.

GOLD-THREAD; CANKER-ROOT [GOLDTHREAD]
  (Coptis trifolia) Crowfoot family [Buttercup family]

Flowers - Small white, solitary, on a slender scape 3 to 6 in. high. Sepals 5 to 7, petal-like, falling early; petals 5 or 6, inconspicuous, like club-shaped columns; stamens numerous carpels few, the stigmatic surfaces curved. Leaves: From the base, long petioled, divided into 3 somewhat fan-shaped, shining, evergreen, sharply toothed leaflets. Rootstock: Thread-like, long, bright yellow, wiry, bitter. Preferred Habitat - Cool mossy bogs, damp woods. Flowering Season - May-August Distribution - Maryland and Minnesota northward to circumpolar regions.

The shining, evergreen, thrice-parted leaves with which this charming little plant carpets its retreats form the best of backgrounds to set off the fragile, tiny white flowers that look like small wood anemones. Why does the gold-thread choose to dwell where bees and butterflies, most flowers' best friends, rarely penetrate? Doubtless because the cool, damp habitat that develops abundant fungi also perfectly suits the fungus gnats and certain fungus-feeding beetles that are its principal benefactors. "The entire flower is constructed with reference to their visits," says Mr. Clarence Moores Weed; "the showy sepals attract their attention; the abnormal petals furnish them food; the many small stamens with white anthers and white pollen furnish a surface to walk upon, and a foreground in which the yellow nectar-cups are distinctly visible; the long-spreading recurved stigmas cover so large a portion of the blossom that it would be difficult even for one of the tiny visitors to take many steps without contact with one of them." On a sunny June day the lens usually reveals at least one tiny gnat making his way from one club-shaped petal to another - for the insignificant petals are mere nectaries - and transferring pollen from flower to flower.

Dig up a plant, and the fine tangled, yellow roots tell why it was given its name. In the good old days when decoctions of any herb that was particularly nauseous were swallowed in the simple faith that virtue resided in them in proportion to their revolting taste, the gold-thread's bitter roots furnished a tea much valued as a spring tonic and as a cure for ulcerated throats and canker-sore mouths of helpless children.

WHITE BANEBERRY
 (Actaea alba) Crowfoot family

Flowers - Small, white, in a terminal oblong raceme. Calyx of 3 to 5 petal-like, early-falling sepals; petals very small, 4 to 10, spatulate, clawed; stamens white, numerous, longer than petals; 1 pistil with a broad stigma. Stem: Erect, bushy, to 2 ft. high. Leaves: Twice or thrice compounded of sharply toothed and pointed, sometimes lobed, leaflets, petioled. Fruit: Clusters of poisonous oval white berries with dark purple spot on end, formed from the pistils. Both pedicels and peduncles much thickened and often red after fruiting. Preferred Habitat - Cool, shady, moist woods. Flowering Season - April-June. Distribution - Nova Scotia to Georgia and far West.

However insignificant the short fuzzy clusters of flowers lifted by this bushy little plant, we cannot fail to name it after it has set those curious white berries with a dark spot on the end, which Mrs. Starr Dana graphically compares to "the china eyes that small children occasionally manage to gouge from their dolls' heads." For genera-tions they have been called "doll's eyes" in Massachusetts. Especial-ly after these poisonous berries fully ripen and the rigid stems which bear them thicken and redden, we cannot fail to notice them. As the sepals fall early, the white stamens and stigmas are the most conspicuous parts of the flowers. A cluster opening its blossoms almost simultaneously, the plant's only hope of cross-fertilization lies in the expectation that the small female bees (Halictus) which come for pollen - no nectar being secreted - will leave some brought from another flower on the stigma as they enter, and before collecting a fresh supply. The time elapsing between the maturity of the stigmas and the anthers is barely perceptible; nevertheless there is a tendency toward the former maturing first.

The RED BANEBERRY, COHOSH, or HERB-CHRISTOPHER (A. rubra; A. spicata, var. rubra of Gray) - a more common species northward, although with a range, habit, and aspect similar to the preceding, may be known by its more ovoid raceme of feathery white flowers, its less sharply pointed leaves, and, above all, by its rigid clusters of oval red berries on slender pedicels, so conspicuous in the woods of late summer.

BLACK COHOSH; BLACK SNAKEROOT; TALL BUGBANE
(Cimicifuga racemosa) Crowfoot family [Buttercup family]

Flowers - Fetid, feathery, white, in an elongated wand-like raceme, 6 in. to 2 ft. long, at the end of a stem 3 to 8 ft. high. Sepals petal-like, falling early; 4 to 8 small stamen-like petals 2-cleft; stamens very numerous, with long filaments; 1 or 2 sessile pistils with broad stigmas. Leaves: Alternate, on long petioles, thrice compounded of oblong, deeply toothed or cleft leaflets, the end leaflet often again compound. Fruit: Dry oval pods, their seeds in 2 rows. Preferred Habitat - Rich woods and woodland borders, hillsides. Flowering Season - June-August. Distribution - Maine to Georgia, and westward from Ontario to Missouri.

Tall white rockets, shooting upward from a mass of large handsome leaves in some heavily shaded midsummer woodland border, cannot fail to impress themselves through more than one sense, for their odor is as disagreeable as the fleecy white blossoms are striking. Obviously such flowers would be most attractive to the carrion and meat flies. Cimicifuga, meaning to drive away bugs, and the old folk-name of bugbane testify to a degree of offensiveness to other insects, where the flies' enjoyment begins. As these are the only insects one is likely to see about the fleecy wands, doubtless they are their benefactors. The countless stamens which feed them generously with pollen willingly left for them alone must also dust them well as they crawl about before flying to another fetid lunch.

The close kinship with the baneberries is detected at once on examining one of these flowers. Were the vigorous plant less offensive to the nostrils, many a garden would be proud to own so decorative an addition to the shrubbery border.

WOOD ANEMONE; WIND FLOWER
(Anemone quinquefolia) Crowfoot family

Flowers - Solitary, about 1 in. broad, white or delicately tinted with blue or pink outside. Calyx of 4 to 9 oval, petal-like sepals; no petals; stamens and carpels numerous, of indefinite number. Stem: Slender, 4 to 9 in. high, from horizontal elongated rootstock. Leaves: On slender petioles, in a whorl of 3 to 5 below the flower, each leaf divided into 3 to 5 variously cut and lobed parts; also a late-appearing leaf from the base. Preferred Habitat - Woodlands, hillsides, light soil, partial shade. Flowering Season - April-June. Distribution - Canada and United States, south to Georgia, west to Rocky Mountains.

According to one poetical Greek tradition, Anemos, the wind, employs these exquisitely delicate little star-like namesakes as heralds of his coming in early spring, while woods and hillsides still lack foliage to break his gust's rude force. Pliny declared that only the wind could open anemones! Another legend utilized by countless poets pictures Venus wandering through the forests grief-stricken over the death of her youthful lover.

"Alas, the Paphian! fair Adonis slain!
Tears plenteous as his blood she pours amain;
But gentle flowers are born and bloom around
From every drop that falls upon the ground:
Where streams his blood, there blushing springs the rose;
And where a tear has dropped, a wind-flower blows."

Indeed, in reading the poets ancient and modern for references to this favorite blossom, one realizes as never before the significance of an anthology, literally a flower gathering.

But it is chiefly the European anemone that is extolled by the poets. Nevertheless our more slender, fragile, paler-leaved, and smaller-flowered species, known, strange to say, by the same scientific name, possesses the greater charm. Doctors, with more prosaic eyes than the poets, find acrid and dangerous juices in the anemone and its kin. Certain European peasants will run past a colony of these pure innocent blossoms in the belief that the very air is tainted by them. Yet the Romans ceremonially picked the first anemone of the year, with an incantation supposed to guard them against fever. The identical plant that blooms in our woods, which may be found also in Asia, is planted on graves by the Chinese, who call it the "death flower."

To leave legend and folk lore, the practical scientist sees in the anemone, trembling and bending before the wind, a perfect adaptation to its environment. Anchored in the light soil by a horizontal rootstock; furnished with a stem so slender and pliable no blast can break it; its pretty leaves whorled where they form a background to set off the fragile beauty of the solitary flower above them; a corolla economically dispensed with, since the white sepals are made to do the advertising for insects; the slightly nodding attitude of the blossom in cloudy weather, that the stigmas may be in the line of the fall of pollen jarred out by the wind in case visitors seeking pollen fail to bring any from other anemones - all these features teach that every plant is what it is for excellent reasons of its own; that it is a sentient being, not to be admired for superficial beauty merely, but also for those same traits which operate in the human race, making it the most interesting of studies.

Note the clusters of tuberous dahlia-like roots, the whorl of thin three-lobed rounded leaflets on long, fine petioles immediately below the smaller pure white or pinkish flowers usually growing in loose clusters, to distinguish the more common RUE-ANEMONE (Syndesmon thalictroides - Thalictrum anemonoides of Gray) from its cousin the solitary flowered wood or true anemone. Generally there are three blossoms of the rue-anemone to a cluster, the central one opening first, the side ones only after it has developed its sta- mens and pistils to prolong the season of bloom and encourage cross-pollination by insects. In the eastern half of the United States, and less abundantly in Canada, these are among the most familiar spring wild flowers. Pick them and they soon wilt miserably; lift the plants early, with a good ball of soil about the roots, and they will unfold their fragile blossoms indoors, bringing with them some- thing of the unspeakable charm of their native woods and hillsides just waking into life.

The TALL or SUMMER ANEMONE (A. Virginiana), called also THIMBLE-WEED from its oblong, thimble-like fruit-head, bears solitary, inconspicuous greenish or white flowers, often over an inch across, and generally with five rounded sepals, on erect, long stalks from June to August. Contrasted with the dainty tremulous little spring anemones, it is a rather coarse, stiff, hairy plant two or three feet tall. Its preference is for woodlands, whereas another summer bloomer, the LONG-FRUITED ANEMONE (A. cylindrica), a smal- ler, silky-hairy plant often confused with it, chooses open places, fields, and roadsides. The leaves of the thimble-weed, which are set in a whorl high up on the stem, and also spring from the root, after the true anemone fashion, are long petioled, three-parted, the divi- sions variously cut, lobed, and saw-edged. The flower-stalks which spring from this whorl continue to rise throughout the summer. The first, or middle of these peduncles, lacks leaves; later ones bear two leaves in the middle, from which more flower-stalks arise, and so on.

VIRGIN'S BOWER; VIRGINIA CLEMATIS; TRAVELLER'S JOY;
OLD MAN'S
BEARD
  (Clematis Virginiana) Crowfoot family

Flowers - White and greenish, about 1 in. across or less, in loose clusters from the axils. Calyx of 4 or 5 petal-like sepals; no petals; stamens and pistils numerous, of indefinite number; the staminate and pistillate flowers on separate plants; the styles feathery, and over 1 in. long in fruit. Stem: Climbing, slightly woody. Leaves:

Opposite, slender petioled, divided into 3 pointed and widely toothed or lobed leaflets. Preferred Habitat - Climbing over woodland borders, thickets, roadside shrubbery, fences, and walls; rich, moist soil. Flowering Season - July-September. Distribution - Georgia and Kansas northward less common beyond the Canadian border.

Fleecy white clusters of wild clematis, festooning woodland and roadside thickets, vary so much in size and attractiveness that one cannot but investigate the reason. Examination shows that comparatively few of the flowers are perfect, that is, few contain both stamens and pistils; the great majority are either male - the more showy ones - or female - the ones so conspicuous in fruit - and, like Quakers in meeting, the sexes are divided. The plant that bears staminate blossoms produces none that are pistillate, and vice versa - another marvelous protection against that horror of the floral race, self-fertilization, and a case of absolute dependence on insect help to perpetuate the race. Since the clematis blooms while insect life is at its height, and after most, if not all, of the Ranunculaceae have withdrawn from the competition for trade; moreover, since its white color, so conspicuous in shady retreats, and its accessible nectar attract hosts of flies and the small, short-tongued bees chiefly, that are compelled to work for it by transferring pollen while they feed, it goes without saying that the vine is a winner in life's race.

Charles Darwin, who made so many interesting studies of the power of movement in various plants, devoted special attention to the clematis clan, of which about one hundred species exist but, alas! none to our traveller's joy, that flings out the right hand of good fellowship to every twig within reach, winds about the sapling in brotherly embrace, drapes a festoon of flowers from shrub to shrub, hooks even its sensitive leafstalks over any available support as it clambers and riots on its lovely way. By rubbing the footstalk of a young leaf with a twig a few times on any side, Darwin found a clematis leaf would bend to that side in the course of a few hours, but return to the straight again if nothing remained on which to hook itself. "To show how sensitive the young petioles are," he wrote, "I may mention that I just touched the undersides of two with a little watercolor which, when dry, formed an excessively thin and minute crust but this sufficed in twenty-four hours to cause both to bend downwards."

In early autumn, when the long, silvery, decorative plumes attached to a ball of seeds form feathery, hoary masses even more fascinating than the flower clusters, the name of old man's beard is most suggestive. These seeds never open, but, when ripe, each is borne on the autumn gales, to sink into the first moist, springy resting place.

The English counterpart of our virgin's bower is fragrant.

TALL MEADOW-RUE
(Thalictrum polyganum; T. Cornuti of Gray) Crowfoot family

Flowers - Greenish white, the calyx of 4 or 5 sepals, falling early; no petals; numerous white, thread-like, green-tipped stamens, spreading in feathery tufts, borne in large, loose, compound terminal clusters 1 ft. long or more. Stem: Stout, erect, 3 to 11 ft. high, leafy, branching above. Leaves: Arranged in threes, compounded of various shaped leaflets, the lobes pointed or rounded, dark above, paler below. Preferred Habitat- Open sunny swamps, beside sluggish water, low meadows. Flowering Season - July-September. Distribution - Quebec to Florida, westward to Ohio.

Masses of these soft, feathery flowers, towering above the ranker growth of midsummer, possess an unseasonable, ethereal, chaste, spring-like beauty. On some plants the flowers are white and exquisite; others, again, are dull and coarser. Why is this? Because these are what botanists term polygamous flowers, i.e., some of them are perfect, containing both stamens and pistils; some are male only others, again, are female. Naturally an insect, like ourselves, is first attracted to the more beautiful male blossoms, the pollen bearers, and of course it transfers the vitalizing dust to the dull pistillate flowers visited later. But the meadow-rue, which produces a superabundance of very light, dry pollen, easily blown by the wind, is often fertilized through that agent also, just as grasses, plantains, sedges, birches, oaks, pines, and all cone-bearing trees are. As might be expected, a plant which has not yet ascended the evolutionary scale high enough to economize its pollen by making insects carry it invariably, overtops surrounding vegetation to take advantage of every breeze that blows.

The EARLY MEADOW-RUE (T. dioicum), found blooming in o-pen, rocky woods during April and May, from Alabama northward to Labrador, and westward to Missouri, grows only one or two feet high, and, like its tall sister, bears fleecy, greenish-white flowers, the staminate and the pistillate ones on different plants. These produce no nectar; they offer no showy corolla advertisement to catch the eye of passing insects; yet so abundant is the dry pollen produced by the male blossoms that insects which come to feed on it must occasionally transfer some, albeit this primitive genus still depends largely on the wind. Not its flower, but the exquisite foliage resembling sprays of a robust maidenhair fern, is this meadow-rue's chief charm.

The PURPLISH MEADOW-RUE (T. purpurascens), so like the tall species in general characteristics that one cannot tell the dried and pressed specimens of these variable plants apart, is easily named afield by the purplish tinge of its green polygamous flowers. Often its stems show color also. Sometimes, not always, the plant is downy, and the comparatively thick leaflets, which are dark green above, are waxy beneath. We look for this meadow-rue in copses and woodlands from Northern Canada to Florida, and far westward after the early meadow-rue has flowered, but before the tall one spreads its fleecy panicles. Quite as decorative as the flower clusters are the compound seed-bearing stars.

TWIN-LEAF; RHEUMATISM ROOT
 (Jeffersonia diphylla) Barberry family

Flowers - White, 1 in. broad, solitary, on a naked scape about 7 in. high in flower, more than twice as tall in fruit. Calyx of 4 petal-like sepals falling early; 8 longer, flat, oblong petals; 8 stamens; 1 pistil. Leaves: From the root, long-petioled, rounded, palmately veined, cleft into 2 divisions. Fruit: A leathery, many-seeded capsule, slit horizontally. Preferred habitat - Rich shady woods. Flowering Season - April-May. Distribution - New York to Virginia, west to Ontario and Tennessee.

Like many little darkies in the United States, this low plant was named for Thomas Jefferson. One suspects from a glance at its solitary white flower and deeply divided leaves that it is not far removed from the May apple, which is characterized by even greater Jeffersonian simplicity of habit, although separated into another genus.

MAY APPLE; HOG APPLE; MANDRAKE; WILD LEMON
(Podophyllum peltatum) Barberry family

Flowers - White, solitary, large, unpleasantly scented, nodding from the fork between a pair of terminal leaves. Calyx of 6 short-lived sepals; 6 to 9 rounded, flat petals stamens as many as petals or (usually) twice as many; 1 pistil, with a thick stigma. Stem: 1 to 1 1/2 ft. high, from a long, running rootstock. Leaves: Of flowerless stems (from separate root-stock), solitary, on a long petiole from base, nearly 1 ft. across, rounded, centrally peltate, umbrella fashion, 5 to 7 lobed, the lobes 2-cleft, dark above, light green below. Leaves of flowering stem 1 to 3, usually a pair, similar to others, but smaller. Fruit: A fleshy, yellowish, egg-shaped, many-seeded fruit about 2 in. long. Preferred habitat - Rich, moist woods. Flowering Season - May. Distribution - Quebec to the Gulf of Mexico, westward to Minnesota and Texas.

In giving this plant its abridged scientific name, Linnaeus seemed to see in its leaves a resemblance to a duck's foot (Anapodophyllum) but equally imaginative American children call them green umbrellas, and declare they unfurl only during April showers. In July, a sweetly mawkish, many-seeded fruit, resembling a yellow egg-tomato, delights the uncritical palates of little people, who should be warned, however, against putting any other part of this poisonous, drastic plant in their mouths. Physicians best know its uses. Dr. Asa Gray's statement about the harmless fruit "eaten by pigs and boys" aroused William Hamilton Gibson, who had happy memories of his own youthful gorges on anything edible that grew.

"Think of it, boys!" he wrote; "and think of what else he says of it: 'Ovary ovoid, stigma sessile, undulate, seeds covering the lateral placenta each enclosed in an aril.' Now it may be safe for pigs and billygoats to tackle such a compound as that, but we boys all like to know what we are eating, and I cannot but feel that the public health officials of every township should require this formula of Dr. Gray's to he printed on every one of these big loaded pills, if that is what they are really made of."

BLOODROOT; INDIAN PAINT; RED PUCCOON
(Sanguinaria Canadensis) Poppy family

Flowers - Pure white, rarely pinkish, golden centered, 1 to 1 1/2 in. across, solitary, at end of a smooth naked scape 6 to 14 in. tall. Calyx of 2 short-lived sepals; corolla of 8 to 12 oblong petals, early falling; stamens numerous; 1 short pistil composed of 2 carpels. Leaves: Rounded, deeply and palmately lobed, the 5 to 9 lobes often cleft. Rootstock: Thick, several inches long, with fibrous roots, and filled with orange-red juice. Preferred habitat - Rich woods and borders; low hillsides. Flowering Season - April-May. Distribution - Nova Scotia to Florida, westward to Nebraska.

Snugly protected in a papery sheath enfolding a silvery-green leaf-cloak, the solitary erect bud slowly rises from its embrace, sheds its sepals, expands into an immaculate golden-centered blossom that, poppy-like, offers but a glimpse of its fleeting loveliness ere it drops its snow-white petals and is gone. But were the flowers less ephemeral, were we always certain of hitting upon the very time its colonies are starring the woodland, would it have so great a charm? Here to-day, if there comes a sudden burst of warm sunshine; gone tomorrow, if the spring winds, rushing through the nearly leafless woods, are too rude to the fragile petals - no blossom has a more evanescent beauty, none is more lovely. After its charms have been displayed, up rises the circular leaf-cloak on its smooth reddish petiole, unrolls, and at length overtops the narrow, oblong seed-vessel. Wound the plant in any part, and there flows an orange-red juice, which old-fashioned mothers used to drop on lumps of sugar and administer when their children had coughs and colds. As this fluid stains whatever it touches - hence its value to the Indians as a war-paint - one should be careful in picking the flower. It has no value for cutting, of course; but in some rich, shady corner of the garden, a clump of the plants will thrive and bring a suggestive picture of the spring woods to our very doors. It will be noticed that plants having thick rootstocks, corms, and bulbs, which store up food during the winter, like the irises, Solomon's seals, bloodroot, adder's tongue, and crocuses, are prepared to rush into blossom far earlier in spring than fibrous-rooted species that must accumulate nourishment after the season has opened.

A newly opened flower which is in the female stage has its anthers tightly closed, and pollen must therefore be carried from distinct plants by the short-tongued bees and flies out collecting it. No nectar rewards their search, although they alight on young blossoms in the expectation of finding some food, and so cross-fertilize them. Late in the afternoon the petals, which have been in a showy horizontal position during the day, rise to the perpendicular before closing to protect the flower's precious contents for the morrow's visitors. In the blossom's staminate stage, abundant pollen is collected by the hive bees chiefly; but, those of the Halictus tribe, the mining bees and the Syrphidae flies also pay profitable visits. Inasmuch as the hive bee is a naturalized foreigner, not a native, the bloodroot probably depended upon the other little bees to fertilize it before her arrival. For ages this bee's small relatives and the flowers they depended upon developed side by side, adapting themselves to each other's wants. Now along comes an immigrant and profits by their centuries of effort.

DUTCHMAN'S BREECHES; WHITE HEARTS; SOLDIER'S CAP; EAR-DROPS (Bicuculla Cucullaria; Dicentra cucullaria of Gray) Poppy family

Flowers - White, tipped with yellow, nodding in a 1-sided raceme. Two scale-like sepals; corolla of 4 petals, in 2 pairs, somewhat cohering into a heart-shaped, flattened, irregular flower, the outer pair of petals extended into 2 widely spread spurs, the small inner petals united above; 6 stamens in 2 sets; style slender, with a 2-lobed stigma. Scape: 5 to 10 in. high, smooth, from a bulbous root. Leaves: Finely cut, thrice compound, pale beneath, on slender petioles, all from base Preferred Habitat - Rich, rocky woods. Flowering Season- - April-May. Distribution - Nova Scotia to the Carolinas, west to Nebraska.

Rich leaf mould, accumulated between crevices of rock, makes the ideal home of this delicate, yet striking, flower, coarse-named, but refined in all its parts. Consistent with the dainty, heart-shaped blossoms that hang trembling along the slender stem like pendants from a lady's ear, are the finely dissected, lace-like leaves, the whole plant repudiating by its femininity its most popular name. It was Thoreau who observed that only those plants which require but little light, and can stand the drip of trees, prefer to dwell in the woods - plants which have commonly more beauty in their leaves than in their pale and almost colorless blossoms. Certainly few woodland dwellers have more delicately beautiful foliage than the fumitory tribe.

Owing to this flower's early season of bloom and to the depth of its spurs, in which nectar is secreted by two long processes of the middle stamens, only the long-tongued female bumblebees then flying are implied by its curious formation. Two canals leading to the sweets invite the visitor to thrust in her tongue, and as she hangs from the white heart and presses forward to drain the luscious drops, first on one side, then on the other, her hairy underside necessarily comes in contact with the pollen of younger flowers and - with the later maturing stigmas of older ones, to which she carries it later. But, as might be expected, this intelligent bee occasionally nips holes through the spurs of the flower that makes dining so difficult for her - holes that lesser fry are not slow to investigate.

According to the Rev. Alexander S. Wilson, bumblebees make holes with jagged edges; wasps make clean-cut, circular openings; and the carpenter bees cut slits, through which they steal nectar from deep flowers. Who has tested this statement about the guilty little pilferers on our side of the Atlantic?

SQUIRREL CORN
(Bicuculla Canadensis) Poppy family

Flowers - Irregular, greenish white tinged with rose, slightly fragrant, heart-shaped, with 2 short rounded spurs, over 1/2 in. long, nodding on a slender scape. Calyx of 2 scale-like sepals; corolla heart-shaped at base, consisting of 4 petals in 2 united pairs, a prominent crest on tips of inner ones; 6 stamens in 2 sets; style with 2-lobed stigma. Scape: Smooth, 6 to 12 in. high, the rootstock bearing many small, round, yellow tubers like kernels of corn. Leaves: All from root, delicate, compounded of 3 very finely dissected divisions. Prferred Habitat - Rich, moist woods. Flowering Season - May-June. Distribution - Nova Scotia to Virginia, and westward to the Mississippi.

Any one familiar with the Bleeding-heart (B. eximia) of old-fashioned gardens, found growing wild in the Alleghanies, and with the exquisite White Mountain Fringe (Adlumia fungosa) often brought from the woods to be planted over shady trellises, or with the Dutchman's breeches, need not be told that the little squirrel corn is next of kin or far removed from the pink corydalis. It is not until we dig up the plant and look at its roots that we see why it received its name. A delicious perfume like hyacinths, only fainter and subtler, rises from the dainty blossoms.

BULBOUS or SPRING CRESS

(Cardamine bulbosa; C. rhomboidea of Gray) Mustard family

Flowers - White, about 1/2 in. across, clustered in a simple terminal raceme. Calyx of four sepals; corolla of 4 petals in form of a cross; 6 stamens; 1 compound pistil with a 2-lobed style. Stem: 6 to 18 in. high, erect, smooth, from a tuberous base. Leaves: Basal ones rounded, on long petioles; upper leaves oblong or lance-shaped, toothed or entire-edged, short petioled or seated on stem. Fruit: Very slender, erect pods about 1 in. long, tapering at each end, tipped with a slender style, the stigma prominent; 1 row of seeds in each cell, the pods rapidly following flowers up the stem and opening suddenly. Preferred Habitat - Wet meadows, low ground, near springs. Flowering Season - April-June. Distribution - Nova Scotia to Florida, west to Minnesota and Texas.

Pretty masses of this flower, that look like borders of garden candy-tuft planted beside some trickling brook, are visited and cross-fertilized by small bees, of the Andrena and Halictus clans chiefly. How well the butterflies understand scientific classification with instinct for their sure guide! The caterpillar of that exquisite little white butterfly with a dark yellow triangular spot across his wings, the fulcate orange-tip (Euchloe genutia), a first-cousin of the common small white cabbage butterfly, feeds on this plant and several of its kin, knowing better than if the books had told it so, that all belong to the same cross-bearing family. The watery, biting juice in the Cruciferae - the radishes, nasturtiums, cabbage, peppergrass, water-cress, mustards, and horseradish - by no means protects them from preying worms and caterpillars; but ants, the worst pilferers of nectar extant, let them alone. Authorities declare that the chloride of potassium and iodine these plants contain increase their food value to mankind.

The PURPLE CRESS (C. purpurea), formerly counted a mere variety of the preceding, has now been ranked as a distinct species. Its purplish-pink flowers, found about cold, springy places northward, appear two or three weeks earlier than those of the white spring cress.\ The MEADOW BITTER-CRESS (or CROSS), LADIES' SMOCK, OR CUCKOO-FLOWER (C. pratensis), an immigrant from Europe and Asia now naturalized here north of New Jersey from coast to coast, lifts its larger and more showy white or purplish-pink flowers, that stand well out from the stem on slender pedicels, in loose clusters above watery low-lying ground in April and May.

"Lady-smocks all silver white"

now paint our meadows with delight, as they do Shakespeare's England; but ours have quite frequently a decided pink tinge. The light and graceful growth, and the pinnately divided foliage, give the plant a special charm. In olden times, when it was counted a valuable remedy in hysteria and epilepsy, Linnaeus gave it its generic name Cardamine from two Greek words signifying heart-strengthening.

More bees, flies, butterflies, and other insects visit the ladies' smock than perhaps any other crucifer found here, since it has showy flowers and so much nectar the long-persistent sepals require little pouches to hold it. No wonder this plant has triumphantly marched around the world, leaving its relatives that take less pains to woo and work insects far behind in the race. Owing to a partial revolution of the tall stamens away from the stigmas, a visitor in sipping nectar must brush off some pollen on his head or tongue, although in stormy weather, when the movement of the stamens is incomplete, self-pollination may occasionally occur, according to Muller.

TWO-LEAVED TOOTHWORT; CRINKLE-ROOT
(Dentaria diphylla) Mustard family

Flowers - White, about 1/2 in. across, in a terminal loose cluster, the formation of each similar to that of bulbous cress. Stem: 8 to 15 in. high. Root stock: Long, crinkled, toothed, fleshy, crisp, edible. Leaves: 2, opposite or nearly so, on the stem, compounded of 3 ovate and toothed leaflets; also larger, broader leaves on larger petioles from the rootstock. Fruit: Flat, lance-shaped pods, 1 in. long or over, tipped with the slender style. Perferred Habitat - Rich leaf mould in woods, sometimes in thickets and meadows. Flowering Season - May. Distribution - Nova Scotia to the Carolinas, west to the Mississippi.

Clusters of these pretty, white, cross-shaped flowers, found near the bloodroot, claytonia, anemones, and a host of other delicate spring blossoms, enter into a short but fierce competition with them for the visits of the small Andrena and Halictus bees then flying to collect nectar and pollen for a generation still unborn. In tunnels underground, or in soft, partially decayed wood, each busy little mother places the pellets of pollen and nectar paste, then when her eggs have been laid on the food supply in separate nurseries and sealed up, she dies from exhaustion, leaving her grub progeny to eat its way through the larva into the chrysalis state, and finally into that of a winged bee that flies away to liberty. These are the little bees so constantly seen about willow catkins.

Country children, on their way to school through the woods, often dig up the curious, long crisp root of the toothwort, which tastes much like the water-cress, to eat with their sandwiches at the noon recess. Then, as they examine the little pointed projections on the rootstock, they see why the plant received its name.

Another toothwort found throughout a similar range, the CUT-LEAVED TOOTHWORT, or PEPPER-ROOT (D. laciniata), has its equally edible rootstock scarcely toothed, but rather constricted in places, giving its little tubers the appearance of beads strung into a necklace. Its white or pale purplish-pink cross-shaped flowers, loosely clustered at the end of an unbranched stem, rise by preference above moist ground in rich woods, often beside a spring, from April to June - a longer season for wooing and working its insect friends than the two-leaved toothwort has attained to - hence it is the commoner plant. Instead of having two leaves on its stem, this species spreads whorls of three leaves, thrice divided, almost to the base, the divisions toothed or lobed, and the side ones sometimes deeply cleft. The larger, longer petioled leaves that rise directly from the rootstock have scarcely developed at flowering time.

SHEPHERD'S PURSE; MOTHER'S HEART
(Bursa Bursa-pastoris; Capsella Bursa-pastoris of Gray)
Mustard family

Flowers - Small, white, in a long loose raceme, followed by triangular and notched (somewhat heart-shaped) pods, the valves boat-shaped and keeled. Sepals and petals 4; stamens 6; 1 pistil. Stem: 6 to 18 in. high, from a deep root. Leaves: Forming a rosette at base, 2 to 5 in. long, more or less cut (pinnatifid), a few pointed, arrow-shaped leaves also scattered along stem and partly clasping it. Preferred Habitat - Fields, roadsides, waste places. Flowering Season - Almost throughout the year. Distribution - Over nearly all parts of the earth.

>From Europe this little low plant found its way, to become the commonest of our weeds, so completing its march around the globe. At a glance one knows it to be related to the alyssum and candy-tuft

of our gardens, albeit a poor relation in spite of its vaunted purses - the tiny, heart-shaped seed-pods that so rapidly succeed the flowers. What is the secret of its successful march over the face of the earth? Like the equally triumphant chickweed, it is easily satisfied with unoccupied wasteland, it avoids the fiercest competition for insect trade by prolonging its season of bloom far beyond that of any native flower, for there is not a month in the year when one may not find it even in New England in sheltered places. Having vanquished in the fiercer struggle for survival in the Old World, it finds life here one long holiday; and finally, by clustering a large number of relatively small flowers together, it attracts the insects that this method of arrangement pleases best, the flies (Syrphidae and Muscidae) which cross-fertilize it in fine weather, transferring enough pollen from plant to plant to save the species from degeneracy through close inbreeding. However, the long stamens standing on a level with the stigma are well calculated to self-pollenize the flowers, the flies failing them.

VERNAL WHITLOW-GRASS
(Draba verna) Mustard family

Flowers - Very small, white, distant, growing on numerous scapes 1 to 5 in. high; in formation each flower is similar to all the mustards, except that the 4 petals are 2-cleft, destroying the cross-like effect. Leaves: 1/2 to 1 in. long, in a tuft or rosette on the ground, oblong or spatulate, covered with stiff hairs. Preferred Habitat - Waste lands, sandy fields, and roadsides. Flowering Season - February-May. Distribution - Throughout our area; naturalized from Europe and Asia.

An insignificantly small plant, too common, however, to be wholly ignored. Although each tiny flower secretes four drops of nectar between the bases of the short stamens and the long ones next them, it would be unreasonable to depend wholly upon insects to carry pollen, since there is so little else to attract them. Therefore the anthers of the four long stamens regularly shed directly upon the stigma below them, leaving to the few visitors, the small bees chiefly, the transferring from flower to flower of pollen from the two short stamens which must be touched if they would reach the nectar. In spite of the persistency with which these little blossoms fertilize themselves, they certainly increase at a prodigious rate; but how much larger and more beautiful might they not be if they possessed more executive ability

A similar but larger plant, with its hairy leaves not only tufted at the base, but also alternating up the stiff stem, is the HAIRY ROCK-CRESS (Arabis hirsuta), whose white or greenish flowers, growing in racemes after the usual mustard fashion, are quickly followed by very narrow, flattened pods two inches long or less. Around the world this small traveler has likewise found its way, choosing rocky places to display its insignificant flowers throughout the entire summer to such small bees and flies as seek the nectar in its two tiny glands. It is not to be confused with the saxifrage or stone-breaker.

ROUND-LEAVED SUNDEW; DEW-PLANT
(Drosera rotundifolia) Sundew family

Flowers - Small, white, growing in a 1-sided, curved raceme of buds chiefly. Calyx usually 5-parted; usually 5 petals, and as many stamens as petals; usually 3 styles, but 2-cleft, thus appearing to be twice as many. Scape: 4 to 10 in. high. Leaves: Growing in an open rosette on the ground; round or broader, clothed with reddish bristly hairs tipped with purple glands, and narrowed into long, flat, hairy petioles; young leaves curled like fern fronds. Preferred Habitat - Bogs, sandy and sunny marshes. Flowering Season - July-August. Distribution - Labrador to the Gulf of Mexico and westward. From Alaska to California. Europe and Asia.

Here is a bloodthirsty little miscreant that lives by reversing the natural order of higher forms of life preying upon lower ones, an anomaly in that the vegetable actually eats the animal! The dogbane, as we have seen, simply catches the flies that dare trespass upon the butterflies' preserves, for excellent reasons of its own; the Silenes and phloxes, among others, spread their calices with a sticky gum that acts as limed twigs do to birds, in order to guard the nectar secreted for flying benefactors from pilfering ants; the honey bee being an imported, not a native, insect, and therefore not perfectly adapted to the milkweed, occasionally gets entrapped by it; the big bumblebee is sometimes fatally imprisoned in the moccasin flower's gorgeous tomb - the punishment of insects that do not benefit the flowers is infinite in its variety. But the local Venus's flytrap (Dionaea muscipula), gathered only from the low savannas in North Carolina to entertain the owners of hothouses as it promptly closes the crushing trap at the end of its sensitive leaves over a hapless fly, and the common sundew that tinges the peat-bogs of three continents with its little reddish leaves, belong to a distinct class of carnivorous plants which actually masticate their animal food, depending upon it for nourishment as men do upon cattle slaughtered in an abattoir. Darwin's luminous account of these two species alone, which occupies over three hundred absorbingly interesting pages of his "Insectivorous Plants" should be read by everyone interested in these freaks of nature.

When we go to some sunny cranberry bog to look for these sundews, nothing could be more innocent looking than the tiny plant, its nodding raceme of buds, usually with only a solitary little blossom (that opens only in the sunshine) at the top of the curve, its leaves glistening with what looks like dew, though the midsummer sun may be high in the heavens. A little fly or gnat, attracted by the bright jewels, alights on a leaf only to find that the clear drops, more sticky than honey, instantly glue his feet, that the pretty reddish hairs about him act like tentacles, reaching inward, to imprison him within their slowly closing embrace. Here is one of the horrors of the Inquisition operating in this land of liberty before our very eyes! Excited by the struggles of the victim, the sensitive hairs close only the faster, working on the same principle that a vine's tendrils do when they come in contact with a trellis. More of the sticky fluid pours upon the hapless fly, plastering over his legs and wings and the pores on his body through which he draws his breath. Slowly, surely, the leaf rolls inward, making a temporary stomach; the cruel hairs bind, the glue suffocates and holds him fast. Death alone releases him. And now the leafs orgy begins: moistening the fly with a fresh peptic fluid, which helps in the assimilation, the plant proceeds to digest its food. Curiously enough, chemical analysis proves that this sundew secretes a complex fluid corresponding almost exactly to the gastric juice in the stomach of animals.

Darwin, who fed these leaves with various articles, found that they could dissolve matter out of pollen, seeds, grass, etc.; yet without a human caterer, how could a leaf turn vegetarian? When a bit of any undesirable substance, such as chalk or wood, was placed on the hairs and excited them, they might embrace it temporarily; but as soon as the mistake was discovered, it would be dropped! He also poisoned the plants by administering acids, and gave them fatal attacks of indigestion by overfeeding them with bits of raw beef!

Other common sundews, the SPATULATE-LEAVED SUNDEW (D. intermedia) and the THREAD-LEAVED SUNDEW (D. filiformis) whose purplish-pink flowers are reared above wet sand along the coast, possess contrivances similar to the round-leaved plant's to pursue their gruesome business. Why should these vegetables turn carnivorous? Doubtless because the soil in which they grow can supply little or no nitrogen. Very small roots testify to the small use they serve. The water sucked up through them from the bog aids in the manufacture of the fluid so freely exuded by the bristly glands, but nitrogen must be obtained by other means, even at the sacrifice of insect victims.

EARLY SAXIFRAGE
(Saxifraga Virginiensis) Saxifrage family

Flowers - White, small, numerous, perfect, spreading into a loose panicle. Calyx 5-lobed; 5 petals; 10 stamens; 1 pistil with 2 styles. Scape: 4 to 12 in. high, naked, sticky-hairy. Leaves: Clustered at the base, rather thick, obovate, toothed, and narrowed into spatulate-margined petioles. Fruit: Widely spread, purplish-brown pods. Preferred Habitat - Rocky woodlands, hillsides. Flowering Season - March-May Distribution - New Brunswick to Georgia, and westward a thousand miles or more.

Rooted in clefts of rock that, therefore, appears to be broken by this vigorous plant, the saxifrage shows rosettes of fresh green leaves in earliest spring, and soon whitens with its blossoms the most forbidding niches. (Saxum = a rock; frango = 1 break.) At first a small ball of green buds nestles in the leafy tuffet, then pushes upward on a bare scape, opening its tiny, white, five-pointed star flowers as it ascends, until, having reached the allotted height, it scatters them in spreading clusters that last a fortnight. Again we see that, however insignificantly small nectar-bearing flowers may be, they are somehow protected from crawling pilferers; in this case by the commonly employed sticky hairs in which ants' feet become ensnared. As the anthers mature before the stigmas are ready to receive pollen, certainly the flowers cannot afford to send empty away the benefactors on whom the perpetuation of their race depends; and must prevent it even with the most heroic measures.

FALSE MITERWORT; COOLWORT; FOAM-FLOWER; NANCY-OVER-THE-GROUND
  (Tiarella cordifolia) Saxifrage family.

  Flowers - White, small, feathery, borne in a close raceme at the top of a scape 6 to 12 in. high. Calyx white, 9-lobed; 5 clawed petals; 10 stamens, long-exserted; 1 pistil with 2 styles. Leaves: Long-petioled from the rootstock or runners, rounded or broadly heart-shaped, 3 to 7-lobed, toothed, often downy along veins beneath. Preferred Habitat - Rich, moist woods, especially along mountains. Flowering Season - April-May. Distribution - Nova Scotia to Georgia, and westward scarcely to the Mississippi.

  Fuzzy, bright white foam-flowers are most conspicuous in the forest when seen against their unevenly colored leaves that carpet the ground. A relative, the TRUE MITERWORT or BISHOP'S CAP (Mitella diphylla), with similar foliage, except that two opposite leaves may be found almost seated near the middle of its hairy stem, has its flowers rather distantly scattered on the raceme, and their fine petals deeply cut like fringe. Both species may be found in bloom at the same time, offering an opportunity for comparison to the confused novice. Now, tiarella, meaning a little tiara, and mitella, a little miter, refer, of course, to the odd forms of their seed-cases; but all of us are not gifted with the imaginative eyes of Linnaeus, who named the plants. Xenophon's assertion that the royal tiara or turban of the Persians was encircled with a crown helps us no more to see what Linnaeus saw in the one case than the fact that the papal miter is encircled by three crowns helps in the other. And as for the lofty, two-peaked cap worn by bishops in the Roman Church, a dozen plants, with equal propriety, might be said to wear it.

CAROLINA GRASS OF PARNASSUS
 (Parnassia Caroliniana) Saxifrage family

Flowers - Creamy white, delicately veined with greenish, solitary, 1 in. broad or over, at the end of a scape 8 in. to 2 ft. high, 1 ovate leaf clasping it. Calyx deeply 5-lobed; corolla of 5 spreading, parallel veined petals; 5 fertile stamens alternating with them, and 3 stout imperfect stamens clustered at base of each petal; 1 very short pistil with 4 stigmas. Leaves: >From the root, on long petioles, broadly oval or rounded, heart-shaped at base, rather thick. Preferred Habitat - Wet ground, low meadows, swamps. Flowering Season - July-September. Distribution - New Brunswick to Virginia, west to Iowa.

What's in a name? Certainly our common grass of Parnassus, which is no grass at all, never starred the meadows round about the home of the Muses, nor sought the steaming savannas of the Carolinas. The European counterpart (P. palustris), fabled to have sprung up on Mount Parnassus, is at home here only in the Canadian border States and northward.

At first analysis one is puzzled by the clusters of filaments at the base of each petal. Of what use are they? We have seen in the case of the beard-tongue and the turtle-head that even imperfect stamens sometimes serve useful ends, or they would doubtless have been abolished. A fly or bee mistaking, as he well may, the abortive anthers for beads of nectar on this flower, alights on one of the white petals, a convenient, spreading landing place; but finding his mistake, and guided by the greenish lines, the pathfinders to the true nectaries situated on the other side of the curious fringy structures, he must, because of their troublesome presence, climb over them into the center of the flower to suck its sweets from the point where he will dust himself with pollen in young blossoms. Of course he will carry some of their vitalizing powder to the late maturing stigmas of older ones. Without the fringe of imperfect stamens, that serves as a harmless trellis easily climbed over, the visitor might stand on the petals and sip nectar without rendering any assistance in cross-fertilizing his entertainers.

NINEBARK (Opulaster opulifolius; Spiraea opulifolia of Gray) Rose family

Flowers - White or pink, small, in numerous rounded terminal clusters to 2 in. broad. Calyx 5-lobed; 5 rounded petals inserted in its throat; 20 to 40 stamens; several pistils. Stem: Shrubby, 3 to 10 ft. high, with long, recurved branches, the loose bark peeling off annually in thin strips. Leaves: Simple, heart-shaped or rounded, 3-lobed, toothed. Fruit: 3 to 5 smooth, shining, reddish, inflated, pointed pods. Preferred Habitat - Rocky banks, riversides. Flowering Season - June. Distribution - Canada to Georgia, west to Kansas.

Whether the nurserymen agree with Dr. Gray or not when he says these balls of white flowers possess "no beauty," the fact remains that numbers of the shrubs are sold for ornament, especially a golden-leaved variety. But the charm certainly lies in their fruit. (Opulus = a wild cranberry tree.) When this is plentifully set at the ends of long branches that curve backward, and the bladder-like pods have taken on a rich purplish or reddish hue, the shrub is undeniably decorative. Even the old flowers, after they have had their pollen carried away by the small bees and flies, show a reddish tint on the ovaries which deepens as the fruit forms; and Ludwig states that this is not only to increase the conspicuousness of the shrubs, but to entice unbidden guests away from the younger flowers. Who will tell us why the old bark should loosen every year and the thin layers separate into not nine, but dozens of ragged strips?

MEADOW-SWEET; QUAKER LADY; QUEEN-OF-THE-MEADOW (Spiraea salicifolia) Rose family

Flowers - Small, white or flesh pink, clustered in dense pyramidal terminal panicles. Calyx 5 cleft; carolla of 5 rounded petals; stamens numerous; pistils 5 to 8. Stem: 2 to 4 ft. high, simple or bushy, smooth, usually reddish. Leaves: Alternate, oval or oblong, saw-edged. Preferred Habitat - Low meadows, swamps, fence-rows, ditches. Flowering Season - June-August. Distribution - Newfoundland to Georgia, west to Rocky Mountains. Europe and Asia.

Fleecy white plumes of meadow-sweet, the "spires of closely clustered bloom" sung by Dora Read Goodale, are surely not frequently found near dusty "waysides scorched with barren heat," even in her Berkshires; their preference is for moister soil, often in the same habitat with a first cousin, the pink steeple-bush. But plants, like humans, are capricious creatures. If the meadow-sweet always elected to grow in damp ground whose rising mists would clog the pores of its leaves, doubtless they would be protected with a woolly absorbent, as its cousins are.

Inasmuch as perfume serves as an attraction to the more highly specialized, aesthetic insects, not required by the spiraeas, our meadow-sweet has none, in spite of its misleading name. Small bees (especially Andrenidae), flies (Syrphidae), and beetles, among other visitors, come in great numbers, seeking the accessible pollen, and, in this case, nectar also, secreted in a conspicuous orange-colored disk. When a floret first opens, or even before, the already mature stigmas overtop the incurved, undeveloped stamens, so that any visitor dusted from other clusters cross-fertilizes it; but as the stigmas remain fresh even after the stamens have risen and shed their abundant pollen, it follows that in long-continued stormy weather, when few insects are flying, the flowers fertilize themselves. Self-fertilization with insect help must often occur in the flower's second stage. The fragrant yellowish-white ENGLISH MEADOW-SWEET (S. ulmaria), often cultivated in old-fashioned gardens here, has escaped locally.

In long, slender, forking spikes the GOAT'S-BEARD (Aruncus Aruncus - Spiraea aruncus of Gray) lifts its graceful panicles of minute whitish flowers in May and June from three to seven feet above the rich soil of its woodland home. The petioled, pinnate leaves are compounded of several leaflets like those on its relative the rose-bush. From New York southward and westward to Missouri, also on the Pacific Coast to Alaska, is its range on this Continent. Very many more beetles than any other visitors transfer pollen from the staminate flowers on one plant to the pistillate ones on another; other plants produce only perfect flowers - the reason different panicles vary so much in appearance.

Another herbaceous perennial once counted a spiraea is the common INDIAN PHYSIC or BOWMAN'S-ROOT (Porteranthus trifoliatus - Gillensia trifoliata of Gray) found blooming in the rich woods during June and July from western New York southward and westward. Two to four feet high, it displays its very loose, pretty clusters of white or pale pink flowers, comparatively few in the whole panicle, each blossom measuring about a half inch across and borne on a slender pedicel. A tubular, 5-toothed calyx has the long slender petals inserted within. Owing to the depth and narrowness of the tube, the small, long-tongued bees cannot reach the nectar without dusting their heads with pollen from the anthers inserted in a ring around the entrance or leaving some on the stigmas of other blossoms. Later, the five carpels make as many hairy, awl-tipped little pods within the reddish cup. The leaves may be compounded of three oblong or ovate, saw-edged leaflets, or merely three-lobed, and with small stipules at their base.

WILD RED RASPBERRY
(Rubus strigosus) Rose family

Flowers - White, about 1/2 in. across, on slender, bristly pedicels, in a loose cluster. Calyx deeply 5-parted, persistent in fruit; 5 erect, short-lived petals, about the length of the sepals; stamens numerous; carpels numerous, inserted on a convex spongy receptacle, and ripening into drupelets. Stem: 3 to 6 ft. high, shrubby, densely covered with bristles; older, woody stems with rigid, hooked prickles. Leaves: Compounded of 3 to 5 ovate, pointed, and irregularly saw-edged leaflets, downy beneath, on bristly petioles. Fruit: A light red, watery, tender, high-flavored, edible berry; ripe July-September. Preferred Habitat - Dry soil, rocky hillsides, fence-rows, hedges. Flowering Season - May-July. Distribution - Labrador to North Carolina, also in Rocky Mountain region.

Who but the bees and such small visitors care about the raspberry blossoms? Notwithstanding the nectar secreted in a fleshy ring for their benefit, comparatively few insects enter the flowers, whose small, erect petals imply no hospitable welcome. Occasionally a visitor laden with pollen from another plant alights in the center of a blossom, and leaves some on the stigmas in bending his head down between them and the stamens to reach the refreshment; but inasmuch as the erect petals allow no room for the stamens to spread out and away from the stigmas, it follows that self-fertilization very commonly occurs.

Of course, men and children, bears and birds, are vastly more interested in the delicious berries; men for the reason that several excellent market varieties, some white or pale red, the Cuthbert and Hansall berries among others, owe their origin to this hardy native. Many superior sorts derived from its European counterpart (R. Idaeus) cannot well endure our rigorous northern climate. As in the case of most berry-bearing species, the raspberry depends upon the birds to drop its undigested seeds over the country, that new colonies may arise under freer conditions. Indeed, one of the best places for the budding ornithologist to take opera-glasses and notebook is to a raspberry patch early in the morning.

The BLACK RASPBERRY, BLACK CAP or SCOTCH CAP or THIMBLE-BERRY (R. occidentalis), common in such situations as the red raspberry chooses, but especially in burned-over districts from Virginia northward and westward, has very long, smooth, cane-like stems, often bending low until they root again at the tips. These are only sparingly armed with small, hooked prickles, no bristles. The flowers, which are similar to the preceding, but clustered more compactly, are sparingly visited by insects; nevertheless when self-fertilized, as they usually are, abundant purplish-black berries, hollow like a thimble where they drop from the spongy receptacle, ripen in July. Numerous garden hybrids have been derived from this prolific species also. Indeed its offspring are the easiest raspberries to grow, since they form new plants at the tips of the branches, yet do not weaken themselves with suckers, and so, even without care, yield immense crops. One need not stir many feet around a good raspberry patch to enjoy a Transcendental feast.

HIGH BUSH BLACKBERRY; BRAMBLE
(Rubus villosus) Rose family

Flowers - White, 1 in. or less across, in terminal raceme-like clusters. Calyx deeply 5-parted, persistent; 5 large petals; stamens and carpels numerous, the latter inserted on a pulpy receptacle. Stem: 3 to 10 ft. high, woody, furrowed, curved, armed with stout, recurved prickles. Leaves: Compounded of 3 to 5 ovate, saw-edged leaflets, the end one stalked, all hairy beneath. Fruit: Firmly attached to the receptacle; nearly black, oblong juicy berries 1 in. long or less, hanging in clusters. Ripe, July-August. Preferred Habitat - Dry soil, thickets, fence-rows, old fields, waysides. Low altitudes. Flowering Season - May-June. Distribution - New England to Florida, and far westward.

"There was a man of our town,
  And he was wondrous wise,
  He jumped into a bramble bush" -

If we must have poetical associations for every flower, Mother Goose furnishes several.

But for the practical mind this plant's chief interest lies in the fact that from its wild varieties the famous Lawton and Kittatinny blackberries have been derived. The late Peter Henderson used to tell how the former came to be introduced. A certain Mr. Secor found an unusually fine blackberry growing wild in a hedge at New Rochelle, New York, and removed it to his garden, where it increased apace. But not even for a gift could he induce a neighbor to relieve him of the superfluous bushes, so little esteemed were blackberries in his day. However, a shrewd lawyer named Lawton at length took hold of it, exhibited the fruit, advertised it cleverly, and succeeded in pocketing a snug little fortune from the sale of the prolific plants. Another fine variety of the common wild blackberry, which was discovered by a clergyman at the edge of the woods on the Kittatinny Mountains in New Jersey, has produced fruit under skilled cultivation that still remains the best of its class. When clusters of blossoms and fruit in various stages of green, red, and black hang on the same bush, few ornaments in Nature's garden are more decorative.

Because bramble flowers show greater executive ability than the raspberries do, they flaunt much larger petals, and spread them out flat to attract insect workers as well as to make room for the stamens to spread away from the stigmas - an arrangement which gives freer access to the nectar secreted in a fleshy ring at the base. Heavy bumblebees, which require a firm support, naturally alight in the center, just as they do in the wild roses, and deposit on the early maturing stigmas some imported pollen. They may therefore be regarded as the truest benefactors, and it will be noticed that for their special benefit the nectar is rather deeply concealed, where short-tongued insects cannot rob them of it. Small bees, which come only to gather pollen from first the outer and then the inner rows of stamens, and a long list of other light-weight visitors, too often a-light on the petals to effect cross-fertilization regularly, but they usually self-fertilize the blossoms. Competition between these flowers and the next is fierce, for their seasons overlap.

The DEWBERRY or LOW RUNNING BLACKBERRY (R. Canadensis), that trails its woody stem by the dusty roadside, in dry fields, and on sterile, rocky hillsides, calls forth maledictions from the bare-footed farmer's boy, except during June and July, when its prickles are freely forgiven it in consideration of the delicious, black, seedy berries it bears. He is the last one in the world to confuse this vine with the SWAMP BLACKBERRY (R. hispidus), a smaller flowered runner, slender and weakly prickly as to its stem, and insignificant and sour as to its fruit. Its greatest charm is when we come upon it in some low meadow in winter, when its still persistent, shining, large leaves, that have taken on rich autumnal reds, glow among the dry, dead weeds and grasses.

CREEPING DALIBARDA
(Dalibarda repens) Rose family

Flowers - White, solitary, or 2 at end of a scape 2 to 5 in. high. Calyx deeply, unevenly 5 or 6 parted, the larger divisions toothed; 5 petals falling early; numerous stamens; 5 to 10 carpels forming as many dry drupelets within the persistent calyx. Stem: Creeping, slender, no prickles. Leaves: Long petioled, in tufts from the runner, almost round, heart-shaped at base, crenate-edged, both sides hairy. Preferred habitat - Woods and wooded hillsides. Flowering Season - June-September. Distribution - Nova Scotia to Pennsylvania, and westward to the Mississippi.

This delicate blossom, which one might mistake for a white violet among a low tuft of violet-like leaves, shows its rose kinship by its rule of five and its numerous stamens. Like the violet again, however, it bears curious little economical flowers near the ground - flowers which never open, and so save pollen. These, requiring no insects to fertilize them, waste no energy in putting forth petals to advertise for visitors. Nevertheless, to save the species from degeneracy from close inbreeding, this little plant needs must display a few showy blossoms to insure cross-fertilized seed; for the offspring of such defeats the offspring of self-fertilized plants in the struggle for existence.

VIRGINIA STRAWBERRY
(Fragaria Virginiana) Rose family

Flowers - White, loosely clustered at summit of an erect hairy scape usually shorter than the leaves. Calyx persistent in fruit, deeply 5-cleft, with 5 bracts between the divisions; 5 petals; stamens and pistils numerous, the latter inserted on a cushion-like receptacle becoming fleshy in fruit. Staminate and pistillate flowers, from separate roots. Stem: Running, and forming new plants. Leaves: Tufted from the root, on hairy petioles 2 to 6 in. tall, compounded of 3 broadly oval, saw-edged leaflets. Fruit. An ovoid, glistening red berry, the minute achenes imbedded in pits on its surface. Ripe,

June-July. (Latin, fragum = fragrant fruit, the strawberry.) Preferred Habitat - Dry fields, banks, roadsides, woodlands. Flowering Season - April-June. Distribution - New Brunswick to the Gulf of Mexico, and westward to Dakota.

"Doubtless God could have made a better berry, but doubtless God never did." Whether one is kneeling in the fields, gathering the sun-kissed, fragrant, luscious, wet scarlet berries nodding among the grass, or eating the huge cultivated fruit smothered with sugar and cream, one fervently quotes Dr. Boteler with dear old lzaak Walton. Shakespeare says : "My lord of Ely, when I was last in Holborn, I saw good strawberries in your garden there." Is not this the first reference to the strawberry under cultivation? Since the time of Henry V, what multitudes of garden varieties past the reckoning have been evolved from the smooth, conic EUROPEAN WOOD STRAWBERRY (F. vesca) now naturalized in our Eastern and Middle States, as well as from our own precious pitted native! Some authorities claim the berry received its name from the straw laid between garden rows to keep the fruit clean, but in earliest Anglo-Saxon it was called streowberie, and later straberry, from the peculiarity of its straying suckers lying as if strewn on the ground; and so, after making due allowance for the erratic, go-as-you-please spelling of early writers, it would seem that there might be two theories as to the origin of the name.

Since the different sexes of these flowers frequently occur on separate plants, good reason have they to woo insect messengers with a showy corolla, a ring of nectar, and abundant pollen to be transferred while they are feasted. Lucky is the gardener who succeeds in keeping birds from pecking their share of the berries which, of course, were primarily intended for them. In English gardens one is almost certain to find a thrush or two imprisoned under the nets so futilely spread over strawberry beds, just as their American cousin, the robin, is caught here in June.

A young botanist may be interested to note the difference in the formation of the raspberry or blackberry and the strawberry: in the former it is the carpels (ovaries) that swell around the spongy receptacle into numerous little fruits (drupelets) united into one berry, whereas it is the cushion-like receptacle itself in the strawberry blossom that swells and reddens into fruit, carrying with it the tiny yellow pistils to the surface.

The NORTHERN WILD STRAWBERRY (F. Canadensis), with clusters of elongated, oblong little berries delightful to three senses, comes over the Canadian border no farther south than the Catskills. Nearly all strawberry plants show the useless but charming eccentricity of bursting into bloom again in autumn, the little white-petaled blossoms coming like unexpected flurries of snow.

No one will confuse our common, fruiting species with the small, yellow-flowered DRY or BARREN STRAWBERRY (Waldsteinia fragarioides), more nearly related to the cinquefoils. Tufts of its pretty trefoliate leaves, sent up from a creeping rootstock, carpet the woods and hillsides from New England and along the Alleghanies to Georgia, and westward a thousand miles or more. Flowers in May and June.

WHITE AVENS
(Geum Canadense; G. album of Gray) Rose family

Flowers - White or pale greenish yellow, about 1/2 in. across, loosely scattered in small clusters on slender peduncles. Calyx persistent, 5-cleft, with little bracts between the reflexed divisions; 5 petals, equaling or shorter than the sepals; stamens and carpels numerous, the latter collected on a short, bristly-hairy receptacle; styles smooth below, hairy above, jointed. Stem: 2 1/2 ft. high or less, slender, branching above. Leaves: Seated on stem or short petioled, of 3 to 5 divisions, or lobed, toothed small stipules; also irregularly divided large root-leaves on long petioles, 3-foliate, usually the terminal leaflet large, broadly ovate side leaflets much smaller, all more or less lobed and toothed. Fruit: A ball of achenes, each ending in an elongated, hooked style. Preferred Habitat - Woodland borders, shady thickets and roadsides. Flowering Season - June-September. Distribution - Nova Scotia to Georgia, west to the Mississippi or beyond.

Small bees and flies attracted to sheltered, shady places by these loosely scattered flowers at the ends of zig-zagged stems, pay for the nectar they sip from the disk where the stamens are inserted, by carrying some of the pollen lunch on their heads from the older to the younger flowers, which mature stigmas first. But saucy bumble-bees, undutiful pilferers from the purple avens, rarely visit blossoms so inconspicuous. Insects failing these, they are well adapted to pollenize themselves. Most of us are all too familiar with the seeds, clinging by barbed styles to any garment passing their way, in the hope that their stolen ride will eventually land them in good colonizing ground. Whoever spends an hour patiently picking off the various seed tramps from his clothes after a walk through the woods and fields in autumn, realizes that the by hook or by crook method of scattering offspring is one of Nature's favorites. Simpler plants than those with hooked achenia produce enormous numbers of spores so light and tiny that the wind and rain distribute them wholesale.

RED CHOKE-BERRY; DOGBERRY TREE
(Aronia arbutifolia; Pyrus arbutifolia of Gray) Apple family

Flowers - White or magenta tinged, 1/2 in. across or less, in terminal, compound cymes, finally overtopped by young sterile shoots. Calyx 5-lobed, hairy; 5 concave, spreading petals; stamens numerous; 3 to 5 styles united at base; ovary woolly. Stem: Shrubby, branching, usually low, rarely 12 ft. high. Leaves: Alternate, petioled, oval to oblong, finely cut-edged, smooth above, matted with woolly hairs underneath. Fruit: Small, round or top-shaped, bright red berries. Preferred Habitat - Swamps, low ground, wet thickets. Flowering Season - March-May. Distribution - Nova Scotia to Gulf of Mexico, westward to the Mississippi.

Another common species often found in the same haunts, the BLACK CHOKE-BERRY (A. nigra), with similar flowers, the berries very dark purple, was formerly confounded with the red choke-berry. But because it sometimes elects to live in dry ground its leaves require no woolly mat on the underside to absorb vapors arising from wet retreats. No wonder that the insipid little berries. related to apples, pears, and other luscious fruits, should share with a cousin, the mountain ash, or rowan, the reproachful name of dog-berry.

JUNEBERRY; SERVICEBERRY; MAY-CHERRY
(Amelanchier Canadensis) Apple family

Flowers - Pure white, over 1 in. across, on long, slender pedicels, in spreading or drooping racemes, with silky, reddish bracts, early falling, among them. Calyx persistent, 5-parted; 5 long, narrow, tapering petals, 3 or 4 times the length of calyx; numerous stamens inserted on calyx throat; 2 to 5 styles, hairy at base. Stem: A large shrub or tree, usually much less than 25 ft. high, rarely twice that height, wood very hard and heavy. Leaves: Alternate, oval, tapering at tip, finely saw-edged, smooth (like the pear tree's), often hairy when young. Fruit. Round, crimson, sweet, edible, seedy berries, ripe in June and July. Preferred Habitat - Woodland borders, pasture thickets, dry soil. Flowering Season - March-May. Distribution - Newfoundland to the Gulf of Mexico, westward over a thousand miles.

Silvery-white chandeliers, hanging from the edges of the woods, light Flora's path in earliest spring, before the trees and shrubbery about them have begun to put forth foliage, much less flowers. Little plants that hug the earth for protection while rude winds rush through the forest and across the hillsides, are already starring her way with fragile, dainty blossoms; but what other shrub, except the serviceberry's twin sister the shadbush, or perhaps the spicebush, has the temerity to burst into bloom while March gusts howl through the naked forests? Little female bees of the Andrena tribe, already at work collecting pollen and nectar for generations yet unborn, buzz their gratitude about the beautiful feathery clusters that lean away from the crowded thicket with a wild, irregular grace. Nesting birds have abundant cause for gratitude also, for the attractive, sweet berries, that ripen providentially early; but, of course, the bees which transfer pollen from flower to flower, and the birds which drop the seeds far and wide, are not the receivers of wholly disinterested favors.

The SHADBUSH or SWAMP SUGAR-PEAR (A. Botryapium), because it was formerly accounted a mere variety (oblongifolia) of the preceding species, still shares with it its popular names; but swamps, river banks, brook sides, and moist thickets are its habitat. Consequently both its inflorescence and pale green, glossy foliage are covered with a sort of whitish cotton, absorbent when young, to prevent the pores from clogging with vapors arising from its damp retreats. Late in the season, when streams narrow or dry up altogether, and the air becomes drier, as the sun rises higher in the heavens, the foliage is usually quite smooth. It will be noticed that, lovely as the shadbush is, its smaller flowers have shorter pedicels than the serviceberry's; consequently its feathery sprays, which are flung outward to the sunshine in April and May, lack something of the grace for which its sister stands preeminent. Under cultivation both species assume conventional form, and lose the wild irregularities of growth that charm us in Nature's garden. Indians believed, what is an obvious fact, that when this bush whitens the swampy river banks, shad are swimming up the stream from the sea to spawn. Then, too, the nighthawk, returning from its winter visit south, booms forth its curious whirring, vibrating, jarring sound as it drops through the air at unseen heights, a dismal, weird noise which the red man thought proceeded from the shad spirits come to warn the schools of fish of their impending fate.

COMMON HAWTHORN: WHITE THORN; SCARLET-FRUITED THORN; RED HAW;
MAYFLOWERS
 (Cratoegus coccinea) Apple family

 Flowers - White, rarely pinkish, usually less than 1 in. across, numerous, in terminal corymbs. Calyx 5-lobed; 5 spreading petals inserted in its throat numerous stamens; styles 3 to 5. Stem: A shrub or small tree, rarely attaining 30 ft. in height (Kratos = strength, in reference to hardness and toughness of the wood); branches spreading, and beset with stout spines (thorns) nearly 2 in. long. Leaves: Alternate, petioled, 2 to 3 in. long, ovate, very sharply cut or lobed, the teeth glandular-tipped. Fruit: Coral red, round or oval; not edible. Preferred Habitat - Thickets, fence-rows, woodland borders. Flowering Season - May. Distribution - Newfoundland and Manitoba southward to the Gulf of Mexico.

"The fair maid who, the first of May,
  Goes to the fields at break of day
  And washes in dew from the hawthorn tree
  Will ever after handsome be."

Here is a popular recipe omitted from that volume of heart-to-heart talks entitled "How to Be Pretty though Plain"!

The sombre-thoughted Scotchman, looking for trouble, tersely observes:

"Mony haws,
  Mony snaws."

But in delicious, blossoming May, when the joy of living fairly intoxicates one, and every bird's throat is swelling with happy music, who but a Calvinist would croak dismal prophecies? In Ireland, old crones tell marvelous tales about the hawthorns, and the banshees which have a predilection for them. So much for folklore.

As one might suspect from the rather disagreeable odor of these blossoms, they are most attractive to flies and beetles, which, carrying pollen from older flowers, leave some on the stigmas that are already mature in newly-opened ones. A concave nectar-secreting disk, not concealed by the filaments in this case, is eagerly pilfered by numerous little short-lipped insects which render no benefit in return; but many others assist in self-pollination after the anthers ripen. The splendid monarch butterfly (Anosia plexippus), the banded purple (Basilarchia arthemis), whose caterpillar feeds on hawthorn foliage, and the light brown hunter's butterfly [American painted lady] (Pyrameis huntera [Vanessa virginiensis]) are, among the visitors seen flitting about this exquisite little tree in early May, when it is fairly white with bloom.

The RED-FRUITED THORN (C. mollis), more hairy on its twigs, petioles, calices, and fruit than the preceding, but so like it in most respects it was formerly accounted a mere variety, is an earlier and even more prolific bloomer, the generous, large clusters of malodorous flowers coming with the leaves in April, and lasting until the common hawthorn starts into lively competition with it for insect trade.

Numerous long, slender thorns, often measuring a finger-length, distinguish the COCKSPUR or NEWCASTLE THORN (C. Crus-Galli), whose abundant small flowers and shining, leathery leaves, dull underneath, are conspicuous in thickets from Quebec to the Gulf. Immense numbers of little bees, among many other visitors, may be noted on a fine day in May and early June about this showy shrub or tree. Because it blooms later than its rival sisters, it has the insect wooers then abroad all to itself.

While most of our beautiful native hawthorns have been introduced to European gardens, it is the WHITE THORN or MAY (C. Oxyacantha) of Europe and Asia which is most commonly cultivated here. Truly a shrub, like a prophet, is not without honor save in its own country.

WHITE SWEET CLOVER; BOKHARA or TREE CLOVER; WHITE MELILOT; HONEY
LOTUS
  (Melilotus alba) Pea family

Flowers - Small, white, fragrant, papilionaceous, the standard petal a trifle longer than the wings; borne in slender racemes.
Stem: 3 to 10 ft. tall, branching. Leaves: Rather distant, petioled, compounded of 3 oblong, saw-edged leaflets; fragrant, especially when dry.
Preferred Habitat - Wastelands, roadsides.
Flowering Season - June-November.
Distribution - United States, Europe, Asia.

Happy must the honeybees have been to find that the sweet clover, one of their dearest delights in the Old World, had preceded them in immigrating to the New. Immense numbers of insects - bees in great variety, wasps, flies, moths, and beetles - visit the little blossoms that provide entertainment so generous and accessible; but honey-bees are ever especially abundant. Slight weight depresses the keel, releasing the stigma and anthers therefore, so soon as a bee alights and opens the flower, he is hit below the belt by the projecting stigma. Pollen carried by him there from other clovers comes off on its sticky surface before his abdomen gets freshly dusted from the anthers, which are necessarily rubbed against while he sips nectar. On the removal of his pressure, the floret springs back to its closed condition, to protect the precious nectar and pollen from rain and pilferers. As the stigma projects too far beyond the anthers to be likely to receive any of the flower's own pollen, good reason is there for the blossoms guarding their attractions for the benefit of their friends, which transfer the vitalizing dust from one floret to another. By clustering its small flowers in spikes, to make them conspicuous, as well as to facilitate dining for its benefactors; by prolonging its season of bloom, to get relief from the fiercest competition for insect trade, and so to insure an abundance of vigorous cross-fertilized seed, this plant reveals at a glance some of the reasons why it has been able to establish itself so quickly throughout our vast area.

Both the white and the yellow sweet clover put their leaves to sleep at night in a remarkable manner: the three leaflets of each leaf twist through an angle of 90 degrees, until one edge of each vertical blade is uppermost. The two side leaflets, Darwin found, always tend to face the north with their upper surface, one facing north-northwest and the other north-northeast, while the terminal leaflet escapes the chilling of its sensitive upper surface through radiation by twisting to a vertical also, but bending to either east or west, until it comes in contact with the vertical upper surface of either of the side leaflets. Thus the upper surface of the terminal and of at least one of the side leaflets is sure to be well protected through the night; one is "left out in the cold."

The dried branches of sweet clover will fill a room with delightful fragrance; but they will not drive away flies, nor protect woolens from the ravages of moths, as old women once taught us to believe.

The ubiquitous WHITE or DUTCH CLOVER (Trifolium repens), whose creeping branches send up solitary round heads of white or pinkish flowers on erect, leafless stems, from May to December, in fields, open waste land, and cultivated places throughout our area, Europe, and Asia, devotes itself to wooing bees, since these are the only insects that effect cross-fertilization regularly, other visitors aiding it only occasionally. When nets are stretched over these flowers to exclude insects, only one-tenth the normal quantity of fertile seed is set. Therefore, for the bee's benefit, does each little floret conceal nectar in a tube so deep that small pilferers cannot reach it; but when a honeybee, for example, depresses the keel of the papilionaceous blossom, abundant reward awaits him in consideration of his services in transferring pollen. After the floret which he has been the means of fertilizing closes over its seed-vessel on his departure, it gradually withers, grows brown, and hangs downward, partly to indicate to the next bee that comes along which fords in the head still contain nectar, and which are done for; partly to hide the precious little vigorous green seed-pod in the center of each withered, papery corolla from the visitation of certain insects whose minute grubs destroy countless millions of the progeny of less careful plants. Thus the erect florets in a head stand awaiting their benefactors; those drooping around the outer edge are engaged in the most serious business of life. Sometimes a solitary old maid remains standing, looking anxiously for a lover, at the end of the season. Usually all the florets are then bent down around the stem in a brown and crumpled mass. But however successfully the clover guards its seeds from annihilation, its foliage is the favorite food of very many species of caterpillars and of all grazing cattle the world around. This is still another plant frequently miscalled shamrock. Good luck or bad attends the finding of the leaves, when compounded of an even or an odd number of leaflets more than the normal count, according to the saying of many simple-minded folk.

The little RABBIT'S-FOOT, PUSSY, OLD-FIELD, or STONE CLOVER (T. arvense) has silky plumed calices to hold its minute whitish florets, giving the dense, oblong heads a charming softness and dove color after it has gone to seed. Like most other clovers, it has come to us from the Old World.

FLOWERING SPURGE
(Euphorbia corollata) Spurge family

Flowers - (Apparently) white, small, borne in forked, long-stalked umbels, subtended by green bracts; but the true flowers are minute, and situated within the white cup-shaped involucre, usually mistaken for a corolla. Staminate flowers scattered over inner surface of involucre, each composed of a single stamen on a thread-like pedicel with a rudimentary calyx or tiny bract below it. A solitary pistillate flower at bottom of involucre, consisting of 3-celled ovary; 3 styles, 2-cleft, at length forming an erect 3-lobed capsule separating into 3 2-valved carpels. Stem: 1 to 3 ft. high, often brightly spotted, simple below, umbellately 5-branched above (usually). Leaves: Linear, lance-shaped or oblong, entire; lower ones alternate, upper ones whorled. Preferred Habitat - Dry soil, gravelly or sandy. Flowering Season - April-October. Distribution - From Kansas and Ontario to the Atlantic.

A very commonplace and uninteresting looking weed is this spurge, which no one but a botanist would suspect of kinship with the brilliant vermilion poinsettia, so commonly grown in American greenhouses. Examination shows that these little bright white cups of the flowering spurge, simulating a five-cleft corolla, are no more the true flowers in the one case than the large red bracts around the poinsettia's globular greenish blossom involucres are in the other. From the milky juice alone one might guess the spurge to be related to the rubber plant. Still another familiar cousin is the stately castor-oil plant; and while the common dull purplish IPECAC SPURGE (E. Ipecacuanhae) also suggests unpleasant doses, it is really a member of quite another family that furnishes the old-fashioned emetic. The flowering spurge, having its staminate and pistillate flowers distinct, depends upon flies, its truest benefactors, to transfer pollen from the former to the latter.

STAGHORN SUMAC; VINEGAR TREE
(Rhus hirta; R. typhina of Gray) Sumac family

Flowers - Greenish or yellowish white, very small, usually 5-parted, and borne in dense upright, terminal, pyramidal clusters. Stem: A shrub or small tree, 6 to 40 ft. high, the ends of branches forked somewhat like a stag's horns. Leaves. Compounded of 11 to 31 lance-shaped, saw-edged leaflets, dark green above, pale below; the petioles and twigs often velvety-hairy. Fruit: Small globules, very thickly covered with crimson hairs. Preferred Habitat - Dry, rough or rocky places, banks, roadsides. Flowering Season - June. Distribution - Nova Scotia to Georgia, and westward 1500 miles.

Painted with glorious scarlet, crimson, and gold, the autumnal foliage of the sumacs, and even the fruit, so far eclipse their inconspicuous flowers in attractiveness that one quite ignores them. Not so the small, short-tongued bees (chiefly Andrenidae) and flies (Dipteria) seeking the freely exposed nectar secreted in five orange-colored glands in the shallow little cups. As some of the flowers are staminate and some pistillate, although others show a tendency to revert to the perfect condition of their ancestors, it behooves them to entertain their little pollen-carrying visitors generously, otherwise no seed can possibly be set. And how the autumnal landscape would suffer from the loss of the decorative, dark-red, velvety panicles! Beware only of the poison sumac's deadly, round grayish-white berries.

Most sumacs contain more or less tannin in their bark and leaves, that are therefore eagerly sought by agents for the leather merchants. The beautiful SMOKE or MIST TREE (R. cotinus), commonly imported from southern Europe to adorn our lawns (although a similar species grows wild in the Southwest), serves a more utilitarian purpose in supplying commerce with a rich orange-yellow dyewood known as young fustic. All this tribe of shrubs and trees contain resinous, milky juice, drying dark like varnish, which in a Japanese species is transformed by the clever native artisans into their famous lacquer. With a commercial instinct worthy of the Hebrew, they guard this process as a national secret.

The SMOOTH, UPLAND, or SCARLET SUMAC (R. glabra), similar to the staghorn, but lacking its velvety down, and usually of much lower growth, is the very common and widely distributed shrub of dry roadsides, railroad banks, and barren fields. Another low-growing, but more or less downy upland sumac, the DWARF, BLACK, or MOUNTAIN SUMAC (R. copallina), may be known by its dark, glossy green foliage, pale on the underside, and by the broadening of the stem into wings between the leaflets. Hungry migrating birds alight to feast on the harmless acid red fruit when the gorgeous autumnal foliage illuminates their route southward. But while they are, of course, the natural agents for distributing the plants over the country, men find that by cutting bits of any sumac root and planting them in good garden soil, strong specimens are secured within a year. An exquisite cut-leaved variety of the smooth sumac adorns many fine lawns.

Everyone should know the POISON SUMAC (R. Vernix - R. venenata of Gray) as the shrub above all others to avoid. Like its cousin, the POISON or THREE-LEAVED IVY (R. radicans), which once had the specific name Toxicodendron, although Linnaeus applied that title to a hairy shrub of the Southern States, the poison sumac causes most painful swelling and irritation to the skin of some people, though they do nothing more than pass it by when the wind is blowing over it. Others may handle both these plants with impunity. In spring they are especially noisome; but when the pores of the skin are opened by perspiration, people who are at all sensitive should give them a wide berth at any season. Usually the poison sumac grows in wet or swampy ground; its bark is gray, its leaf-stalks are red; the leaves are compounded, of fewer leaflets than those of the innocent sumacs - that is, of from seven to thirteen - which are green on both sides; the flowers, which are dull whitish-green, grow in loose panicles from the axils of the leaves, and naturally the berries follow them in the same unusual situation. "By their fruits ye shall know them:" all the harmless sumacs have red fruit clusters at the ends of the branches, whereas both the poison sumac's and the poison ivy's axillary clusters are dull grayish-white.

AMERICAN HOLLY
(Ilex opaca) Holly family

Flowers - Very small, greenish or yellowish white, from 3 to 10 staminate ones in a short cyme; fertile flowers usually solitary, scattered. Stem: A small tree of very slow growth, rarely attaining any great height. Leaves: Evergreen, thick, rigid, glossy, elliptical, scalloped edged, spiny-tipped. Fruit: Round, red berries. Preferred Habitat - Moist woods and thickets. Flowering Season - April-June. Distribution - Maine to the Gulf of Mexico, west to Texas, chiefly near the coast and south of New York.

Happily we continue to borrow all the beautiful Old World associations, poetical and legendary, that cluster about the holly at Christmas time, although our native tree furnishes most of our holiday decorations. So far back as Pliny's day, the European holly had all manner of supernatural qualities attributed to it: its insignificant little flowers caused water to freeze, he tells us; because it was believed to repel lightning, the Romans planted it near their houses; and a branch of it thrown after any refractory animal, even if it did not hit him, would subdue him instantly, and cause him to lie down meekly beside the stick! Can it be that the Italian peasants, who still believe cattle kneel in their stalls at midnight on the anniversary of Jesus' birth, decorate the mangers on Christmas eve with holly, among other plants, because of a survival of this old pagan notion about its subduing effect on animals?

Would that the beautiful holly of English gardens (I. Aquifolium), more glossy and spiny of leaf and redder of berry than our own, might live here; but it is too tender to withstand New England winters, and the hot, dry summers farther south soon prove fatal. Ilex was the ancient name, not of these plants, but of the holly oak.

The MOUNTAIN HOLLY (Ilicioides mucronata - Nemopanthes Canadensis of Gray) a shrub of the northern swamps, about six feet high, and by no means confined to mountainous regions, since it is also abundant in the middle West, has smooth-edged, elliptic, petioled leaves, ash-colored bark, small, solitary, narrow-petalled staminate and pistillate flowers on long, threadlike pedicels from the leaf-axils in May. In August dull pale-red berries appear. Darwin proved that seed set with the help of pollen brought from distinct plants produces offspring that vanquishes the offspring of seed set with pollen brought from another flower on the same plant in the struggle for existence. Thus we see, in very many ambitious plants besides those of the holly tribe, a tendency to separate the male and the female flowers as widely as possible.

BLACK ALDER; WINTERBERRY FEVER-BUSH
(Ilex verticillata) Holly family

Flowers - Small, greenish white, the staminate clusters 2 to 10 flowered the fertile ones 1 to 3 flowered. Stem: A shrub 6 to 25 ft. high. Leaves: Oval, tapering to a point, about 1 in. wide, saw-edged, dark green, smooth above, hairy, especially along veins underneath. Fruit: Bright red berries, about the size of a pea, apparently whorled around the twigs. Preferred Habitat - Swamps, ditches, fencerows, and low thickets. Flowering Season - June-July. Distribution - Nova Scotia to Florida, west to Missouri.

Beautiful bright red berries, dotted or clustered along the naked twigs of the black alder, add an indispensable cheeriness to the somber winter landscape. Bunches of them, commonly sold in the city streets for household decoration, bring twenty-five cents each; hence the shrubs within a large radius of each market get ample pruning every autumn. The leaves turn black before dropping off.

The SMOOTH WINTERBERRY (I. laevigata), a similar species, but of more restricted range, ripens its larger, orange-red berries earlier than the preceding, and before its leaves, which turn yellow, not black, in autumn, have fallen. Another distinguishing feature is that its small, greenish-white staminate flowers grow on long, very slender pedicels; whereas the solitary fertile flowers are much nearer the stern.

BITTERSWEET; WAX-WORK; STAFF-TREE
(Celastrus scandens) Staff-tree family

Flowers - Small, greenish-white, 5-parted, some staminate, some pistillate only; in terminal compound racemes 4 in. long or less. Stem: Woody, twining. Leaves: Alternate, oval, tapering, finely toothed, thin, with a tendency to show white variations. Fruit: A yellow-orange berry-like capsule, splitting at maturity and curling back to display the scarlet, pulpy coating of the seeds within. Preferred Habitat - Rich soil of thickets, fence rows, and wayslde tangles. Flowering Season - June. Distribution - North Carolina, New Mexico, and far north.

Not to be hung above mirror and picture frames in farmhouse parlors, as we have been wont to think, do the brilliant clusters of orange-red wax-work berries attract the eye, where they brighten old walls, copses, and fence rows in autumn; but to advertise their charming wares to hungry migrating birds, which will drop the seeds concealed within the red berry perhaps a thousand miles away, and so plant new colonies. On the smaller, less specialized bees and flies the vine depends in June to carry pollen from its staminate flowers to the fertile ones, whose thick, erect pistil would wither without fruiting without their help.

But the best laid plans of other creatures than mice and men "gang aft a-gley." What mean the little cottony tufts all along the stems of so very many bittersweet vines, but that these have foes as well as friends? Curious little parasitic tree-hoppers (Membracis binotata), which spend their entire lives on the stems, sucking the juices through their little beaks, just as the aphids moor themselves to the tender rose-twigs, might be mistaken for thorns during one of their protective masquerades. Again they look like diminutive flocks of fowl, their heads ever pointing in one direction, no matter how the vine may twist and turn - always toward the top of the branch, that they may the better siphon the sap down their tiny throats. Toward the end of summer the females, which have a sharp instrument at the rear of their bodies, cut deeply into the juicy food-store, the cambium layer of bark, and there deposit their eggs. Presently, a nest being filled, the mother emits a substantial froth at the end of her ovipositor, and proceeds to construct the cottony, corrugated dome over her nursery which first attracted our attention. This is especially skilful work, for she works behind her, evidently not from sight, but from instinct only. Inasmuch as the young hoppers will not come forth until the following summer, some such snug protection is required during winter's cold and snows. With hordes of little parasites constantly preying on its juices, is it any wonder the vine is often too enfeebled to produce seed, or that

the leaves lose part of their color and become, as we say, variegated? Occasionally one finds the cottony nursery domes of this little
hopper on the locust tree - the favorite home of its big, noisy relative, the so-called locust, or cicada.

NEW JERSEY TEA; WILD SNOWBALL; RED-ROOT
(Ceanothus Americanus) Buckthorn family

Flowers - Small, white, on white pedicels, crowded in dense, oblong, terminal clusters. Calyx white, hemispheric, 5-lobed; petals,
hooded and long-clawed; 5 stamens with long filaments; style short,
3-cleft. Stems: Shrubby, 1 to 3 ft. high, usually several, from a deep
reddish root. Leaves: Alternate, ovate-oblong, acute at tip, finely
saw-edged, 3-nerved, on short petioles. Preferred Habitat - Dry,
open woods and thickets. Flowering Season - May-July. Distribution
- Ontario south and west to the Gulf of Mexico.

Light, feathery clusters of white little flowers crowded on the
twigs of this low shrub interested thrifty colonial housewives of
Revolutionary days not at all; the tender, young, rusty, downy
leaves were what they sought to dry as a substitute for imported
tea. Doubtless the thought that they were thereby evading George
the Third's tax and brewing patriotism in every kettleful added a
sweetness to the homemade beverage that sugar itself could not
impart. The American troops were glad enough to use New Jersey
tea throughout the war. A nankeen or cinnamon-colored dye is
made from the reddish root.

NORTHERN, WILD, FOX, or PLUM GRAPE
(Vitis Labrusca) Grape family

Flowers - Greenish, small, deliciously fragrant, some staminate,
some pistillate, rarely perfect; the fertile flowers in more compact
panicles than the sterile ones. Stem: Climbing with the help of tendrils; woody, bark loose. Leaves: Large, rounded or lobed, toothed,
rusty-hairy underneath, especially when young, each leathery leaf

opposite a tendril or a flower cluster. Fruit: Clusters containing a few brownish, purple, musky-scented grapes, 3/4 in. across. Ripe, August-September. Preferred Habitat - Sunny thickets, loamy or gravelly soil. Flowering Season - June. Distribution - New England to Georgia, west to Minnesota and Tennessee.

Aesop's fox may never have touched the grapes of fable, but this, our wild species, certainly retains a strong foxy odor, which at least suggests that he came very near them. Tough pulp and thick skin by no means deter birds and beasts from feasting on this fruit, and so dispersing the seeds; but mankind prefers the tender, delightful flavored Isabella, Catawba, and Concord grapes derived from it. The Massachusetts man who produced the Concord variety in the town whose name he gave it, declares he would be a millionaire had he received only a penny royalty on every Concord grapevine planted.

What fragrance is more delicious than that of the blossoming grape? To swing in a loop made by some strong old vine, when the air almost intoxicates one with its sweetness on a June evening, is many a country child's idea of perfect bliss. Not until about nine o'clock do the leaves "go to sleep" by becoming depressed in the center like saucers. This was the signal for bedtime that one child, at least, used to wait for. We have seen in the clematis how its sensitive leafstalks hook themselves over any support they rub against; but the grapevine has gone a step farther, and by discarding an occasional flower cluster and prolonging the flower stalk into a coiling, forking tendril it moors itself to the thicket. We know that all tendrils are either transformed leaves, as in the case of the pea vine, where each branch of its tendril represents a modified leaflet; or they are transformed flower stalks or other organs. Occasionally the tendril of a grapevine reveals its ancestry by bearing a blossom or a cluster of flowers, and sometimes even fruit, about midway on the coil, which attempts to fill all offices at once like Pooh Bah.

The phylloxera having destroyed many of the finest vineyards in Europe, it would seem that Americans have the best of chances to supply the world with high-class wines, for there is not a State in the Union where the vine will not flourish. Here its worst enemy is mildew, a parasitical fungus which attacks the leaves, revealing itself in yellowish-brown patches on the upper side, and thin, frosty patches underneath. Soon the leaves become sere, and then they fall. The microscope reveals a miniature forest of growth in each leaf, with the threadlike roots of the fungi searching about the leaf cells for food. To burn old leaves, and to blow sulphur over the vine while it is wet, are efficacious remedies. Bees and wasps which puncture grapes to feast on them, are the innocent means of destroying quantities.

Both the RIVERSIDE or SWEET-SCENTED GRAPE (V. vulpina; formerly V. cordifolia, var. riparia) - whose bluish-black, bloom-covered fruit begins to ripen in July; and the FROST, CHICKEN, POSSUM, or WINTER GRAPE (V. cordifolia), whose smaller, shining black berries are not at their best till after frost, grow along streams and preferably in rocky situations. The shining, light green, thin leaves of the sweet-scented species are sharply lobed, the three to seven lobes have acute teeth, and the tendrils are intermittent. The frost grape's leaves, which are commonly three or four inches wide, are deeply heart-shaped, entire (rarely slightly three-lobed), tapering to a long point and acutely toothed.

Another familiar member of the Grape family, the VIRGINIA CREEPER, FALSE GRAPE, AMERICAN or FIVE-LEAVED IVY, also erroneously called WOODBINE (Parthenocissus quinquefolia; formerly Ampelopsis quinquefolia) - is far more charming in its glorious autumnal foliage, when its small dark blue berries hang from red peduncles, than when its insignificant greenish flower clusters appear in July. The leaves, compounded of five leaflets, should sufficiently distinguish the harmless vine from the three-leaved poison ivy, sometimes confounded with it. From Manitoba and Mexico to the Atlantic Ocean, and even in Cuba, the Virginia creeper rambles over thickets, fences, and walls, ascends trees, festoons rocky woodlands, drapes our verandas, making its way with the help of modified flower stalks that are now branching tendrils, each branch bearing an adhesive disk at the end. "In the course of about two days after a tendril has arranged its branches so as to press upon any surface," says Darwin, "its curved tips swell, become bright red, and form on their undersides little disks or cushions with which they adhere firmly." It is supposed that these disks secrete a cement. At any rate, we know that they have a very tenacious hold, because often one contracting tendril, as elastic as a steel spring, supports, by means of these little disks, the entire weight of the branch it lifts up. Darwin concluded that a tendril with five disk-bearing branches, on which he experimented, would stand a strain of ten pounds, even after ten years' exposure to high winds and softening rains.

WHITE VIOLETS
(Viola) Violet family

Three small-flowered, white, purple-veined, and almost beardless species which prefer to dwell in moist meadows, damp, mossy places, and along the borders of streams, are the LANCE-LEAVED VIOLET (V. lanceolata), the PRIMROSE-LEAVED VIOLET (V. prirnulaefolia), and the SWEET WHITE VIOLET (V. blanda), whose leaves show successive gradations from the narrow, tapering, smooth, long-petioled blades of the first to the oval form of the second and the almost circular, cordate leaf of the delicately fragrant, little white blanda, the dearest violet of all. Inasmuch as these are short-spurred species, requiring no effort for bees to drain their nectaries, no footholds in the form of beards on the side petals are provided for them. The purple veinings show the stupidest visitor the path to the sweets.

The sprightly CANADA VIOLET (V. Canadensis), widely distributed in woodlands, chiefly in hilly and mountainous regions, rears tall, leafy stems terminated by faintly fragrant white or pale lavender blossoms, purple-tinged without and purple veined, the side petals bearded, the long sepals tapering to sharp points. Here we see a violet in the process of changing from the white ancestral type to the purple color which Sir John Lubbock, among other scientists, considers the highest step in chromatic evolution. This species has heart-shaped, saw-edged leaves which taper acutely. From May even to July is its regular blooming season; but the delightful family eccentricity of flowering again in autumn appears to be a confirmed habit with the Canada violet.

ENCHANTER'S NIGHTSHADE
(Circaea Lutetiana) Evening Primrose family

Flowers - Very small, white, slender pedicelled, in terminal and lateral racemes. Calyx 2-parted, hairy 2 petals, 2 alternate stamens. Stem: 1 to 2 ft. high, slender, branching, swollen at nodes. Leaves: Opposite, tapering to a point, distantly toothed, 2 to 4 in. long, slender petioled. Fruit: Pear-shaped, 2-celled, densely covered with stiff, hooked hairs. Preferred Habitat - Woods; shady roadsides. Flowering Season - June-August. Distribution - Nova Scotia to Georgia, westward to Nebraska. Europe and Asia.

Why Circe, the enchantress, skilled in the use of poisonous herbs, should have had her name applied to this innocent and insignificant looking little plant is not now obvious; neither is the title of nightshade any more appropriate.

Each tiny flower having a hairy calyx, that acts as a stockade against ants and other such crawling pilferers, we suspect there are abundant sweets secreted in the fleshy ring at the base of the styles for the benefit of the numerous flies seen hovering about. Among other visitors, watch the common housefly alighting on the knobby stigma, a most convenient landing place, where he leaves some pollen carried on his underside from other nightshade blossoms. In clasping the bases of the two pliable stamens, his only available supports as he sucks, he will surely get well dusted again, that he may fertilize the next blossom he flies to for refreshment. The nightshade's little pear-shaped seed vessels, armed with hooked bristles by which they steal a ride on any passing petticoat or trouser leg, reveal at a glance how this plant has contrived to travel around the globe.

A smaller, weaker species (Circaea alpina), found in cool, moist woods, chiefly north, has thin, shining leaves and soft, hooked hairs on its vagabond seeds. Less dependence seems to be placed on these ineffective hooks to help perpetuate the plant than on the tiny pink bulblets growing at the end of an exceedingly slender thread sent out by the parent roots.

AMERICAN SPIKENARD; INDIAN ROOT; SPIGNET
(Aralia racemosa) Ginseng family

Flowers - Greenish white, small, 5-parted, mostly imperfect, in a drooping compound raceme of rounded clusters. Stem: 3 to 6 ft. high, branches spreading. Roots: Large, thick, fragrant. Leaves: Compounded of heart-shaped, sharply tapering, saw-edged leaflets from 2 to 5 in. long, often downy underneath. Lower leaves often enormous. Fruit: Dark reddish-brown berries. Preferred Habitat - Rich open woods, wayside thickets, light soil. Flowering Season - July-August. Distribution - New Brunswick to Georgia, west to the Mississippi.

A striking, decorative plant, once much sought after for its medicinal virtues - still another herb with which old women delight to dose their victims for any malady from a cold to a carbuncle. Quite a different plant, but a relative, is the one with hairy, spike-like shoots from its fragrant roots, from which the "very precious" ointment poured by Mary upon the Saviour's head was made. The nard, an Indian product from that plant, which is still found growing on the distant Himalayas, could then be imported into Palestine only by the rich.

The wild spikenard, or false Solomon's seal, has not the remotest connection with this tribe of plants. Inasmuch as some of the American spikenard's tiny flowers are staminate and some pistillate, while others again are perfect, they depend upon flies chiefly - but on some wasps and beetles, too - to transfer pollen and enable the fertile ones to set seed. How certain of the winter birds gormandize on the resinous, spicy little berries! A flock of juncos will strip the fruit from every spikenard in the neighborhood the first day it arrives from the North.

The WILD or FALSE SARSAPARILLA (A. nudicaulis), so common in woods, hillsides, and thickets, shelters its three spreading umbels of greenish-white flowers in May and June beneath a canopy formed by a large, solitary, compound leaf. The aromatic roots, which run horizontally sometimes three feet or more through the soil, send up a very short, smooth proper stem which lifts a tall leafstalk and a shorter, naked flower stalk. The single large leaf, of exquisite bronzy tints when young, is compounded of from three to five oval, toothed leaflets on each of its three divisions. The tiny five-parted flowers have their petals curved backward over the calyx to make their refreshments more accessible for the flies, on which they chiefly rely for aid in producing those close clusters of dark-purple berries on which migrating birds feast in early autumn. By these agents the plant has been distributed from Newfoundland to the Carolinas, westward from Manitoba to Missouri, which is not surprising when we remember that certain birds travel from the Gulf of Mexico to the Great Lakes in a single night. While the true sarsaparilla of medicine should come from a quite different herb that flourishes in Mexico and South America, this one furnishes a commercial substitute enormously used as a blood purifier and cooling summer drink. Burrowing rabbits delight to nibble the long, slender, fragrant roots.

The GINSENG (Panax quinquefolium; Aralia quinquefolia of Gray) found in rich woods from Quebec to Alabama, and westward to Nebraska - that is, where found at all, for much hunting has all but exterminated it in many regions - bears a solitary umbel of small yellowish-green, five-parted, polygamous flowers in July and August at the end of a smooth stem about a foot high. Bright crimson berries follow the clusters on the female plants in early autumn. Three long-petioled leaves, which grow in a whorl at the top of the low stem, are palmately divided into five thin, ovate, pointed, and irregularly toothed leaflets. But it is the deep fusiform root, simple or branched, about which the Americanized Chinese, at least, are most concerned. For centuries Chinese physicians have ascribed miraculous virtues to the Manchurian ginseng. Not only can it remove fatigue and restore lost powers, but by its use veterans became frisky youths again according to these wise men of the East. In short, they consider it the panacea for all ills (Panax: pan = all, akos = remedy) - the source of immortality. Naturally the roots were and are in great demand, especially such as branch so as to resemble the human form. (Both the Chinese name Schin-sen, and Garan-toguen, the Indian one, are said to mean like a man. Here is an interesting clue for the ethnologists to follow !) Imperial edict prohibited the Chinese from digging up their native plant lest it be exterminated.

So Jesuit missionaries, who discovered our similar ginseng, were not slow in exporting it to China when it was literally worth its weight in gold. Indeed, it is always sold by weight - a fact on which the heathen Chinee "with ways that are dark and tricks that are vain" not infrequently relies. Chinamen, who gather large quantities in our Western States to sell to the wholesale druggists for export, sometimes drill holes into the largest roots, pour in melted lead, and plug up the drills so ingeniously that druggists refuse to pay for a Chinaman's diggings until they have handled and weighed each root separately.

The DWARF GINSENG, OR GROUND NUT (P. trifolium; Aralia trifolia of Gray) whose little white flowers are clustered in feathery, fluffy balls above the whorl of three compound leaves in April and May, chooses low thickets and moist woods for its habitat - often in the same neighborhood with its larger relative. Yellowish berries follow the fragrant white pompons. One must burrow deep, like the rabbits, to find its round, pungent, sweet, nut-like root, measuring about half an inch across, which few have ever seen.

WILD CARROT; QUEEN ANNE'S LACE; BIRD'S-NEST
  (Daucus Carota) Carrot family

Flowers - Small, of unequal sizes (polygamous), white, rarely pinkish gray, 5-parted, in a compound, flat, circular umbel, the central floret often dark crimson; the umbels very concave in fruit. An involucre of narrow, pinnately cut bracts. Stem: 1 to 3 ft. high, with stiff hairs; from a deep, fleshy, conic root. Leaves: Cut into fine, fringy divisions; upper ones smaller and less dissected. Preferred Habitat - Wastelands, fields, roadsides. Flowering Season - June-September. Distribution - Eastern half of United States and Canada. Europe and Asia.

A pest to farmers, a joy to the flower lover, and a welcome signal for refreshment to hosts of flies, beetles, bees, and wasps, especially to the paper-nest builders, the sprangly wild carrot lifts its fringy foliage and exquisite lacy, blossoms above the dry soil of three continents. From Europe it has come to spread its delicate wheels over our summer landscape, until whole fields are whitened by them east of the Mississippi. Having proved fittest in the struggle for survival in the fiercer competition of plants in the over-cultivated Old World, it takes its course of empire westward year by year, Finding most favorable conditions for colonizing in our vast, uncultivated area; and the less aggressive, native occupants of our soil are only too readily crowded out. Would that the advocates of unrestricted immigration of foreign peasants studied the parallel examples among floral invaders!

What is the secret of the wild carrots' triumphal march? As usual, it is to be sought chiefly in the flower's scheme to attract and utilize visitors. Nectar being secreted in open disks near to one another, the shortest-tongued insects can lick it up from the Umbelliferae with even less loss of time than from the tubular florets of the Cornpositae. Over sixty distinct species of insects may be taken on the wild carrot by any amateur, since it blooms while insect life is at its height but, as might be expected, the long-tongued and color-loving, specialized bees and butterflies do not often waste time on florets so easily drained by the mob. Ants find the stiff hairs on the stem disagreeable obstacles to pilfering; but no visitors seem to object to the flowers' suffocating odor.

One of these lacy, white umbels must be examined under a lens before its delicate structure and perfection of detail can be appreciated. Naturally a visitor is attracted first by the largest, most showy florets situated around the outer edge of the wheel, on which he leaves pollen, brought from another umbel; and any vitalizing dust remaining on his under side may be left on the less conspicuous hermaphrodite blossoms as he makes his way toward the center, where the tiny, pollen-bearing florets are grouped. From the latter, as he flies away, he will carry fresh pollen to the outer row of florets on another umbel, and so on - at least this is the usual and highly advantageous method. After general fertilization, the slender flower-stalks curl inward, and the umbel forms a hollow nest that gradually contracts as it dries, almost, if not quite, closing at the top, albeit the fiction that bees and spiders make their home in the seeding umbels circulates freely.

Still another fiction is that the cultivated carrot, introduced to England by the Dutch in Queen Elizabeth's reign, was derived from this wild species. Miller, the celebrated English botanist and gardener, among many others, has disproved this statement by utterly failing again and again to produce an edible vegetable from this wild root. When cultivation of the garden carrot lapses for a few generations, it reverts to the ancestral type -a species quite distinct from Daucus Carota.

SMOOTHER SWEET CICELY
(Washingtonia longistylis; Osmorrhiza longistylis of Gray)
Carrot family

Flowers - Small, white, 5-parted; in few rayed, long-peduncled umbels, with small bracts below them. Stem: 1 1/2 to 3 ft. high, branching, from thick, fleshy, fragrant, edible roots. Leaves: Lower ones often very large, long-petioled, thrice-compound, and again divided, the leaflets ovate, pointed, deeply toothed, slightly downy; upper leaves less compound, nearly sessile. Preferred Habitat - Rich, moist woods and thickets. Flowering Season - May-June. Distribution - Nova Scotia to the Carolinas, westward to Dakota.

Graceful in gesture, with delicate, fernlike leaves and anise-scented roots that children, like rabbits, delight to nibble, the sweet cicely attracts attention by its fragrance, however insignificant its flowers. In wooded places, such as it prefers to dwell in, white blossoms, which are far more noticeable in a dim light than colored ones, and finely cut leaves that can best withstand the drip from trees, abound. These white umbels bear a large proportion of male, or pollen-bearing, florets to the number of hermaphrodite, or two-sexed, florets; but as the latter mature their pollen before their stigmas become susceptible to it, self-fertilization is well guarded against, and cross-fertilization is effected with the help of as many flies as small bees, which come in numbers to lick up the nectar so freely exposed in consideration of their short tongues. We have to thank these little creatures for the long, slender seeds, armed with short bristles along the ribs, that they may snatch rides on our garments, together with the beggar-ticks, burdock, cleavers, and other vagabond colonists in search of unoccupied ground. Be sure you know the difference between sweet cicely and the poisonous water hemlock before tasting the former's spicy root.

Was there no more important genus - containing, if possible, red, white, and blue flowers - to have named in honor of the Father of his Country?

Another member of the Carrot family, the SANICLE or BLACK SNAKEROOT (Sanicula Marylandica), found blooming from May to July in such rich, moist woodlands and shrubbery as the sweet cicely prefers, lifts spreading, two to four rayed umbels of insignificant-looking but interesting little greenish-white florets. At first the tips of the five petals are tucked into the center of each little flower; underneath them the stamens are now imprisoned while any danger of self-fertilizing the stigma remains. The few hermaphrodite florets have their styles protruding from the start, and incoming insects leave pollen brought from staminate florets on the early-maturing stigmas. After cross-fertilization has been effected, it is the pistil's turn to keep out of the way, and give the imprisoned stamens a chance: the styles curve until the stigmas are pressed against the sides of the ovary, that not a grain of pollen may touch them; the petals spread and release the stamens; but so great is the flower's zeal not to be fertilized with its own pollen that it sometimes holds the anthers tightly between the petals until all the vitalizing dust has been shed! Around the hermaphrodite florets are a large number of male florets in each hemispheric cluster. Hooked bristles and slender, curved styles protrude from the little burr-like seeds, that any creature passing by may give them a lift to fresh colonizing land! The firm bluish-green leaves, palmately divided into from five to seven oblong, irregularly saw-edged segments, the upper leaves seated on the stem, the lower ones long-petioled, help us to identify this common weed.

With splendid, vigorous gesture the COW-PARSNIP (Heracleum lanatum) rears itself from four to eight feet above moist, rich soil from ocean to ocean in circumpolar regions as in temperate climes. A perfect Hercules for coarseness and strength does it appear when contrasted with some of the dainty members of the carrot tribe. In June and July, when a myriad of winged creatures are flying, large, compound, many-rayed umbels of both hermaphrodite and male white flowers are spread to attract their benefactors the flies, of which twenty-one species visit them regularly, besides small bees, wasps, and other short-tongued insects, which have no difficulty in licking up the freely exposed nectar. The anthers, maturing first, compel cross-fertilization which accounts for the plant's vigor and its aggressive march across the continent. A very stout, ridged, hairy stem, the petioled leaves compounded of three broadly ovate, lobed and saw-edged divisions, downy on the underside, and the great umbels, which sometimes measure a foot across, all bear out the general impression of a Hercules of the fields.

FOOL'S PARSLEY, or CICELY, or DOG-POISON (Aethusa cynapium), a European immigrant found in waste ground and rubbish heaps from Nova Scotia to New Jersey and westward to the Mississippi, should be known only to be avoided. The dark bluish-green, finely divided, rather glossy leaves when bruised do not give out the familiar fragrance of true parsley; the little narrow bracts, turned downward around each separate flower-cluster, give it a bearded appearance, otherwise the white umbel suggests a small wild carrot head of bloom. Cows have died from eating this innocent-looking little plant among the herbage; but most creatures know by instinct that it must not be touched.

Strange that a family which furnishes the carrot, parsnip, parsley, fennel, caraway, coriander, and celery to mankind, should contain many members with deadly properties. Fortunately the large, coarse WATER HEMLOCK, SPOTTED COWBANE, MUSQUASH ROOT, or BEAVER-POISON (Cicuta maculata) has been branded as a murderer. Purple streaks along its erect branching stem correspond to the marks on Cain's brow. Above swamps and low ground it towers. Twice or thrice pinnate leaves, the lower ones long-stalked and often enormous, the leaflets' conspicuous veins apparently ending in the notches of the coarse, sharp teeth, help to distinguish it from its innocent relations sometimes confounded with it. Its several tuberiform fleshy roots contain an especially deadly poison; nevertheless, some highly intelligent animals, beavers, rabbits, and the omnivorous small boy among others have mistaken it for sweet-cicely with fatal results. Indeed, the potion drunk by Socrates and other philosophers and criminals at Athens, is thought to have been a decoction made from the roots of this very hemlock. Many little white flowers in each cluster make up a large umbel; and many umbels to a plant attract great numbers of flies, small bees, and wasps, which sip the freely exposed nectar apparently with only the happiest consequences, as they transfer pollen from the male to the proterandrous hermaphrodite flowers. Just as the cow-parsnip shows a preponderance of flies among its visitors, so the water hemlock seems to attract far more bees and wasps than any of the umbel-bearing carrot tribe. It blooms from the end of June through August.

Still another poisonous species is the HEMLOCK WATER-PARSNIP (Sium cicutaefolium), found in swampy places throughout Canada and the United States from ocean to ocean. The compound, long-rayed umbels of small white flowers, fringy-bracted below, which measure two or three inches across; the extremely variable pinnate leaves, which may be divided into from three to six pairs of narrow and sharply toothed leaflets (or perhaps the lower long-stalked ones as finely dissected as a wild carrot leaf where they grow in water), and the stout, grooved, branching stem, from two to six feet tall, are its distinguishing characteristics. In these umbels it will be noticed there are far more hermaphrodite, or two-sexed, florets (maturing their anthers first), than there are male ones; consequently quantities of unwelcome seed are set with the help of small bees, wasps, and flies, which receive generous entertainment from July to October.

The MOCK BISHOP'S-WEED (Ptilimnium capillaceum), a slender, delicate, dainty weed found chiefly in saltwater meadows from Massachusetts to Florida and around the Gulf coast to Texas, has very finely dissected, fringy leaves and compound umbels two to four inches across, of tiny white florets, with threadlike bracts below. It blooms throughout the summer.

FLOWERING DOGWOOD
 (Cornus florida) Dogwood family

Flowers - (Apparently) large, white or pinkish, the four conspicuous parts simulating petals, notched at the top, being really bracts of an involucre below the true flowers, clustered, in the center, which are very small, greenish yellow, 4-parted, perfect. Stem: A large shrub or small tree, wood hard, bark rough. Leaves: Opposite, oval, entire-edged, petioled, paler underneath. Fruit: Clusters of egg-shaped scarlet berries, tipped with the persistent calyx. Preferred Habitat - Woodlands rocky thickets, wooded roadsides. Flowering Season - April-June. Distribution - Maine to Florida, west to Ontario and Texas.

Has Nature's garden a more decorative ornament than the flowering dogwood, whose spreading flattened branches whiten the woodland borders in May as if an untimely snowstorm had come down upon them, and in autumn paint the landscape with glorious crimson, scarlet, and gold, dulled by comparison only with the clusters of vivid red berries among the foliage? Little wonder that nurserymen sell enormous numbers of these small trees to be planted on lawns. The horrors of pompous monuments, urns, busts, shafts, angels, lambs, and long-drawn-out eulogies in stone in many a cemetery are mercifully concealed in part by these boughs, laden with blossoms of heavenly purity.

"Let dead names be eternized in dead stone,
But living names by living shafts be known.
Plant thou a tree whose leaves shall sing
Thy deeds and thee each fresh, recurrent spring."

Fit symbol of immortality! Even before the dogwood's leaves fall in autumn, the round buds for next year's bloom appear on the twigs, to remain in consoling evidence all winter with the scarlet fruit. When the buds begin to swell in spring, the four reddish-purple, scale-like bracts expand, revealing a dozen or more tiny green flowers clustered within for the large, white, petal-like parts, with notched, tinted, and puckered lips, into which these reddish bracts speedily develop, and which some of us have mistaken for a corolla, are not petals at all - not the true flowers - merely appendages around the real ones, placed there, like showy advertisements, to attract customers. Nectar, secreted in a disk on each minute ovary, is eagerly sought by little Andrena and other bees, besides flies and butterflies. Insects crawling about these clusters, whose florets are all of one kind, get their heads and undersides dusted with pollen, which they transfer as they suck. Hungry winter birds, which bolt the red fruit only when they can get no choicer fare, distribute the smooth, indigestible stones far and wide.

When the Massachusetts farmers think they hear the first brown thrasher in April advising them to plant their Indian corn, reassuringly calling, "Drop it, drop it - cover it up, cover it up - pull it up, pull it up, pull it up" (Thoreau), they look to the dogwood flowers to confirm the thrasher's advice before taking it.

The LOW or DWARF CORNEL, or BUNCHBERRY (C. canadensus) whose scaly stem does its best to attain a height of nine inches, bears a whorl of from four to six oval, pointed, smooth leaves at the summit. From the midst of this whorl comes a cluster of minute greenish florets, encircled by four to six large, showy, white petal-like bracts, quite like a small edition of the flowering dogwood blossom. Tight clusters of round berries, that are lifted upward on a gradually lengthened peduncle after the flowers fade (May-July), brighten with vivid touches of scarlet shadowy, mossy places in cool, rich woods, where the dwarf cornels, with the partridge vine, twin flower, gold thread, and fern, form the most charming of carpets.

Other common dogwoods there are - shrubs from three to ten feet in height - which bear flat clusters of small white flowers without the showy petal-like bracts, imitating a corolla, as in the two preceding species, but each little four-parted blossom attracting its miscellaneous crowd of benefactors by association with dozens of its counterparts in a showy cyme. Because these flowers expand farther than the minute florets of the dwarf cornel or the flowering dogwood, and the sweets are therefore more accessible, all the insects which fertilize them come to the shrub dogwoods too, and in addition very many beetles, to which their odor seems especially attractive. ("Odore carabico o scarabeo" - Delpino.) The ROUND-LEAVED CORNEL or DOGWOOD [now ROUNDLEAF DOGWOOD] (C. circinata), found on shady hillsides, in open woodlands, and roadside thickets - especially in rocky districts -

from Nova Scotia to Virginia, and westward to Iowa, may be known by its greenish, warty twigs; its broadly ovate, or round petioled, opposite leaves, short-tapering to a point, and downy beneath; and, in May and June, by its small, flat, white flower-clusters about two inches across, that are followed by light-blue (not edible) berries.

Even more abundant is the SILKY CORNEL, KINNIKINNICK, or SWAMP DOGWOOD [now SILKY DOGWOOD] (C. amonum; C. sericca of Gray) found in low, wet ground, and beside streams, from Nebraska to the Atlantic Ocean, south to Florida and north to New Brunswick. Its dull-reddish twigs, oval or oblong leaves, rounded at the base but tapering to a point at the apex, and usually silky-downy with fine, brownish hairs underneath (to prevent the pores from clogging with vapors arising from its damp habitat); its rather compact, flat clusters of white flowers from May to July, and its bluish berries are its distinguishing features. The Indians loved to smoke its bark for its alleged tonic effect.

The RED-OSIER CORNEL or DOGWOOD (C. stolonifera), which has spread, with the help of running shoots, through the soft soil of its moist retreats, over the British Possessions north of us and throughout the United States from ocean to ocean, except at the extreme south, may be known by its bright purplish-red twigs; its opposite, slender, petioled leaves, rather abruptly pointed at the apex, roughish on both sides, but white or nearly so beneath; its small, flat-topped white flower-clusters in June or July; and finally, by its white or lead-colored fruit.

In good, rich, moist soil another white-fruited species, the PA-NICLED CORNEL or DOGWOOD (C. candidissima; C. paniculata of Gray) rears its much-branched, smooth, gray stems. In May or June the shrub is beautiful with numerous convex, loose clusters of white flowers at the ends of the twigs. So far do the stamens diverge from the pistil that self-pollination is not likely; but an especially large number of the less specialized insects, seeking the freely expo-sed nectar, do all the necessary work as they crawl about and fly from shrub to shrub. This species bears comparatively long and narrow leaves, pale underneath. Its range is from Maine to the Carolinas and westward to Nebraska.

WHITE ALDER; SWEET PEPPERBUSH; ALDER-LEAVED CLETHRA
(Clethra alnifolia) White Alder family

Flowers - Very fragrant, white, about 1/3 in. across, borne in long, narrow, upright, clustered spikes, with awl-shaped bracts. Calyx of 5 sepals; 5 longer petals; 10 protruding stamens, the style longest. Stem: A much-branched shrub, 3 to 10 ft. high. Leaves: Alternate, oblong or ovate, finely saw-edged above the middle at least, green on both sides, tapering at base into short petioles. Preferred Habitat - Low, wet woodland and roadside thickets; swamps; beside slow streams; meadows. Flowering Season - July-August. Distribution - Chiefly near the coast, in States bordering the Atlantic Ocean.

Like many another neglected native plant, the beautiful sweet pepperbush improves under cultivation; and when the departed lilacs, syringa, snowball, and blossoming almond, found with almost monotonous frequency in every American garden, leave a blank in the shrubbery at midsummer, these fleecy white spikes should exhale their spicy breath about our homes. But wild flowers, like a prophet, may remain long without honor in their own country. This and a similar but more hairy species found in the Alleghany region, the MOUNTAIN SWEET PEPPERBUSH (C. acuminata), with pointed leaves, pale beneath, and spreading or drooping flower-spikes, go abroad to be appreciated. Planted beside lakes and streams on noblemen's estates, how overpowering must their fragrance be in the heavy, moisture-laden air of England! Even in our drier atmosphere, it hangs about the thickets like incense.

ROUND-LEAVED PYROLA; PEAR-LEAVED, or FALSE WINTER-GREEN; INDIAN or CANKER LETTUCE
(Pyrola rolundifolia) Wintergreen family

Flowers - Very fragrant, white, in a spike; 6 to 20, nodding from an erect, bracted scape 6 to 20 in. high. Calyx 5-parted corolla, over

1/2 in. across, of 5 concave, obtuse petals 10 stamens, protruding pistil, style curved, stigma 5-lobed. Leaves: All spreading from the base by margined petioles; shining leathery green, round or broadly oval, obtuse, 1 1/2 to 3 in. long, persistent through the winter. Preferred Habitat - Open woods. Flowering Season - June-July. Distribution - Nova Scotia to Georgia, west to Ohio and Minnesota.

Deliciously fragrant little flowers, nodding from an erect, slender stalk, when seen at a distance are often mistaken for lilies-of-the-valley growing wild. But closer inspection of the rounded, pearlike leaves in a cluster from the running root, and the concave, not bell-shaped, white, waxen blossoms, with the pistil protruding and curved, indicate the commonest of the pyrolas. Some of its kin dwell in bogs and wet places, but this plant and the shin-leaf carpet drier woodland where dwarf cornels, partridge vines, pipsissewa, and goldthread weave their charming patterns too. Certain of the lovely pyrola clan, whose blossoms range from greenish white, flesh-color, and pink to deep purplish rose, have so many features in common they were once counted mere varieties of this round-leaved wintergreen - an easygoing classification broken up by later-day systematists, who now rank the varieties as distinct species. It will be noticed that all these flowers have their anthers erect in the bud but reversed at flowering time, each of the two sacs opening by a pore which, in reality, is at the base of the sac, though by reversion it appears to be at the top. To these pores small bees and flies fasten their short lips to feed on pollen, some of which will be necessarily .jarred out on them as they struggle for a foothold on the stamens, and will be carried by them to another flower's protruding stigma, which impedes their entrance purposely to receive the imported pollen.

By reason of the old custom of clapping on a so-called "shinplaster" to every bruise, regardless of its location on the human body, a lovely little plant, whose leaves were once counted a first aid to the injured, still suffers instead under an unlovely name. The SHINLEAF (P. elliptica) sends up a naked flower-stalk, scaly at the base, often with a bract midway, and bearing at the top from seven to fifteen very fragrant, nodding, waxen, greenish-white blossoms, similar to the round-leaved wintergreen's. But on the thinner, dull, dark-green, upright leaves, with slight wavy indentations, scarcely to be called teeth, on the margins, their shorter leaf-stalks often reddish, one chiefly depends to name this common plant. It is usually found, in company with a few or many of its fellows, in rich woodlands so far west as the Rocky Mountains, blooming from June to August, according to the climate of its wide range.

When the little SERRATED or ONE-SIDED WINTERGREEN (P. secunda) first sends up its slender raceme in June or July, it is erect but presently the small, greenish-white flowers, opening irregularly along one side, appear to weigh it downward into a curve. Usually several bracted scapes rise from a running, branched rootstock, to a height of from three to (rarely) ten inches above a cluster of basal evergreen leaves. These latter are rather thin, oval, slightly pointed, wavy or slightly saw-edged, the midrib prominent above and below. A peculiarity of the flowers is, that their petals are partially welded together into little bells, with the clapper (alias the straight green pistil) protruding, and the stamens united around its base. After the blossoms have been fertilized, the tiny, round, five-scalloped seed capsules, with the pistil still protruding, remain in evidence for months, as is usual in the pyrola clan. Small as the plant is, it has managed to distribute itself over Europe, Asia, and the woods and thickets of our own land from Labrador to Alaska, southward to California, Mexico, and the District of Columbia.

Another little globe-trotter, so insignificant in size that one is apt to overlook it until its surprisingly large blossom appears in June or July, is the ONE-FLOWERED WINTERGREEN (Moneses uniflora), found in cool northern woods, especially about the roots of pines, in such yielding soil as will enable its long stem to run just below the surface. ONE-FLOWERED PYROLA, it is often called, although it belongs to a genus all its own. A boldly curved stalk, like a miniature Bo-peep crook, enables the solitary white or pink widely open flower to droop from the tip, thus protecting its precious contents from rain, and from crawling pilferers, to whom a pendent blossom is as inaccessible as a hanging bird's nest is to snakes. This five-petalled waxen flower, half an inch across or over, with its ten white, yellow-tipped stamens, and green, club-shaped pistil projecting from a conspicuous round ovary, never nods more than six inches above the ground, often at only half that height. When there is no longer need for the stalk to crook, that is to say, after the flower has begun to fruit, it gradually straightens itself out so that the little seed capsule, with the style and its five-lobed stigma still persistent, is held erect. The thin, rounded, finely notched leaves, measuring barely an inch in length, are clustered in whorls next the ground. Whether one comes upon colonies of this gregarious little plant, or upon a lonely straggler, the "single delight" (moneses), as Dr. Gray called the solitary flower, is one of the joys of a tramp through the summer woods.

INDIAN PIPE; ICE-PLANT; GHOST-FLOWER; CORPSE-PLANT
(Monotropa uniflora) Indian-pipe family

Flowers - Solitary, smooth, waxy, white (rarely pink), oblong-bell shaped, nodding from the tip of a fleshy, white, scaly scape 4 to 10 in. tall. Calyx of 2 to 4 early-falling white sepals; 4 or 5 oblong, scale-like petals; 8 or 10 tawny, hairy stamens; a 5-celled, egg-shaped ovary, narrowed into the short, thick style. Leaves: None. Roots: A mass of brittle fibers, from which usually a cluster of several white scapes arises. Fruit: A 5-valved, many-seeded, erect capsule. Preferred Habitat - Heavily shaded, moist, rich woods, especially under oak and pine trees. Flowering Season - June-August. Distribution - Almost throughout temperate North America.

Colorless in every part, waxy, cold, and clammy, Indian pipes rise like a company of wraiths in the dim forest that suits them well. Ghoulish parasites, uncanny saprophytes, for their matted roots prey either on the juices of living plants or on the decaying matter of dead ones, how weirdly beautiful and decorative, they are! The strange plant grows also in Japan, and one can readily imagine how fascinated the native artists must be by its chaste charms.

Yet to one who can read the faces of flowers, as it were, it stands a branded sinner. Doubtless its ancestors were industrious, honest creatures, seeking their food in the soil, and digesting it with the help of leaves filled with good green matter (chlorophyll) on which virtuous vegetable life depends; but some ancestral knave elected to live by piracy, to drain the already digested food of its neighbors; so the Indian pipe gradually lost the use of parts for which it had need no longer, until we find it today without color and its leaves degenerated into mere scaly bracts. Nature has manifold ways of illustrating the parable of the ten pieces of money. Spiritual law is natural law: "From him that hath not, even that he hath shall be taken away." Among plants as among souls, there are all degrees of backsliders. The foxglove, which is guilty of only sly, petty larceny, wears not the equivalent of the striped suit and the shaved head; nor does the mistletoe, which steals crude food from the tree, but still digests it itself, and is therefore only a dingy yellowish green. Such plants, however, as the broomrape, pinesap, beechdrops, the Indian pipe, and the dodder - which marks the lowest stage of degradation of them all - appear among their race branded with the mark of crime as surely as was Cain.

No wonder this degenerate hangs its head; no wonder it grows black with shame on being picked, as if its wickedness were only just then discovered! To think that a plant related on one side to many of the loveliest flowers in Nature's garden- - the azaleas, laurels, rhododendrons, and the bonny heather - and on the other side to the modest but no less charming wintergreen tribe, should have fallen from grace to such a depth! Its scientific name, meaning a flower once turned, describes it during only a part of its career. When the minute, innumerable seeds begin to form, it proudly raises its head erect, as if conscious that it had performed the one righteous act of its life.

LABRADOR TEA
(Ledum Groenlandicum; L. latifolium of Gray) Heath family

Flowers - White, 5-parted, 1/2 in. across or less, numerous, borne in terminal, umbellate clusters rising from scaly, sticky bud-bracts. Stem: A compact shrub 1 to 4 ft. high, resinous, the twigs woolly-hairy. Leaves: Alternate, thick, evergreen, oblong, obtuse, small, dull above, rusty-woolly beneath, the margins curled. Preferred Habitat - Swamps, bogs, wet mountain woods. Flowering Season - May-June. Distribution - Greenland to Pennsylvania, west to Wisconsin.

Whoever has used the homeopathic lotion distilled from the leaves of Ledum palustre, a similar species found at the far North, knows the tea-like fragrance given forth by the leaves of this common shrub when crushed in a warm hand. But because the homeopathists claim that like is cured by like, are we to assume that these little bushes, both of which afford a soothing lotion, also irritate and poison? It may be; for they are next of kin to the azaleas, laurels, and rhododendrons, known to be injurious since Xenophon's day. At the end of May, when the Labrador tea is white with abundant flower clusters, one cannot but wonder why so desirable an acquisition is never seen in men's gardens here among its relatives. Over a hundred years ago the dense, compact little shrub was taken to England to adorn sunny bog gardens on fine estates. Doubtless the leaves have woolly mats underneath for the reason given in reference to the Steeple-bush.

WILD ROSEMARY; MARCH HOLY ROSE; WATER ANDROMEDA; MOORWORT
(Andromeda Polifolia) Heath family

Flowers - White or pink-tinted, small, round, tubular, 5-toothed at the tip; drooping from curved footstalks in few-flowered terminal umbels. Calyx deeply 5-parted; 10 bearded stamens; style like a column. Stem: A sparingly branched, dwarf shrub, 6 in. to 3 ft. tall. Leaves: Linear to lance-shape, evergreen, dark and glossy above, with a prominent white bloom underneath, the margins curled. Preferred Habitat - Cool bogs, wet places. Flowering Season - May-June. Distribution - Pennsylvania and Michigan, far northward.

Only a delightfully imaginative optimist like Linnaeus could feel the enthusiasm he expended on this dwarf shrub, with its little, white, heath-like flowers, which most of us consider rather insignificant, if the truth be told. But then the blossoms he found in Lapland must have been much pinker than any seen in American swamps, since they reminded him of "a fine female complexion."

"This plant is always fixed on some little turfy hillock in the midst of the swamps," he wrote, "just as Andromeda herself was chained to a rock in the sea, which bathed her feet as the fresh water does the roots of this plant.... As the distressed virgin cast down her blushing face through excessive affliction, so does this rosy-colored flower hang its head, growing paler and paler till it withers away." Under the old go-as-you-please method of applying scientific names, most of this shrub's relatives shared with it the name of the fair maid whom Perseus rescued from the dragons.

The beautiful, low-growing STAGGERBUSH (Pieris Mariana) has its small, cylindric, five-parted, white or pink-tinted flowers clustered at intervals along one side of the upright, nearly leafless, smooth, dark-dotted branches of the preceding year. When the glossy oval leaves, black dotted beneath, are freshly put forth in early summer - for the shrub is not strictly an evergreen, however late the old leaves may cling - it is said that stupid sheep and calves, which find them irresistibly attractive, stagger about from their poisonous effect just as they do after feeding on this shrub's relative the Lamb-kill (q.v.). In sandy soil from southern New England to Florida, rarely far inland, one finds the staggerbush in bloom from May to July. On the dry plains of Long Island, where it is common indeed, it appears a not unworthy relative of the FETTERBUSH (Pieris fioribunda), that exquisite little evergreen with quantities of small white urns drooping along its twigs, which nurserymen acquire from the mountains of our Southern States to adorn garden shrubbery at home and abroad. Mr. William Robinson, in his delightful book, "The English Flower Garden" (a book, by the way, that Rudyard Kipling reads as the Puritan read his Bible), counts this fetterbush among the "indispensables."

Much taller than the preceding dwarfs is the COMMON PRIVET ANDROMEDA found in swamps and low ground from New England to the Gulf and in the southwest (Xolisma ligustrina). Whoever has seen the privet almost universally grown in hedges is familiar with the general aspect of this much-branched shrub. Most farmers' boys know the Andromeda's mock May-apple, a hollow, stringy growth of insect origin, which they are not likely to confuse with the pulpy, juicy apple found on the closely related azaleas (q.v.). Abundant terminal spike-like or branched clusters of white, globular, four or five parted flowers in close array, attract quantities of bees from the end of May to early July, notwithstanding each individual flower measures barely an eighth of an inch across. We have seen the fine hair-triggers which other members of this same family, the beautiful pink laurels (q.v.), have set to be sprung by an incoming visitor. Now this Andromeda, and similarly several of its immediate kin, have a quite different, but equally effective, method of throwing pollen on its friends who come to call. When one of the little banded bees clings, as he must, to the tiny flower scarce half his size, thrusting his tongue obliquely through the globe's narrow opening to reach the nectar, suddenly a shower of pollen is inhospitably thrown upon him from within. In probing between the ring of anthers (that are pressed against the style by the S-shaped curvature of the filaments so as to retain the pollen), he needs must displace some of them and release the vitalizing dust through the large terminal pores in the anther-sacs. Is he discouraged by such rough treatment? Not at all. Off he flies to another Andromeda blossom, and leaves some of the dust with which he is powdered on the sticky stigma that impedes his entrance, before precipitating a fresh shower as he sips another reward. The straight column-like pistil, stigmatic on its tip only, allows the flower's own pollen to slide harmlessly down its sides. How exquisite are the most minute adjustments of floral mechanism! Is it possible for one to remain an agnostic after the evidences even the flowers show us of infinite wisdom and love?

Another denizen of swamps and low ground, next of kin to the trailing arbutus, is the LEATHERLEAF, or DWARF CASSANDRA (Chamaedaphne calyculata), a modest little shrub, its stiff, slender branches plentifully set with thick oblong leaves that grow gradually smaller the higher they go, and when young are densely covered with minute scurfy scales. Sometimes before the snow has melted in April, the leafy terminal shoots are hung with multitudes of little waxy-white, cylindric, typical heath flowers only about a quarter of an inch long, each nodding from a leaf axil, and the whole forming one-sided racemes. But as the shrub ranges from Newfoundland to Georgia, and westward to Illinois, British Columbia, and Alaska, some people find it blooming even in July. Mythological names were evidently in high favor among the botanists who labeled the genuses comprising the heath family: Phyllodoce, the sea-nymph; Cassiope, mother of Andromeda; Leucothoe; Andromeda herself;

Pieris, a name sometimes applied to the Muses from their supposed abode at Pieria, Thessaly; and Cassandra, daughter of Priam, the prophetess who was shut up in a mad-house because she prophesied the ruin of Troy - these names are as familiar to the student of this group of shrubs today as they were to the devout Greeks in the brave days of old.

CREEPING WINTERGREEN; CHECKERBERRY; PARTRIDGE-BERRY; MOUNTAIN
TEA; GROUND TEA; DEER, BOX, or SPICE BERRY
  (Gaultheria procumbens) Heath family

Flowers - White, small, usually solitary, nodding from a leaf axil. Corolla rounded bell-shape, 5-toothed; calyx 5-parted, persistent; 10 included stamens, their anther-sacs opening by a pore at the top. Stem: Creeping above or below ground, its branches 2 to 6 in. high. Leaves: Mostly clustered at top of branches; alternate, glossy, leathery, evergreen, much darker above than underneath, oval to oblong, very finely saw-edged; the entire plant aromatic. Fruit: Bright red, mealy, spicy, berry-like; ripe in October. Preferred Habitat - Cool woods, especially under evergreens. Flowering Season - June-September. Distribution - Newfoundland to Georgia, westward to Michigan and Manitoba.

However truly the poets may make us feel the spirit of Nature in their verse, can many be trusted when it comes to the letter of natural science? "Where camels arch their cool, dark boughs o'er beds of wintergreen," wrote Bryant; yet it is safe to say that nine colonies of this hardy little plant out of every ten he saw were under evergreen trees, not dogwoods. When the July sun melts the fragrance out of the pines high overhead, and the dim, cool forest aisles are more fragrant with commingled incense from a hundred natural censers than any stone cathedral's, the wintergreen's little waxy bells hang among the glossy leaves that form their aromatic carpet. On such a day, in such a resting place, how one thrills with the consciousness that it is good to be alive!

Omnivorous children who are addicted to birch-chewing, prefer these tender yellow-green leaves tinged with red, when newly put forth in June - "Youngsters" rural New Englanders call them then. In some sections a kind of tea is steeped from the leaves, which also furnish the old-fashioned embrocation, wintergreen oil. Late in the year the glossy bronze carpet of old leaves dotted over with vivid red "berries" invites much trampling by hungry birds and beasts, especially deer and bears, not to mention well-fed humans. Coveys of Bob Whites and packs of grouse will plunge beneath the snow for fare so delicious as this spicy, mealy fruit that hangs on the plant till spring, of course for the benefit of just such colonizing agents as they. Quite a different species, belonging to another family, bears the true Partridgeberry, albeit the wintergreen shares with it a number of popular names. In a strict sense neither of these plants produces a berry; for the fruit of the true partridge[berry] vine (Mitchella repens) is a double drupe, or stone bearer, each half containing four hard, seed-like nutlets; while the wintergreen's so called berry is merely the calyx grown thick, fleshy, and gaily colored - only a coating for the five-celled ovary that contains the minute seeds. Little baskets of wintergreen berries bring none too high prices in the fancy fruit and grocery shops when we calculate how many charming plants such unnatural use of them sacrifices.

Closely allied to the wintergreen is the RED BEARBERRY, KINNIKINIC, BEAR'S GRAPE, FOXBERRY or MEALBERRY, as it is variously called (Arctostaphylos-uva-ursi = bearberry). Trailing its spreading branches over sandy ground, rocky hillsides and steeps until it sometimes forms luxuriant mats, it closely resembles its cousin the arbutus in its manner of growth, and has been mistaken for it by at least one poet. But its tiny, rounded, urn-shaped flowers, which come in May and June, are white, not salver form and pink; the entire plant is not rusty-hairy; the dark little leathery evergreen leaves are spatulate, and, moreover, it bears small but abundant clusters of round, berry-like fruit, an attainment the arbutus still struggles for, but cannot yet reach. Bumblebees are the flower's chief benefactors. Game fowl, especially grouse, but many other birds too, and various animals which are glad to add the clusters of smooth red bearberries to their scanty winter menu, however insipid and dry they may be, have distributed the seed from Labrador across Arctic America to Alaska, southward to Pennsylvania, Illinois, Nebraska, and California. How plants do compel insects, birds, and beasts to work for them! The entire plant is astringent, and has been used in medicine; also by leather dressers.

BLACK or HIGH-BUSH HUCKLEBERRY; WHORTLEBERRY [now TALL HUCKLEBERRY]
(Gaylussacia resinosa) Huckleberry family

Flowers - White and pink, pale or deep, small, cylindric, bell-shaped. 5-parted, borne in 1-sided racemes from the sides of the stiff, grayish branches. Stem: A shrub to 3 ft. high. Leaves: Alternate, oval to oblong, firm, entire edged, green on both sides, dotted underneath with resinous spots, especially when young. Fruit: A round, black, bloomless, sweet, berry-like drupe, containing 10 seed-like nutlets, in each of which is a solitary seed. Ripe, July-August. Preferred Habitat - Moist, sandy soil, thickets, open woods. Flowering Season - May-June. Distribution - Newfoundland to Georgia, west to Manitoba and Kentucky.

This common huckleberry, oftener found in pies and muffins by the average observer than in its native thickets, unfortunately ripens in fly-time, when the squeamish boarder in the summer hotel does well to carefully scrutinize each mouthful. For the abundant fruit set on huckleberry bushes, as on so many others, we are indebted chiefly to the lesser bees, which, receiving the pollen jarred out from the terminal chinks in the anther-sacs on their undersides as they cling, transfer it to the protruding stigmas of the next blossom visited. After fertilization, when the now useless corolla falls, the ten-celled ovary is protected by the encircling calyx, that grows rapidly, swells, fills with juice, and takes on color until it and the ovary together become a so-called berry, whose seeds are dropped far and wide by birds and beasts. "The name huckleberry, which is applied indiscriminately to several species of Vaccinium and Gaylussacia," says Professor L. H. Bailey, "is evidently a corruption of whortleberry. Whortleberry is in turn a corruption of myrtleberry. In the Middle Ages, the true myrtleberry was largely used in cookery and medicine, but the European bilberry or Vaccinium so closely resembled it that the name was transferred to the latter plant, a circumstance commemorated by Linnaeus in the giving of the name Vaccinium Myrtillus to the bilberry. From the European whortleberry the name was transferred to the similar American plants."

A common little bushy shrub, not a true blueberry, found in moist woods, especially beside streams, from New England to the Gulf States, and westward to Ohio, is the BLUE TANGLE, TANGLE-BERRY, or DANGLEBERRY [now TALL HUCKLEBERRY (G. frondosa). It bears a few tiny greenish-pink flowers dangling from pedicels in loose racemes, and corresponding clusters of most delicious, sweet, dark-blue berries, covered with hoary bloom in midsummer. The abundant resinous leaves on its slender gray branches are pale and hoary beneath. The caterpillars of several species of sulphur butterflies (Colias) feed on huckleberry leaves.

To a genus quite distinct from the huckleberries belong the true blueberries, however interchangeably these names are misused. Perhaps the first species to send its fruit to market in June and July is the DWARF, SUGAR, or LOW-BUSH BLUEBERRY (Vaccinium Pennsylvanicum), sometimes six inches tall, never more than twenty inches. It prefers sandy or rocky soil from southern New Jersey far northward, and west to Illinois. Shortly after the small, bell-shaped, white or pink flowers, that grow in racemes on the ends or sides of the angular, green, warty branches of nearly all blueberry bushes, have been fertilized by bees, this species forms an especially sweet berry with a bloom on its blue surface. The alternate oblong leaves, smooth and green on both sides, are very finely and sharply saw-edged.

Another, and perhaps the commonest, as it is the finest, species, whose immature fruit is still green or red when the dwarf's is ripe, is the HIGH-BUSH, TALL, or SWAMP BLUEBERRY (V. corymbosum), found in low wet ground from Virginia westward to the Mississippi, and very far north. Only the bees and their kind concern themselves with the little cylindric, five-parted, nectar-bearing flowers. These appear with the oblong, entire leaves, paler below than above. But thousands of fruit sellers and housekeepers depend on the sweet blueberries (with a pleasant acid flavor) as a market staple. In July and August, even in early September, the berries arrive in the cities. One picker in New Jersey claims to have filled an entire crate with the fruit of a single bush.

The DEERBERRY, BUCKBERRY, or SQUAW HUCKLEBERRY (V. stainineum), common in dry woods and thickets from Maine and Minnesota to the Gulf States, puts forth quantities of small greenish-white, yellow, or purplish-green, open bell-shaped, five-cleft flowers, nodding from hair-like pedicels in graceful, leafy-bracted racemes. Both the tips of the stamens and the style protrude like a fringe. No creature, unless hard pressed by hunger, could relish the greenish or yellowish berries. This is a low-growing, spreading shrub, with firm oval or oblong tapering leaves, dull above, and pale, sometimes even hoary, underneath.

CREEPING SNOWBERRY
(Chiogenes hispidula) Huckleberry family

Flowers - Very small, white, few, solitary, nodding on short, curved peduncles from the leaf axils. Calyx 2-bracted, 4-cleft; corolla a short 4-cleft bell; 8 short stamens, each anther sac opening by a slit to the middle; 1 pistil, the ovary 4-celled. Stem: Creeping along the ground, the slender, leafy, hairy branches 3 to 12 in. long. Leaves: Evergreen, alternate, 2-ranked, oval, very small, dark and glossy above, coated with stiff, rusty hairs underneath, the edges curled. Fruit: A snow-white, round or oval, mealy, aromatic berry; ripe August-September. Preferred Habitat - Cool bogs; low, moist, mossy woods. Flowering Season - May-June. Distribution - North Carolina and Michigan northward to the British Possessions.

Allied on the one hand to the cranberry, so often found with it in the cool northern peat bogs, and on the other to the delicious blueberries, this "snow-born" berry, which appears on no dining table, nevertheless furnishes many a good meal to hungry birds and fagged pedestrians. Both the pretty foliage and the fruit have the refreshing flavor of sweet birch.

PYXIE; FLOWERING MOSS; PINE-BARREN BEAUTY
(Pyxidanthera barbulata) Diapensia family

Flowers - Abundant, white, or sometimes pink, about 1/4 in. across, 5-parted, solitary, seated at tips of branches. Stem: Prostrate, creeping, much branched, the main branches often 1 ft. long, very leafy, growing in mat-like patches. Leaves: Moss-like, very narrow, pointed, seated on stem, and overlapping like scales, on upper part of branches. Preferred Habitat - Dry sandy soil; pine barrens. Flowering Season - March-May. Distribution - New Jersey, south to North Carolina.

Curiously enough, this creeping, tufted, mat-like little plant is botanically known as a shrub, yet it is lower than many mosses, and would seem to the untrained eye to be certainly of their kin. In earliest spring, when Lenten penitents, jaded with the winter's frivolities in the large cities, seek the salubrious pine lands of southern New Jersey and beyond, they are amazed and delighted to find the abundant little evergreen mounds of pyxie already starred with blossoms. The dense mossy cushions, plentifully sprinkled with pink buds and white flowers, are so beautiful, one cannot resist taking a few tuffets home to naturalize in the rock garden. Planted in a mixture of clear sand and leaf-mould, with exposure to the morning sun, pyxie will smile up at us from under our very windows, spring after spring, with increased charms; whereas the arbutus, that untamable wildling, carried home from the pinewoods at the same time, soon sulks itself to death.

STARFLOWER; CHICKWEED-WINTERGREEN; STAR ANEMONE
(Trientalis Americana) Primrose family

Flowers - White, solitary, or a few rising on slender, wiry footstalks above a whorl of leaves. Calyx of 5 to 9 (usually 7) narrow sepals. Corolla wheel-shaped, 1/2 in. across or less, deeply cut into (usually) 7 tapering, spreading, petal-like segments. Stem: A long horizontal rootstock, sending up smooth stem-like branches 3 to 9 in. high, usually with a scale or two below. (Trientalis = one-third of a foot, the usual height of a plant.) Leaves: 5 to 10, in a whorl at summit; thin, tapering at both ends, of unequal size, 1 1/2 to 4 in. long. Preferred Habitat - Moist shade of woods and thickets. Flowering Season - May-June. Distribution - From Virginia and Illinois far north.

Is any other blossom poised quite so airily above its whorl of leaves as the delicate, frosty-white little starflower? It is none of the anemone kin, of course, in spite of one of its misleading folk names; but only the wind-flower has a similar lightness and grace. No nectar rewards the small bee and fly visitors; they get pollen only. Those coming from older blossoms to a newly opened one leave some of the vitalizing dust clinging to them on the moist and sticky stigma, which will wither to prevent self-fertilization before the flower's own curved anthers mature and shed their grains. Sometimes, when the blossoms do not run on schedule time, or the insects are not flying in stormy weather, this well laid plan may gang a-gley. An occasional lapse matters little; it is perpetual self-fertilization that Nature abhors.

INDIAN HEMP: AMY-ROOT
 (Apocynum cannabinum) Dogbane family

Flowers - Greenish white, about 1/4 in. across, on short pedicels, in dense clusters at ends of branches and from the axils. Calyx of 5 segments; corolla nearly erect, bell-shaped, 5-lobed, with 5 small triangular appendages alternating with the stamens within its tube. Stem: 1 to 4 ft. high, branching, smooth, often dull reddish, from a deep, vertical root. Leaves: Opposite, entire, 2 to 6 in. long, mostly oblong, abruptly pointed, variable. Fruit: A pair of slender pods, the numerous seeds tipped with tufts of hairs. Preferred Habitat - Gravelly soil, banks of streams, low fields. Flowering Season - June-August. Distribution - Almost throughout the United States and British Possessions.

Instead of setting a trap to catch flies and hold them by the tongue in a vise-like grip until death alone releases them, as its heartless sister the spreading dogbane does (q.v.), this awkward, rank herb lifts clusters of smaller, less conspicuous, but innocent, flowers, with nectar secreted in rather shallow receptacles, that even short-tongued insects may feast without harm. Honey and mining bees, among others; wasps and flies in variety, and great numbers of the spangled fritillary (Argynnis cybele) and the banded hair-streak (Thecla calanus) among the butterfly tribe; destructive bugs and beetles attracted by the white color, a faint odor, and liberal entertainment, may be seen about the clusters. Many visitors are useless pilferers, no doubt; but certainly the bees which depart with pollen masses cemented to their lips or tongues, to leave them in the stigmatic cavities of the next blossoms their heads enter, pay a fair price for all they get.

>From the fact that Indians used to substitute this very common plant's tough fiber for hemp in making their fishnets, mats, baskets, and clothing, came its popular name; and from their use of the juices to poison mangy old dogs about their camps, its scientific one.

WHORLED or GREEN-FLOWERED MILKWEED
(Asclepias verticillata) Milkweed family

Flowers - White or greenish, on short pedicels, in several small terminal clusters. Calyx inferior; corolla deeply 5-parted, the oblong segments turned back; a 5-parted, erect crown of hooded nectaries between them and the stamens, each shorter than the incurved horn within. Stem: 1 to 2 1/2 ft. tall, simple or sparingly branched, hairy, leafy to summit, containing milky juice. Leaves: In upright groups, very narrow, almost thread-like, from 3 to 7 in each whorl. Fruit: 2 smooth, narrow, spindle-shaped, upright pods, the seeds attached to silky fluff; 1 pod usually abortive. Preferred Habitat - Dry fields, hills, uplands. Flowering Season - July-September. Distribution - Maine and far westward, south to Florida and Mexico.

In describing the common milkweed (q.v.), so many statements were made that apply quite as truly to this far daintier and more ethereal species, the reader is referred back to the pink and magenta section. Compared with some of its rank-growing, heavy relatives, how exquisite is this little denizen of the uplands, with its whorls of needle-like leaves set at intervals along a slender swaying stem! The entire plant, with its delicate foliage and greenish-white umbels of flowers, rather suggests a member of the carrot tribe; and much the same class of small-sized, short-tongued visitors come to seek its accessible nectar as we find about the parsnips, for example. When little bees alight - and these are the truest benefactors, however frequently larger bees, wasps, flies, and even the almost useless butterflies come around - their feet slip about within the low crown to find a secure lodging. As they rise to fly away after sucking, the pollen masses which have attached themselves to the hairs on the lower part of their legs are drawn out, to be transferred to other blossoms, perhaps today, perhaps not for a fortnight. Annoying as they may be, it is very rarely, indeed, that an insect can rid itself of the pollen masses carried from either orchids or milkweeds, except by the method Nature intended; and it is not until the long-suffering bee is outrageously loaded that he attains his greatest usefulness to milkweed blossoms. "Of ninety-two specimens bearing corpuscula of Asclepias verticillata," says Professor Robertson, "eighty-eight have them on hairs alone, and four on the hairs and claws." And again: "As far as the mere application of pollen to an insect is concerned, a flower with loose pollen has the advantage. But the advantage is on the side of Asclepias after the insect is loaded with it. It is only a general rule that insects keep to flowers of a particular species on their honey and pollen gathering expeditions. If a bee dusted with loose pollen visits flowers of another species, it will not long retain pollen in sufficient quantity to effectually fertilize flowers of the original species. On the other hand, if an insect returns at any time during the day, or even after a few days, to the species of Asclepias from which it got a load of pollinia, it may bring with it all or most of the pollinia which it has carried from the first plants visited. The firmness with which the pollinia keep their hold on the insect is one of the best adaptations for cross-fertilization."

Ants, the worst pilferers of nectar extant, find the hairy stem of the whorled milkweed, as well as its sticky juice, most discouraging, if not fatal, obstacles to climbing. How daintily the goldfinch picks at the milkweed pods and sets adrift the seeds attached to silky aeronautic fluff!

WILD POTATO-VINE; MAN-OF-THE-EARTH; MECHA-MECK
(Ipomoea pandurata) Morning-glory family

Flowers - Funnel form, wide-spread, 2 to 3 in. long, pure white or pinkish purple inside the throat; the peduncles 1 to 5 flowered. Stem: Trailing over the ground or weakly twining, 2 to 12 ft. long. Leaves: Heart, fiddle, or halbert shaped (rarely 3-lobed), on slender petioles. Root: Enormous, fleshy. Preferred Habitat - Dry soil, sandy or gravelly fields or hills. Flowering Season - May-September. Distribution - Ontario, Michigan, and Texas, east to the Atlantic Ocean.

No one need be told that this flaring, trumpet-shaped flower is next of kin to the morning-glory that clambers over the trellises of countless kitchen porches, and escapes back to Nature's garden whenever it can. When the ancestors of these blossoms welded their five petals into a solid deep bell, which still shows on its edges the trace of five once separate parts, they did much to protect their precious contents from rain; but some additional protection was surely needed against the little interlopers not adapted to fertilize the flower, which could so easily crawl down its tube. Doubtless the hairs on the base of the filaments, between which certain bumblebees and other long-tongued benefactors can easily penetrate to suck the nectar secreted in a fleshy disk below, act as a stockade to little would-be pilferers. The color in the throat serves as a pathfinder to the deep-hidden sweets. How pleasant the way is made for such insects as a flower must needs encourage! For these the perennial wild potato vine keeps open house far later in the day than its annual relatives. Professor Robertson says it is dependent mainly upon two bees, Entechnia taurea and Xenoglossa ipomoeae, the latter its namesake.

One has to dig deep to find the huge, fleshy, potato-like root from which the vine derived its name of man-of-the-earth. Such a storehouse of juices is surely necessary in the dry soil where the wild potato lives.

Happily, the COMMON MORNING-GLORY (I. purpurea) - the Convolvulus major of seedsmen's catalogues - has so commonly escaped from cultivation in the eastern half of the United States and Canada as now to deserve counting among our wild flowers, albeit

South America is its true home. Surely no description of this com-monest of all garden climbers is needed; everyone has an opportu-nity to watch how the bees cross-fertilize it.

The vine has a special interest because of Darwin's illuminating experiments upon it when he planted six self-fertilized seeds and six seeds fertilized with the pollen brought from flowers on a diffe-rent vine, on opposite sides of the same pot. Vines produced by the former reached an average height of five feet four inches, whereas the cross-pollenized seed sent its stems up two feet higher, and produced very many more flowers. If so marked a benefit from imported pollen may be observed in a single generation, is it any wonder that ambitious plants employ every sort of ingenious device to compel insects to bring them pollen from distant flowers of the same species? How punctually the MOON-FLOWER (I. grandiflo-ra), next of kin to the morning-glory, opens its immense, pure whi-te, sweet-scented flowers at night to attract night-flying moths, be-cause their long tongues, which only can drain the nectar, may not be withdrawn until they are dusted with vitalizing powder for ex-port to some waiting sister.

GRONOVIUS' or COMMON DODDER; STRANGLE-WEED; LOVE VINE; ANGEL'S
HAIR
(Cuscuta gronovii) Dodder family

Flowers - Dull white, minute, numerous, in dense clusters. Calyx inferior, greenish white, 5-parted; corolla bell-shaped, the 5 lobes spreading, 5 fringed scales within; 5 stamens, each inserted on corolla throat above a scale; 2 slender styles. Stem: Bright orange yellow, thread-like, twining high, leafless. Preferred Habitat - Moist soil, meadows, ditches, beside streams. Flowering Season - July-September. Distribution - Nova Scotia and Manitoba, south to the Gulf States.

Like tangled yellow yarn wound spirally about the herbage and shrubbery in moist thickets, the dodder grows, its beautiful bright threads plentifully studded with small flowers tightly bunched. Try to loosen its hold on the support it is climbing up, and the secret of its guilt is out at once; for no honest vine is this, but a parasite, a degenerate of the lowest type, with numerous sharp suckers (haustoria) penetrating the bark of its victim, and spreading in the softer tissues beneath to steal all their nourishment. So firmly are these suckers attached, that the golden thread-like stem will break before they can be torn from their hold.

Not a leaf now remains on the vine to tell of virtue in its remote ancestors; the absence of green matter (chlorophyll) testifies to dishonest methods of gaining a living (see Indian pipe); not even a root is left after the seedling is old enough to twine about its hardworking, respectable neighbors. Starting out in life with apparently the best intentions, suddenly the tender young twiner develops an appetite for strong drink and murder combined, such as would terrify any budding criminal in Five Points or Seven Dials! No sooner has it laid hold of its victim and tapped it, than the now useless root and lower portion wither away, leaving the dodder in mid-air, without any connection with the soil below, but abundantly nourished with juices already stored up, and even assimilated, at its host's expense. By rapidly lengthening the cells on the outer side of its stem more than on the inner side, the former becomes convex, the latter concave; that is to say, a section of spiral is formed by the new shoot, which, twining upward, devitalizes its benefactor as it goes. Abundant, globular seed vessels, which develop rapidly, while the blossoming continues unabated, soon sink into the soft soil to begin their piratical careers close beside the criminals which bore them; or better still, from their point of view, float downstream to found new colonies afar. When the beautiful jewelweed - a conspicuous sufferer - is hung about with dodder, one must be grateful for at least such symphony of yellows.

VIRGINIA WATERLEAF
 (Hydrophyllum Virginicum) Waterleaf family

Flowers - White or purplish tinged, in a single or forking cluster on a long peduncle. Calyx deeply 5-parted, the spreading segments very narrow, bristly hairy. Corolla erect, bell-shaped, deeply 5-

lobed; 5 protruding stamens, with soft hairs about their middle; 2 styles united to almost the summit. Stem: Slender, rather weak, to 3 ft. long, leafy, sparingly branched, from a scaly rootstock. Leaves: Alternate, lower ones on long petioles, 6 to 10 in. long, pinnately divided into 5 to 7 oblong, sharply toothed, acute leaflets or segments; upper leaves similar, but smaller, and with fewer divisions. Preferred Habitat - Rich, moist woods. Flowering season - May-August. Distribution - Quebec to South Carolina, west to Kansas and Washington.

So very many flowers especially adapted to the bumblebee are in bloom when the cymes of the waterleaf uncoil, like the borages, from their immature roll, that some special inducement to attract this benefactor were surely needed. In high altitudes the clusters became deeper hued; but much as the more specialized bees love color, food appeals to them far more. Accordingly the five lobes of each little flower stand erect to increase the difficulty a short-tongued insect would have to drain its precious stores; the stamens are provided with hairs for the same reason; and even the calyx is bristly, to discourage crawling ants, the worst pilferers out. By these precautions against theft, plenty of nectar remains for the large bees. To prevent self-fertilization, pollen is shed on visitors, which remove it from a newly opened flower before the stigmas become receptive to any; but in any case these are elevated in maturity above the anthers, well out of harm's way.

Early in spring the large lower leaves are calculated to hold the drip from the trees overhead, hence the plant's scientific and popular names.

JIMSONWEED; JAMESTOWN WEED; THORN APPLE; STRAMONIUM; DEVIL'S TRUMPET
 (Datura stramonium) Potato family

 Flowers - Showy, large, about 4 in. high, solitary, erect, growing from the forks of branches. Calyx tubular, nearly half as long as the corolla, 5-toothed, prismatic; corolla funnel-form, deep-throated, the spreading limb 2 in. across or less, plaited, 5-pointed; stamens 5; 1 pistil. Stem: Stout, branching, smooth, 1 to 5 ft. high. Leaves: Alternate, large, rather thin, petioled, egg-shaped in outline, the edges irregularly wavy-toothed or angled, rank-scented. Fruit: A densely prickly, egg-shaped capsule, the lower prickles smallest. The seeds and stems contain a powerful narcotic poison. Preferred Habitat - Light soil, fields, waste land near dwellings, rubbish heaps. Flowering Season - June-September. Distribution - Nova Scotia to the Gulf of Mexico, westward beyond the Mississippi.

When we consider that there are over five million Gypsies wandering about the globe, and that the narcotic seeds of the thorn apple, which apparently heal, as well as poison, have been a favorite medicine of theirs for ages, we can understand at least one means of the weed reaching these shores from tropical Asia. (Hindoo, dhatura). Our Indians, who call it "white man's plant," associate it with the Jamestown settlement - a plausible connection, for Raleigh's colonists would have been likely to carry with them to the New World the seeds of an herb yielding an alkaloid more esteemed in the England of their day than the alkaloid of opium known as morphine. Daturina, the narcotic, and another product, known in medicine as stramonium, smoked by asthmatics, are by no means despised by up-to-date practitioners. Were it not for the rank odor of its leaves, the vigorous weed, coarse as it is, would be welcome in men's gardens. Indeed, many of its similar relatives adorn them. The fragrant petunia and tobacco plants of the flower beds, the potato, tomato, and egg-plant in the kitchen garden, call it cousin.

Late in the afternoon the plaited corolla of this long trumpet-shaped flower expands to welcome the sphinx moths. So deep a tube implies their tongues; not that these are the benefactors to which the blossom originally adapted itself - they were doubtless left behind in Asia - but apparently our moths make excellent substitutes, for there is no abatement of the weed's vigor here, as there surely would be did it habitually fertilize itself. Any time after four o'clock in the afternoon, according to the light, the sphinx moth, a creature of the gloaming, begins its rounds, to be mistaken for a hummingbird seven times out of ten. Hovering about its chosen white or yellow flowers, that open for it at the approach of twilight, it remains poised above one a second, as if motionless - although the faint hum of its wings, while sucking, indicates that no magic suspends it - then darts swift as thought to another deep tube to feast again, of course transferring pollen as it goes. But what if the Jamestown weed miscalculate the hour of her lover's call and open too soon? Mischievous bees, quick to seize so golden an opportunity, squeeze into the flower when it begins to unfold (flies and beetles following them), to steal pollen, which will sometimes be entirely removed before the moth's arrival.

The THORN-APPLE [now PURPLE THORN-APPLE, considered a variant of JIMSONWEED]; PURPLE STRAMONIUM (D. tatula), a similar species, usually with darker leaves, and pale lavender or violet flowers, or with its long, slender tube white, has become at home in so many fields and waste lands east of Minnesota and Texas that no one thinks of it as belonging to tropical America.

Only sphinx moths can reach its deep well of nectar, from which bees are literally barred out by an inward turn of the stamens toward the center of the tube. Caterpillars of our commonest member of the sphinx tribe conceal themselves on the tomato vine by a mimicry of its color so faultless that a bright eye only may detect their presence. In the South the caterpillar of another of these moths (Sphinx Carolina) does fearful havoc under its appropriate alias of "tobacco worm."

CULVER'S-ROOT; CULVER'S PHYSIC
(Leptandra Virginica; Veronica Virginica of Gray) Figwort family

Flowers - Small, white or rarely bluish, crowded in dense spike-like racemes 3 to 9 in. long, usually several spikes at top of stem or from upper axils. Calyx 4-parted, very small; corolla tubular, 4-lobed; 2 stamens protruding; pistil. Stem: Straight, erect, usually unbranched, 2 to 7 ft. tall. Leaves: Whorled, from 3 to 9 in a cluster, lance-shaped or oblong, and long-tapering, sharply saw-edged. Preferred Habitat - Rich, moist woods, thickets, meadows. Flowering Season - June-September. Distribution - Nova Scotia to Alabama, west to Nebraska.

Slender, erect white wands make conspicuous advertisements in shady retreats at midsummer, when insect life is at its height and floral competition for insect favors at its fiercest. Next of kin to the tiny blue speedwell, these minute, pallid blossoms could have little hope of winning wooers were they not living examples of the adage, "In union there is strength.' Great numbers crowded together on a single spike, and several spikes in a cluster that towers above the woodland undergrowth, cannot well be overlooked by the dullest insects, especially as nectar rewards the search of those having mid-length or long tongues. Simply by crawling over the spikes, of which the terminal one usually matures first, they fertilize the little flowers. The pollen thrust far out of each tube in the early stage of bloom, has usually all been brushed off on the underside of bees, wasps, butterflies, flies, and beetles before the stigma matures; nevertheless, when it becomes susceptible, the anthers spread apart to keep out of its way lest any leftover pollen should touch it.

"The leaves of the herbage at our feet," says Ruskin, "take all kinds of strange shapes, as if to invite us to examine them. Star-shaped. heart-shaped, spear-shaped, arrow-shaped, fretted, fringed, cleft, furrowed, serrated, in whorls, in tufts, in wreaths, in spires, endlessly expressive, deceptive, fantastic, never the same from footstalks to blossom, they seem perpetually to tempt our watchfulness, and take delight in outstripping our wonder." Doubtless light is the factor with the greatest effect in determining the position of the leaves on the stem, if not their shape. After plenty of light has been secured, any aid they may render the flowers in increasing their attractiveness is gladly rendered. Who shall deny that the brilliant foliage of the sumacs, the dogwood, and the pokeweed in autumn does not greatly help them in attracting the attention of migrating birds to their fruit, whose seeds they wish distributed? Or that the clustered leaves of the dwarf cornel and Culver's-root, among others, do not set off to great advantage their white flowers which, when seen by an insect flying overhead, are made doubly conspicuous by the leafy background formed by the whorl?

BUTTONBUSH; HONEY-BALLS; GLOBE-FLOWER; BUTTON-BALL SHRUB;

RIVER-BUSH
(Cephalanthus occidentalis) Madder family

Flowers - Fragrant, white, small, tubular, hairy within, 4-parted, the long, yellow-tipped style far protruding; the florets clustered on a fleshy receptacle, in round heads (about 1 in. across), elevated on long peduncles from leaf-axils or ends of branches. Stem: A shrub 3 to 12 ft. high. Leaves: Opposite or in small whorls, petioled, oval, tapering at the tip, entire. Preferred Habitat - Beside streams and ponds; swamps, low ground. Flowering Season - June-September. Distribution - New Brunswick to Florida and Cuba, westward to Arizona and California.

Delicious fragrance, faintly suggesting jasmine, leads one over marshy ground to where the buttonbush displays dense, creamy-white globes of bloom, heads that Miss Lounsberry aptly likens to "little cushions full of pins." Not far away the sweet breath of the white-spiked clethra comes at the same season, and one cannot but wonder why these two bushes, which are so beautiful when most garden shrubbery is out of flower, should be left to waste their sweetness, if not on desert air exactly, on air that blows far from the homes of men. Partially shaded and sheltered positions near a house, if possible, suit these water lovers admirably. Cultivation only increases their charms. We have not so many fragrant wild flowers that any can be neglected. John Burroughs, who included the blossoms of several trees in his list of fragrant ones, found only thirty-odd species in New England and New York.

Examine a well-developed ball of bloom on the button-bush under a magnifying glass to appreciate its perfection of detail. After counting two hundred and fifty minute florets, tightly clustered, one's tired eyes give out. A honey-ball, with a well of nectar in each of these narrow tubes, invites hosts of insects to its hospitable feast; but only visitors long and slender of tongue can drain the last drop, therefore the vicinity of this bush is an excellent place for a butterfly collector to carry his net. Butterflies are by far the most abundant visitors; honey-bees also abound, bumblebees, carpenter and mining bees, wasps, a horde of flies, and some destructive beetles; but the short tongues can reach little nectar. Why do the pistils of the florets protrude so far? Even before each minute bud opened, all its pollen had been shed on the tip of the style, to be in a position to be removed by the first visitor alighting on the ball of bloom. After the removal of the pollen from the still immature stigma, it becomes sticky, to receive the importation from other blossoms. Did not the floret pass through two distinct stages, first male, then female, self-fertilization, not cross-fertilization, would be the inevitable result. The dull red and green seed-balls, which take on brown and bronze tints after frost, make beautiful additions to an autumn bouquet. The bush is next of kin to the coffee.

PARTRIDGE VINE; TWIN-BERRY; MITCHELLA-VINE; SQUAW-BERRY
(Mitchella repens) Madder family

Flowers - Waxy, white (pink in bud), fragrant, growing in pairs at ends of the branches. Calyx usually 4-lobed; corolla funnel-form, about 1/2 in. long, the 4 spreading lobes bearded within; 4 stamens inserted on corolla throat; style with 4 stigmas; the ovaries of the twin flowers united. (The style is long when the stamens are short, or vice versa). Stem: Slender, trailing, rooting at joints, 6 to 12 in. long, with numerous erect branches. Leaves: Opposite, entire, short petioled, oval or rounded, evergreen, dark, sometimes white veined. Fruit: A small, red, edible, double berry-like drupe. Preferred Habitat - Woods; usually, but not always, dry ones. Flowering Season - April-June. Sometimes again in autumn. Distribution - Nova Scotia to the Gulf States, westward to Minnesota and Texas.

A carpet of these dark, shining, little evergreen leaves, spread at the foot of forest trees, whether sprinkled over in June with pairs of waxy, cream-white, pink-tipped, velvety, lilac-scented flowers that suggest attenuated arbutus blossoms, or with coral-red "berries" in autumn and winter, is surely one of the loveliest sights in the woods. Transplanted to the home garden in closely packed, generous clumps, with plenty of leaf-mould, or, better still, chopped sphagnum, about them, they soon spread into thick mats in the rockery, the hardy fernery, or about the roots of rhododendrons and the taller shrubs that permit some sunlight to reach them. No woodland creeper rewards our care with greater luxuriance of growth. Growing near our homes, the partridge vine offers an excellent opportunity for study.

The two flowers at the tip of a branch may grow distinct down to their united ovaries, or their tubes may be partly united, like Siamese twins - a union which in either case accounts for the odd shape of the so-called berry, that shows further traces of consolidation in its "two eyes," the remnants of eight calyx teeth. Experiment proves that when only one of the twin flowers is pollenized by insects (excluded from the other one by a net), fruit is rarely set; but when both are, a healthy seeded berry follows. To secure cross-fertilization, the partridge flower, like the bluets (q.v.), occurs in two different forms on distinct plants, seed from either producing after its kind. In one form the style is low within the tube, and the stamens protrude; in the other form the stamens are concealed, and the style, with its four spreading stigmas, is exserted. No single flower matures both its reproductive organs. Short-tongued small bees and flies cannot reach the nectar reserved for the blossom's benefactors because of the hairs inside the tube, which nearly close it; but larger bees and butterflies coming to suck a flower with tall stamens receive pollen on the precise spot on their long tongues that will come in contact with the sticky stigmas of the long-styled form visited later, and there rub the pollen off. The lobes' velvety surface keeps insect feet from slipping.

What endless confusion arises through giving the same popular folk names to different species! The Bob White, which is called quail in New England or wherever the ruffed grouse is known as partridge, is called partridge in the Middle and Southern States, where the ruffed grouse is known as pheasant. But as both these distributing agents, like most winter rovers, whether bird or beast, are inordinately fond of this tasteless partridge berry, as well as of the spicy fruit of quite another species, the aromatic wintergreen (q.v.), which shares with it a number of common names, every one may associate whatever bird and berry that best suit him. The delicious little twin-flower, beloved of Linnaeus, also comes in for a share of lost identity through confusion with the partridge vine.

CLEAVERS; GOOSE-GRASS; BEDSTRAW
(Galium Aparine) Madder family

Flowers - Small, white, 4-parted, inconspicuous, in clusters of 1 to 3 on peduncles from the axils of upper leaves. Stem: 2 to 5 ft. long, scrambling, weak, square; bristly on the angles. Leaves: in whorls of 6 or 8, narrow, midrib and edges very rough. Fruit: Rounded, twin seed-vessels, beset with many hooked bristles.
Preferred Habitat - Shady ground.
Flowering Season - May-September.
Distribution - Eastern half of United States and Canada.

Among some seventy other English folk names by which cleavers are known are the following, taken from Britton and Brown's "Illustrated Flora": "CATCHWEED, BEGGAR-LICE, BURHEAD, CLOVER-GRASS, CLING-RASCAL, SCRATCH-GRASS, WILD HEDGE-BURS, HAIRIF or AIRIF, STICK-A-BACK or STICKLE-BACK, GOSLING-GRASS or GOSLING-WEED, TURKEY-GRASS, PIGTAIL, GRIP or GRIP-GRASS, LOVEMAN, SWEETHEARTS." From these it will be seen that the insignificant little white flowers impress not the popular mind. But the twin burs which steal a ride on every passing animal, whether man or beast, in the hope of reaching new colonizing ground far from the parent plant, rarely fail to make an impression on one who has to pick trailing sprays beset with them off woollen clothing.

Several other similar bur-bearing relatives there are, common in various parts of America as they are in Europe. The SWEET-SCENTED BEDSTRAW (G. trifolium), always with three little greenish flowers at the end of a footstalk, or branched into three pedicels that are one to three flowered, and with narrowly oval, one-nerved leaves arranged in whorls of six on its square stem, ranges from ocean to ocean on this continent, over northern Europe, and in Asia from Japan to the Himalayas. It will be noticed that plants depending upon the by hook or by crook method of travel are among the best of globe trotters. This species becomes increasingly fragrant as it dries.

COMMON ELDER; BLACK-BERRIED, AMERICAN or SWEET
ELDER; ELDERBERRY
(Sambucus Canadensis) Honeysuckle family

Flowers - Small, creamy, white, numerous, odorous, in large, flat-topped, or convex cymes at ends of branches. Calyx tubular, minute; corolla of 5 spreading lobes; 5 stamens; style short, 3-parted. Stem: A shrub 4 to 10 ft. high, smooth, pithy, with little wood. Leaves: Opposite, pinnately compounded of 5 to 11 (usually 7) oval, pointed, and saw-edged leaflets, heavy-scented when crushed. Fruit: Reddish-black, juicy "berries" (drupes). Preferred Habitat - Rich, moist soil; open situation. Flowering Season - June-July. Distribution - Nova Scotia to the Gulf of Mexico, and westward 2,000 miles.

Flowers far less beautiful than these flat-spread, misty clusters, that are borne in such profusion along the country lane and meadow hedgerows in June, are brought from the ends of the earth to adorn our over-conventional gardens. Certain European relatives, with golden or otherwise variegated foliage that looks sickly after the first resplendent outburst in spring, receive places of honor with monotonous frequency in American shrubbery borders.

Like the wild carrot among all the umbel-bearers, and the daisy among the horde of composites, the elder flower has massed its minute florets together, knowing that there was no hope of attracting insect friends, except in such union. Where clumps of elder grow - and society it ever prefers to solitude - few shrubs, looked at from above, which, of course, is the winged insect's point of view, offer a better advertisement. There are people who object to the honey-like odor of the flowers. Doubtless this is what most attracts the flies and beetles, while the lesser bees, that frequent them also, are more strongly appealed to through the eye. No nectar rewards visitors, consequently butterflies rarely stop on the flat clusters; but there is an abundant lunch of pollen for such as like it. Each minute floret has its five anthers so widely spread away from the stigmas that self-pollination is impossible; but with the help of small, winged pollen carriers plenty of cross-fertilized fruit forms. With the help of migrating birds, the minute nutlets within the "berries" are distributed far and wide.

When clusters of dark, juicy fruit make the bush top-heavy, it is, of course, no part of their plan to be gathered into pails, crushed and boiled and fermented into the spicy elderberry wine that is still as regularly made in some old-fashioned kitchens as currant jelly and pickled peaches. Both flowers and fruit have strong medicinal properties. Snuffling children are not loath to swallow sugar pills moistened with the homeopathic tincture of Sambucus. The common European species (S. nigra), a mystic plant, was once employed to cure every ill that flesh is heir to; not only that, but, when used as a switch, it was believed to check a lad's growth. Very likely! Every whittling schoolboy knows how easy it is to remove the white pith from an elder stem. An ancient musical instrument, the sambuca, was doubtless made from many such hollow reed-like sticks properly attuned.

A more woody species than the common elder, whose stems are so green it is scarcely like a true shrub, is the very beautiful RED-BERRIED or MOUNTAIN ELDER (S. pubens), found in rocky places, especially in uplands and high altitudes, from the British Possessions north of us to Georgia on the Atlantic Coast, and to California on the Pacific. Coming into bloom in April or May, it produces numerous flower clusters which are longer than broad, pyramidal rather than flat-topped. They turn brown when drying. In young twigs the pith is reddish-brown, not white as in the common elder. Birds with increased families to feed in June are naturally attracted by the bright red fruit; and while they may not distribute the stones over so vast an area as autumn migrants do those of the fall berries, they nevertheless have enabled the shrub to travel across our continent.

HOBBLE-BUSH; AMERICAN WAYFARING TREE (Viburnum alnifolium; V. lantanoides of Gray) Honeysuckle family

Flowers - In loose, compound, flat, terminal clusters, 3 to 5 in. across; the outer, showy, white flowers each about 1 in. across, neutral; inner ones very much smaller, perfect. Calyx 5-parted;

corolla 5-lobed; 5 stamens; 3 stigmas. Stem: A widely and irregularly branching shrub, sometimes 10 ft. high; the young twigs rusty scurfy. Leaves: Opposite, rounded or broadly ovate, pointed at the tip, finely saw-edged, unevenly divided by midrib, scurfy on veins beneath. Fruit: Not edible, berry-like, at first coral-red, afterward darker. Preferred Habitat - Cool, low, moist woods. Flowering Season - May-June. Distribution - North Carolina and Michigan, far northward.

Widespread, irregular clusters of white bloom, that suggest heads of hydrangea whose plan has somehow miscarried, form a very decorative feature of the woods in May, when the shrubbery in Nature's garden, as in men's, is in its glory. For what reason are there two sizes and kinds of flowers in each cluster? Around the outer margin are large showy shams: they lack the essential organs, the stamens and pistil; therefore what use are they? Undoubtedly they are mere advertisements to catch the eye of passing insects - no small service, however. It is the inconspicuous little flowers grouped within their circle that attend to the serious business of life. The shrub found it good economy to increase the size of the outer row of flowers, even at the expense of their reproductive organs, simply to add to the conspicuousness of the clusters, when so many blossoms enter into fierce competition with them for insect trade. Many beetles, attracted by the white color, come to feed on pollen, and often destroy the anthers in their greed. But the lesser bees (Andrena chiefly), and more flies, whose short tongues easily obtain the accessible nectar, render constant service. These welcome guests we have to thank for the clusters of coral-red berries that make the shrub even more beautiful in September than in May.

Because it sometimes sends its straggling branches downward in loops that touch the ground and trip up the unwary pedestrian, who presumably hobbles off in pain, the bush received a name with which the stumbler will be the last to find fault. From the bark of the Wayfaring Tree of the Old World (V. lantana), the tips of whose procumbent branches often take root as they lie on the ground, is obtained bird-lime. No warm, sticky scales enclose the buds of our hardy hobble-bush; the only protection for its tender baby foliage is in the scurfy coat on its twigs; yet with this thin covering, or without it, the young leaves safely withstand the intense cold of northern winters.

The chief beauty of the HIGH BUSH-CRANBERRY, CRANBER-RY TREE, or WILD GUELDER-ROSE (V. Opulus) lies in its clusters of bright red, oval, very acid "berries" (drupes), that are commonly used by country people as a substitute for the fruit they so closely resemble. This is a symmetrical, erect, tall, smooth shrub, found in moist, low ground. Among the Berkshires it grows in perfection. From New Jersey, Michigan, and Oregon far northward is its range; also in Europe and Asia. The broadly ovate, saw-edged, three-lobed leaves are more or less hairy along the veins on the underside. Like the hobble-bush, this one produces an outer circle of showy, neutral flowers, as advertisements, on its peduncled, flat cluster; and small, perfect ones, to reproduce the species, in June or July. As the flies and small pollen-collecting bees move rapidly over a corymb to feast on the layer of nectar freely exposed for their benefit, they usually cross-fertilize the flowers; for, as Muller pointed out, the anthers and stigmas of each come in contact with different parts of the insect's feet or tongue. Beetles, which visit the clusters in great numbers, often prove destructive visitors. Kerner claims that nectar is secreted in the leaves of this species, whether in the two glands that appear at the top of the petioles or not, he does not say. Of what possible advantage to the plant could such an arrangement be? Plants, as well as humans, are not in business for philanthropy.

No garden is complete - was garden ever complete? - without the beautiful SNOWBALL BUSH, a sterile variety of this shrub, with whose abundant balls of white flowers everyone is familiar. When various members of the viburnum and the hydrangea tribes are cultivated, the corollas of both the small interior flowers and those in the showy exterior circle become largely developed, while the reproductive organs of the former gradually become abortive. The snowball bush rather overdoes its advertising business; for however attractive its round white masses of sterile bloom, the effect is of no advantage to itself.

In light, dry, rocky woods, from North Carolina and Minnesota, far northward, grows the common MAPLE-LEAVED ARROW-WOOD or DOCKMACKIE (V. acerifolium), which one might easily mistake for a maple sapling when it is not in flower or fruit. All the blossoms in its slender peduncled, flat-topped, white clusters are perfect; none are sterile for advertising purposes merely, as in the cases of so many of its relatives. The five stamens protrude from each five-lobed little flower for plain reasons. The opposite leaves are broadly ovate, three-ribbed, three-lobed, coarsely toothed, acute at the tip, and, except for their soft hairiness underneath, are too like maple leaves to be mistaken. In autumn, when they take on rich tints, and the clusters of "berries" become first crimson, then nearly black, the shrub is a delight to see.

To become familiar with one of the Viburnum bushes is to recognize any member of the tribe when in blossom or fruit, for all spread more or less flattened, compound cymes of white flowers in late spring or early summer, followed by red or very dark "berries" (drupes); but it is on the leaves that we depend to name a species. The opposite, slender petioled, pale leaves of the ARROW-WOOD or MEALY-TREE (V. dentalum), have no lobes; but are ovate, coarsely toothed, pointed at the tip, prominently pinnately veined. All the flowers in a cyme are perfect; and the drupes, which are at first blue, become nearly black when fully ripe. In moist, or even wet, ground, from the Georgia mountains, western New York, and Minnesota far northward, this smooth, slender, gray shrub is found. Its wood once furnished the Indians with arrows.

A much lower growing, but similar, bush, the DOWNY-LEAVED ARROW-WOOD (V. pubescens), formerly counted a mere variety of the preceding, may be known by the velvety down on the under side of its leaves. It grows in rocky, wooded places, often on some high bank above a stream. Beetles and the less specialized bees visit the flat-topped flower clusters abundantly in May. Short-tongued visitors quickly lick up the abundant nectar secreted at the base of each little style, cross-fertilizing their entertainers as they journey across the cyme. So widely do the anthers diverge, that pollen must often drop on the stigma of a neighboring floret, and quite as often a flower is likely to be self-fertilized through the curvature of the filaments.

The WITHE-ROD OR APPALACHIAN TEA (V. cassinoides; V. nudum of Gray) is found in swamps and wet ground from North

Carolina and Minnesota northward, flowering in May or June. Its dense clusters of perfect, small white flowers, on a rather short peduncle, are followed by oval "berries" that, although pink at first, soon turn a dark blue, with a bloom like the huckleberry's. The opposite, oval to oblong, rather thick, smooth leaves and the somewhat scurfy twigs help the novice to name this common shrub, whose tough, pliable branches make excellent binders for farmer's bundles, but whose leaves cannot be recommended as a substitute for tea.

Beautiful enough for any gentleman's lawn is the SWEET VIBURNUM, NANNY-BERRY, SHEEP-BERRY, or NANNY-BUSH, as it is variously called (V. Lentago). Indeed, its name appears in many nurserymen's catalogues. From Georgia, Indiana, and Missouri far northward it grows in rich, moist soil, sometimes attaining the height of a tree, more frequently that of a good-sized shrub. A profusion of dense white, broad flower clusters, seated among the rich green terminal leaves in May, indicate a feast for migrating birds and hungry beasts, including the omnivorous small boy in October, when the bluish-black, bloom-covered, sweet, edible "berries" ripen. A peculiarity of the ovate, long-tapering, and finely saw-edged leaves is that their long petioles often broaden out and become wavy margined.

Another Viburnum, with smooth, bluish-black, sweet, and edible fruit, that ripens a month earlier than the nanny-berry's, is the similar BLACK HAW, STAG-BUSH or SLOE (V. prunifolium). As its Latin name indicates, the leaves suggest those of a plum tree. It is a very early bloomer; the flat-topped white clusters appearing in April, and lasting through June, in various parts of its range from the Gulf States to southern New England and Michigan. Unlike the hobble-bush and the withe-rod, both the nanny-berry and the black haw have conspicuous winter buds, the latter bush often clothing its tender undeveloped foliage with warm-looking reddish down, although few of its naked kin have so southerly a range.

ONE-SEEDED, BUR- or STAR CUCUMBER; NIMBLE KATE

(Sicyos angulatus) Gourd family

Flowers - Small, greenish-white, 5-parted, of 2 kinds: staminate ones in a loose raceme on a very long peduncle; fertile ones clustered in a little head on a short peduncle. Stem: A climbing vine with branched tendrils; more or less sticky-hairy. Leaves: Broad, 5-angled or 5-lobed, heart-shaped at base, rough, sometimes enormous, on stout petioles. Fruit: From 3 to 10 bur-like, yellowish, prickly seed-vessels in a star-shaped cluster, each containing one seed. Preferred Habitat - Moist, shady waste ground; banks of streams. Flowering Season - June-September. Distribution - Quebec to the Gulf States, and westward beyond the Mississippi.

In a damp, shady, waste corner, perhaps the first weed to take possession is the star cucumber, a poor relation of the musk and water melons, the squash, cucumber, pumpkin, and gourd of the garden. Its sole use yet discovered is to screen ugly fences and rubbish heaps by climbing and trailing luxuriantly over everything within reach. That it thinks more highly of its own importance in the world than men do of it, is shown by the precaution it takes to insure a continuance of its species. By separating the sexes of its flowers, like Quakers at meeting, it prevents self-fertilization, and compels its small-winged visitors to carry the smooth-banded, rough pollen from the staminate to the tiny pistillate group. By roughening its angled stem and leaves, it discourages pilfering ants and other crawlers from reaching the sweets reserved for legitimate benefactors. So extremely sensitive are the tips of the tendrils that by rubbing them with the finger they will coil up perceptibly; then straighten out again if they find they have been deceived, and that there is no stick for them to twine around. Give them a stick, however, and the coils remain fixed.

RATTLESNAKE-ROOT; WHITE LETTUCE or CANKER-WEED;
LION'S-FOOT
 (Nabalus albus) Chickory family

Flower-heads - Composite, numerous, greenish or cream white, or tinged with lilac, fragrant, nodding; borne in loose, open, narrow terminal, and axillary clusters. Each bell-like flowerhead only about 1/4 in. across, composed of 8 to 15 ray flowers, drooping from a cup-like involucre consisting of 8 principal, colored bracts. Stem: 2 to 5 ft. high, smooth, green or dark purplish red, leafy, from a tuberous, bitter root. Leaves: Alternate, variable, sometimes very large, broad, hastate, ovate, or heart-shaped, wavy-toothed, lobed, or palmately cleft; upper leaves smaller, lance-shaped, entire. Preferred Habitat - Woods; rich, moist borders; roadsides. Flowering Season - August-September. Distribution - Southern Canada to Georgia and Kentucky.

Nodding in graceful, open clusters from the top of a shining colored stalk, the inconspicuous little bell-like flowers of this common plant spread their rays to release the branching styles for contact with pollen-laden visitors. These styles presently become a bunch of cinnamon-colored hairs, a seed-tassel resembling a sable paint brush - the principal feature that distinguishes this species from the smaller-flowered TALL WHITE LETTUCE (N. altissimus), whose pappus is a light straw color. Both these plants are most easily recognized when their fluffy, plumed seeds are waiting for a stiff breeze to waft them to fresh colonizing ground.

BONESET; COMMON THOROUGHWORT; AGUE-WEED; INDIAN SAGE

(Eupatorium perfoliatum) Thistle family

Flower-heads - Composite, the numerous, small, dull, white heads of tubular florets only, crowded in a scaly involucre and borne in spreading, flat-topped terminal cymes. Stem: Stout, tall, branching above, hairy, leafy. Leaves: Opposite, often united at their bases, or clasping, lance-shaped, saw-edged, wrinkled. Preferred Habitat - Wet ground, low meadows, roadsides. Flowering Season - July-September. Distribution - From the Gulf States north to Nebraska, Manitoba, and New Brunswick.

Frequently, in just such situations as its sister the Joe-Pye weed selects (q.v.), and with similar intent, the boneset spreads its soft, leaden-white bloom; but it will be noticed that the butterflies, which love color, especially deep pinks and magenta, let this plant alone, whereas beetles, that do not find the butterfly's favorite, fragrant Joe-Pye weed at all to their liking, prefer these dull, odorous flowers. Many flies, wasps, and bees also, get generous entertainment in these tiny florets, where they feast with the minimum loss of time, each head in a cluster containing, as it does, from ten to six-teen restaurants. An ant crawling up the stem is usually discoura-ged by its hairs long before reaching the sweets. Sometimes the stem appears to run through the center of one large leaf that is kinky in the middle and taper-pointed at both ends, rather than between a pair of leaves.

An old-fashioned illness known as break-bone fever - doubtless paralleled to-day by the grippe - once had its terrors for a patient increased a hundredfold by the certainty he felt of taking nauseous doses of boneset tea, administered by zealous old women outside the "regular practice." Children who had to have their noses held before they would - or, indeed, could - swallow the decoction, cheerfully munched boneset taffy instead.

The bright white, wide-spread inflorescence of the WHITE SNA-KEROOT, WHITE or INDIAN SANICLE, or DEERWORT BONE-SET (E. ageratoides) is displayed from July to November in the hope of getting relief from the fiercest competition for the visits of butter-flies, honey and other small bees, wasps, and flies. From July to September the vast army of composites appear in such hopeless predominance that prolonged bloom on the part of any of their number is surely an advantage. In the rich, moist woods, or by shady roadsides, where it prefers to dwell, the white sanicle makes a fine show. Above its fringy bloom how often one sees the exquisi-te little lavender-blue butterflies (Lycaena pseudargiolus) pausing an instant to drain the tiny cups of nectar, and usually transferring pollen from the protruding styles (q.v.) as they flit to another clus-ter.

The opposite, petioled leaves, broadly oval at the base, taper-pointed, coarsely toothed, three-nerved, and veiny, are thin and easily skeletonized by the insects that enjoy the leaves of all this clan of plants. From one to four feet high, the White Snakeroot grows in the United States and Canada as far west as Nebraska.

Closely allied to the eupatoriums, and with similar inflorescence, is the CLIMBING BONESET or HEMPWEED (Willughbaeaa scan-dens; Mikania scandens of Gray.) Straggling over bushes in swamps, by the brookside thicket, or in moist, shady roadsides, the vine reveals its kinship to the boneset instantly it comes into bloom in midsummer, although its flower clusters are occasionally pinkish. The opposite, petioled leaves are quite different from the boneset's, however, being heart-shaped at the base, and taper-pointed, somewhat triangular, two to four inches long, and one or two inches wide. From Massachusetts and the Middle States even to South America and the West Indies is its range.

WHITE ASTERS or STARWORTS
  (Aster = a star) Thistle family

In dry, open woodlands, thickets, and roadsides, from August to October, we find the dainty WHITE WOOD ASTER (A. divaricatus; A. corymbosus of Gray) its brittle zig-zag stem two feet high or less, branching at the top, and repeatedly forked where loose clusters of flower-heads spread in a broad, rather flat corymb. Only a few white rays - usually from six to nine - surround the yellow disk, whose forets soon turn brown. Range from Canada southward to Tennessee.

First to bloom among the white species, beginning in July, is the UPLAND WHITE ASTER (A. ptarmicoides), which elects to grow in the rocky or dry soil of high ground in the northern United States westward to Colorado. The leaves, which resemble grayish-green shining grass-blades, arranged alternately up the rigid stem, and diminishing in size near the top until they become mere bracts among the flowers, enable us to name the plant. The heads, in a branching cluster, are not numerous; each measures barely an inch across its ten to twenty snow-white rays; the center is of a pale yellow-green, turning a light brown in maturity.

The TALL WHITE or PANICLED ASTER (A. paniculatus), in bloom from August to October in different parts of its wide range, attracts great numbers of beetles, which do it more harm than good; but many more butterflies (some of whose caterpillars feed on aster foliage as a staple), quantities of flies, some moths, swarms of bees, wasps, and miscellaneous winged visitors. Professor Robertson found several thousand callers, representing ninety-eight distinct species, on this one aster during four October days. Such popularity as the asters have attained finds its just reward in the triumphant progress of the lovely tribe (q.v.). For the amateur to name each member of such a horde is quite hopeless. In branching, raceme-like clusters, from August to October, this aster displays its numerous flower-heads, less than an inch across, each with a green cup formed of four or five series of overlapping bracts, and many white rays, occasionally violet tipped. The smooth stem, which rises from two to eight feet above moist soil, is plentifully set with alternate, pointed-tipped, lance-shaped leaves, tapering to a sessile or partly clasping base, and sparingly saw-edged. Its range is from Montana east to Virginia, south to Louisiana, north to Ontario and New England.

The bushy little WHITE HEATH ASTER (A. ericoides) every one must know, possibly, as MICHAELMAS DAISY, FAREWELL SUMMER, WHITE ROSEMARY, or FROSTWEED; for none is commoner in dry soil, throughout the eastern United States at least. Its smooth, much branched stem rarely reaches three feet in height, usually it is not over a foot tall, and its very numerous flower-heads, white or pink tinged, barely half an inch across, appear in such profusion from September even to December as to transform it into a feathery mass of bloom.

Growing like branching wands of golden rod, the DENSE-FLOWERED, WHITE-WREATHED, or STARRY ASTER (A. multiflorus) bears its minute flower-heads crowded close along the branches, where many small, stiff leaves, like miniature pine needles, follow them. Each flower measures only about a quarter of an inch across. From Maine to Georgia and Texas westward to Arizona and British Columbia the common bushy plant lifts its rather erect, curving, feathery branches perhaps only a foot, sometimes above a man's head, from August till November, in such dry, open, sterile ground as the white heath aster also chooses.

No one not a latter-day, structural botanist could see why the TALL, FLAT-TOP WHITE ASTER (Doellingeria umbella) is now an outcast from the aster tribe into a separate genus. This common species of moist soil and swamps has its numerous small heads (containing ten to fifteen rays each) arranged in large, terminal, compound clusters (corymbs). The stem, which rises from two to eight feet, has its long-tapering, alternate leaves, hairy on the veins beneath and rough margined.

Late in the fall you may hear the rich tone of a Bombilius, one of the commonest flies seen about flowers, as he darts rapidly among the white asters. Unless you have been initiated, you may mistake this fly for a bee. He sings a very similar song and wears a similar dress; but he is not a very good imitation, after all, and a little familiarity with him will give you courage to catch him in your hand if you are quick enough, for he is incapable of stinging or biting: he can merely make a noise out of all proportion to his size. He is simply living from hour to hour, and lays up no store for the winter, enjoying more or less security from his resemblance to the industrious and dangerous insect which he imitates.

DAISY FLEABANE; SWEET SCABIOUS
(Erigeron annus) Thistle family

Flower-heads - Numerous, daisy-like, about 1/2 in. across; from 40 to 70 long, fine, white rays (or purple- or pink-tinged), arranged around yellow disk florets in a rough, hemispheric cup whose bracts overlap. Stem: Erect, to 4 ft. high, branching above, with spreading, rough hairs. Leaves: Thin, lower ones ovate, coarsely toothed, petioled; upper ones sessile, becoming smaller, lance-shaped. Preferred Habitat: Fields, wasteland, roadsides. Flowering Season: May-November. Distribution: Nova Scotia to Virginia, westward to Missouri.

At a glance one knows this flower to be akin to Robin's plantain (q.v.) the the asters and daisy. A smaller, more delicate species, with mostly entire leaves and appressed hairs (E. ramosus; E. strigosum of Gray) has a similar range and season of bloom. Both soon grow hoary-headed after they have been fertilized by countless insects crawling over them (Erigeron = early old). That either of these plants, or the pinkish, small-flowered, strong-scented SALT-MARSH FLEABANE (Pluchea camphorata), drive away fleas, is believed only by those who have not used them dried, reduced to powder, and sprinkled in kennels, from which, however, they have been known to drive away dogs.

GROUNDSEL-BUSH or -TREE; PENCIL-TREE
(Baccharis halimifolia) Thistle family

Flower-heads: White or yellowish tubular florets, 1 to 5 in peduncled clusters. Staminate and pistillate clusters on different shrubs;
the former almost round at first, the latter conspicuous only when
seeding; then their pappus is white, and about 1/3 in. long. Stem: A
smooth, branching shrub, 3 to 10 ft. high. Leaves: Thick, lower ones
ovate to wedge-shaped, coarsely angular-toothed; upper ones smaller, few-toothed or entire. Preferred Habitat: Salt marshes, tidewater
streams, often far from the coast. Flowering Season: September-
November Distribution: The Atlantic and Gulf coasts from Maine to
Texas.

When the little bright white, silky cockades, clustered at the ends
of the branches, appear on a female groundsel-bush in autumn, our
eyes are attracted to the shrub for the first time. But had not small
pollen carriers discovered it weeks before, the scaly, glutinous cups
would hold no charming, plumed seeds ready to ride on autumn
gales. Self-fertilization has been guarded against by precarious means, but the safest of all devices - separation of the sexes on distinct
plants. These are absolutely dependent, of course, on insect messengers - not visitors merely. Bees, which always show less inclina-
tion to dally from one species of flower to another than any other
guests, and more intelligent directness of purpose when out for
business are the groundsel-bush's truest benefactors. This is the
only shrub among the multitudinous composite clan that most of us
are ever likely to see.

PEARLY or LARGE-FLOWERED EVERLASTING; IMMORTELLE;
SILVER LEAF;
MOONSHINE; COTTON-WEED; NONE-SO-PRETTY

(Anaphalis margaritacea; Antennaria margaritacea of Gray)
Thistle family

Flower-heads - Numerous pearly-white scales of the involucre
holding tubular florets only; borne in broad, rather flat, compound
corymbs at the summit. Stem: Cottony, to 3 ft. high, leafy to the top.

Leaves: Upper ones small, narrow, linear; lower ones broader, lance-shaped, rolled backward, more or less woolly beneath. Preferred Habitat - Dry fields, hillsides, open woods, uplands. Flowering Season - July-September. Distribution - North Carolina, Kansas, and California, far north.

When the small, white, overlapping scales of an everlasting's oblong involucre expand stiff and straight, each pert little flower-head resembles nothing so much as a miniature pond lily, only what would be a lily's yellow stamens are in this case the true flowers, which become brown in drying. It will be noticed that these tiny florets, so well protected in the center, are of two different kinds, separated on distinct heads: the female florets with a tubular, five-cleft corolla, a two-cleft style, and a copious pappus of hairy bristles; the staminate, or male, florets more slender, the anthers tailed at the base. Self-fertilization being, of course, impossible under such an arrangement, the florets are absolutely dependent upon little winged pollen carriers, whose sweet reward is well protected for them from pilfering ants by the cottony substance on the wiry stem, a device successfully employed by thistles also (q.v.).

An imaginary blossom that never fades has been the dream of poets from Milton's day; but seeing one, who loves it? Our amaranth has the aspect of an artificial flower - stiff, dry, soulless, quite in keeping with the decorations on the average farmhouse mantelpiece. Here it forms the most uncheering of winter bouquets, or a wreath about flowers made from the lifeless hair of some dear departed.

In open, rocky places, moist or dry, the CLAMMY EVERLASTING, SWEET BALSAM, OR WINGED CUDWEED (Gnaphalium decurrens) prefers to dwell. A wholesome fragrance, usually mingled with that of sweet fern, pervades its neighborhood. Its yellowish-white little flower-heads clustered at the top of an erect stem, and its pale sage-green leaves, densely woolly beneath, the lower ones seeming to run along the stem, need no further description: every one knows the common everlasting. Its right to the Greek generic name, meaning a lock of wool, no one will dispute. From Pennsylvania and Arizona, north to Nova Scotia and British Columbia, its amaranthine flowers are displayed from July to September, the staminate and the pistillate heads on distinct plants. Many insect visitors approach the flowers; some, like the bees, are working for them in transferring pollen; others, like the ants, which are trying to steal nectar, usually getting killed on the sticky, cottony stem; and, hovering near, ever conspicuous among the larger visitors, is the beautiful hunter's butterfly (Pyrameis huntera), to be distinguished from its sister the painted lady, always seen about thistles, by the two large eye-like spots on the under side of the hind wings. What are these butterflies doing about their chosen plants? Certainly the minute florets of the everlasting offer no great inducements to a creature that lives only on nectar. But that cocoon, compactly woven with silk and petals, which hangs from the stem, tells the story of the hunter's butterfly's presence. A brownish-drab chrysalis, or a slate-colored and black-banded little caterpillar with tufts of hairs on its back, and pretty red and white dots on the dark stripes, shows our butterfly in the earlier stages of its existence, when the everlastings form its staple diet.

When the hepatica, arbutus, saxifrage, and adder's tongue are running for first place among the earliest spring flowers, another modest little competitor joins the race - the DWARF EVERLASTING (Antennaria plantaginifolia), also known as PLANTAIN-LEAVED, MOUSE-EAR, SPRING or EARLY EVERLASTING, WHITE PLANTAIN, PUSSY-TOES and LADIES' TOBACCO. From March to June, in different parts of its wide range, rocky fields, hillsides, and dry, open woods are whitened with broad patches of it, formed by runners; the fertile plants from six to eighteen inches high; the male plants, in distinct patches, smaller throughout. At the base the tufted leaves, which are green on the upper side, but silvery beneath, often woolly when young, are broadly oval or spatulate, the upper leaves oblong to lance-shaped, seated on the woolly stem. Charming little rosettes remain all winter, ready to send up the first flowers displayed by the vast host of composites. Several little heads of fertile florets, resembling tufts of silvery-white silk, are set in pale-greenish cups in a broad cluster at the top of the stem; the staminate florets in whiter cups with more rounded scales. Small bees, chiefly those of the Andrena and Halictus tribe, and many flies, attend to transferring pollen. Our friend, the hunter's butterfly, also hovers near. Range from Labrador to the Gulf of Mexico, westward to Nebraska.

YARROW; MILFOIL; OLD MAN'S PEPPER; NOSEBLEED
(Achillea Millefolium) Thistle family

Flower-heads - Grayish-white, rarely pinkish, in a hard, close, flat-topped, compound cluster. Ray florets 4 to 6, pistillate, fertile; disk florets yellow, afterward brown, perfect, fertile. Stem: Erect, from horizontal rootstalk, 1 to 2 ft. high, leafy, sometimes hairy. Leaves: Very finely dissected (Millefolium = thousand leaf), narrowly oblong in outline. Preferred Habitat - Waste land, dry fields, banks, roadsides. Flowering Season - June-November. Distribution - Naturalized from Europe and Asia throughout North America.

Everywhere this commonest of common weeds confronts us; the compact, dusty-looking clusters appearing not by waysides only, around the world, but in the mythology, folklore, medicine, and literature of many peoples. Chiron, the centaur, who taught its virtues to Achilles that he might make an ointment to heal his Myrmidons wounded in the siege of Troy, named the plant for this favorite pupil, giving his own to the beautiful blue corn-flower (Centaurea Cyanus). As a love-charm; as an herb-tea brewed by crones to cure divers ailments, from loss of hair to the ague; as an inducement to nosebleed for the relief of congestive headache; as an ingredient of an especially intoxicating beer made by the Swedes, it is mentioned in old books. Nowadays we are satisfied merely to admire the feathery masses of lace-like foliage formed by young plants, to whiff the wholesome, nutty, autumnal odor of its flowers, or to wonder at the marvelous scheme it employs to overrun the earth.

Like the daisy, each small flower in a cluster, as symmetrically arranged as brain coral, is made up of a large number of minute but perfect florets, suited to attract insects by making a better show than each could do alone, and by offering them accessible feeding places close together, where they may feast with minimum loss of time. Simultaneous cross-fertilization of many florets must be effected by every visitor crawling over a cluster. The florets in each disk open in regular array toward the centers. At the expense of stamens, which are absent in the grayish-white ray florets, they have attained their development, another instance of "progress by loss" from the evolutionary standpoint. By prolonging its season of bloom to get relief from the fierce competition for insect visitors in midsummer; by increase through seeds, and runners too; by contenting itself with neglected corners of the earth, the yarrow gives us many valuable lessons on how to succeed.

DOG'S or FETID CAMOMILE; MAYWEED; PIG-STY DAISY; DILLWEED;
DOG-FENNEL

(Anthemis Cotula; Maruta Cotula of Gray) Thistle family

Flower-heads - Like smaller daisies, about 1 in. broad; 10 to 18 white, notched, neutral ray florets around a convex or conical yellow disk, whose florets are fertile, containing both stamens and pistil, their tubular corollas 5-cleft. Stem: Smooth, much branched, 1 to 2 ft. high, leafy, with unpleasant odor and acrid taste. Leaves: Very finely dissected into slender segments. Preferred Habitat - Roadsides, dry wasteland, sandy fields. Flowering Season - June-November. Distribution - Throughout North America, except in circumpolar regions.

"Naturalized from Europe, and widely distributed as a weed in Asia, Africa, and Australasia" (Britton and Brown's "Flora"). Little wonder the camomile encompasses the earth, for it imitates the triumphant daisy, putting into practice those business methods of the modern department store, by which the composite horde have become the most successful strugglers for survival.

The unpleasant odor given forth by this bushy little plant repels bees and other highly organized insects; not so flies, which, far from objecting to a fetid smell, are rather attracted by it. They visit the camomile in such numbers as to be the chief fertilizers. As the development of bloom proceeds toward the center, the disk becomes conical, to present the newly opened florets, where a fly alighting on it must receive pollen, to be transferred as he crawls and flies to another head. After fertilization the white rays droop. Dog, used as a prefix by several of the plant's folk names, implies contempt for its worthlessness. It is quite another species, the GARDEN CAMOMILE (A. nobilis) which furnishes the apothecary with those flowers which, when steeped into a bitter aromatic tea, have been supposed for generations to make a superior tonic and blood purifier.

Not so common a plant here, but almost as widespread as the preceding species, is the similar, but not fetid, CORN or FIELD CAMOMILE (A. arvensis), a pest to European farmers. Both are closely related to the garden FEVERFEW, FEATHERFEW, OR PELLITORY (Chrysanthemum Parthenium), which escapes from cultivation whenever it can into waste fields and roadsides.

COMMON DAISY; WHITE-WEED; WHITE OR OX-EYE DAISY; LOVE-ME, LOVE-ME-NOT

(Chrysanthemum Leucanthemum) Thistle family

Flower-heads - Disk florets yellow, tubular, 4 or 5 toothed, containing stamens and pistil; surrounded by white ray florets, which are pistillate, fertile. Stem: Smooth, rarely branched, to 3 ft. high. Leaves: Mostly oblong in outline, coarsely toothed and divided. Preferred Habitat - Meadows, pastures, roadsides, wasteland. Flowering Season - May-November. Distribution - Throughout the United States and Canada; not so common in the South and West.

Myriads and myriads of daisies, whitening our fields as if a belated blizzard had covered them with a snowy mantle in June, fill the farmer with dismay, the flower-lover with rapture. When vacation days have come; when chains and white-capped old women are to be made of daisies by happy children turned out of schoolrooms into meadows; when pretty maids, like Goethe's Marguerite, tell their fortunes by the daisy "petals;" when music bubbles up in a cascade of ecstasy from the throats of bobolinks nesting among the daisies, timothy, and clover; when the blue sky arches over the fairest scenes the year can show, and all the world is full of sunshine and happy promises of fruition, must we Americans always go to English literature for a song to fit our joyous mood?

"When daisies pied, and violets blue,
    And lady-smocks all silver white,
    And cuckoo-buds of yellow hue,

Do paint the meadows with delight-"

sang Shakespeare. His lovely suggestion of an English spring recalls no familiar picture to American minds. No more does Burns's

"Wee, modest crimson-tippit flower."

Shakespeare, Burns, Chaucer, Wordsworth, and all the British poets who have written familiar lines about the daisy, extolled a quite different flower from ours - Bellis perennis, the little pink and white blossom that hugs English turf as if it loved it - the true day's-eye, for it closes at nightfall and opens with the dawn.

Now, what is the secret of the large, white daisy's triumphal conquest of our territory? A naturalized immigrant from Europe and Asia, how could it so quickly take possession? In the overcultivated Old World no weed can have half the chance for unrestricted colonizing that it has in our vast unoccupied area. Most of our weeds are naturalized foreigners, not natives. Once released from the harder conditions of struggle at home (the seeds being safely smuggled in among the ballast of freight ships, or hay used in packing), they find life here easy, pleasant; as if to make up for lost time, they increase a thousandfold. If we look closely at a daisy - and a lens is necessary for any but the most superficial acquaintance - we shall see that, far from being a single flower, it is literally a host in itself. Each of the so-called white "petals" is a female floret, whose open corolla has grown large, white, and showy, to aid its sisters in advertising for insect visitors - a prominence gained only by the loss of its stamens. The yellow center is composed of hundreds of minute tubular florets huddled together in a green cup as closely as they can be packed. Inside each of these tiny yellow tubes stand the stamens, literally putting their heads together. As the pistil within the ring of stamens develops and rises through their midst, two little hair brushes on its tip sweep the pollen from their anthers as a rounded brush would remove the soot from a lamp chimney. Now the pollen is elevated to a point where any insect crawling over the floret must remove it. The pollen gone, the pistil now spreads its two arms, that were kept tightly closed together while any danger of self-fertilization lasted. Their surfaces become sticky, that pollen brought from another flower may adhere to them. Notice that the pistils in the white ray florets have no hairbrushes on their tips, because, no stamens being there, there is no pollen to be swept out. Because daisies are among the most conspicuous of flowers, and have facilitated dining for their visitors by offering them countless cups of refreshment that may be drained with a minimum loss of time, almost every insect on wings alights on them sooner or later. In short, they run their business on the principle of a cooperative department store. Immense quantities of the most vigorous, because cross-fertilized, seed being set in every patch, small wonder that our fields are white with daisies - a long and a merry life to them!

Since all flowers must once have passed through a white stage before attaining gay colors, so evolution teaches, it is not surprising that occasional reversions to the white type should be found even among the brightest-hued species. Again, some white flowers which are in a transition state show aspirations after color, often so marked in individuals as to mislead one into believing them products of a far advanced colored type. Also, pale colors blanch under a summer sun. These facts must be borne in mind, and the blue, pink, and yellow blossoms should be investigated before the reader despairs of identifying a flower not found in the white group.

YELLOW AND ORANGE FLOWERS

"All variations which render the blossoms more attractive, either by scent, color, size of corolla, or quantity of nectar, make the insect visit more sure, and therefore the production of seed more likely. Thus, the conspicuous blossoms secure descendants which inherit the special variations of their parents, and so, generation after generation, we have selections in favor of conspicuous flowers, where insects are at work. Their appreciation of color, because it has brought the blossom possessing it more immediately into their view, and more surely under their attention, has enabled them, through the ages, to be preparing the specimens upon which man now operates, he taking up the work where they have left it, selecting, inoculating, and hybridizing, according to his own rules of taste, and developing a beauty which insects alone could never have evolved. His are the finishing touches, his the apparent effects, yet no less is it true, that the results of his floriculture would never have been attainable without insect helpers. It is equally certain, that the beautiful perfume, and the nectar also, are, in their present development, the outcome of repeated insect selection, and here, it seems to me, we get an inkling of a deep mystery: Why is life, in all its forms, so dependent upon the fusion of two individual elements? Is it not, that thus the door of progress has been opened? If each alone had reproduced, itself all-in-all, advance would have been impossible, the insect and human florists and pomologists, like the improvers of animal races, would have had no platform for their operation, and not only the forms of life, but life itself would have been stereotyped unalterably, ever mechanically giving repetition to identical phenomena." - Frank R. Cheshire in "Bees and Beekeeping."

YELLOW AND ORANGE FLOWERS
GOLDEN CLUB
 (Orontium aquaticum) Arum family

Flowers - Bright yellow, minute, perfect, crowded on a spadix (club) 1 to 2 in. long; the scape, 6 in. to 2 ft. tall, flattened just below it; the club much thickened in fruit. Leaves: All from root, petioled, oblong-elliptic, dull green above, pale underneath, 5 to 12 in. long, floating or erect. Preferred Habitat - Shallow ponds, standing water, swamps. Flowering Season - April-May. Distribution - New England to the Gulf States, mostly near the coast.

A first cousin of cruel Jack-in-the-pulpit, the skunk cabbage, and the water-arum (q.v.), a poor relation also of the calla lily, the golden club seems to be denied part of its tribal inheritance - the spathe, corresponding to the pulpit in which Jack preaches, or to the lily's showy white skirt. In the tropics, where the lily grows, where insect life teems in myriads and myriads, and competition among the flowers for their visits is infinitely more keen than here, she has greater need to flaunt showy clothes to attract benefactors than her northern relatives. But the golden club, which looks something like a calla stripped of her lovely white robe, has not lacked protection for its little buds from the cold spring winds while any was needed. By the time we notice the plant in bloom, however, its bract-like spathe has usually fallen away, as if conscious that the pretty mosaic club of golden florets, so attractive in itself, was quite able to draw all the visitors needed without further help. Merely by crawling over the clubs, flies and midges cross-fertilize them.

PERFOLIATE BELLWORT; STRAW BELL
 (Uvularia perfoliala) Bunch-flower family

Flowers - Fragrant, pale yellow, about 1 in. long, drooping singly (rarely 2) from tips of branches; perianth narrow, bell-shaped, of 6 petal-like segments, rough within, spreading at the tip; 6 stamens; 3 styles united to the middle. Stem: 6 to 20 in. high, smooth, shining, forking about half way. Leaves: Apparently strung on the slender stem, oval, tapering at tip. Preferred Habitat - Moist, rich woods; thickets. Flowering Season - May-June. Distribution - Quebec to the Gulf of Mexico, west to Mississippi.

Hanging like a palate (uvula) from the roof of a mouth, according to imaginative Linnaeus, the little bellwort droops, and so modestly hides behind the leaf its footstalk pierces that the eye often fails to find it when so many more showy blossoms arrest attention in the May woods. Slight fragrance helps to guide the keen bumblebee to the pale yellow bell. The tips spreading apart very little and the flower being pendent, how is she to reach the nectar secreted at the base of each of its six divisions? Is it not more than probable that the inner surface is rough, as if dusted with yellow meal, to provide a foothold for her as she clings? Now securely hanging from within the inhospitable flower, her long tongue can easily drain the sweets, and in doing so she will receive pollen, to be deposited, in all probability, on the stigmatic style branches of the next bellwort entered.

With a more westerly range than the perfoliate species, the similar LARGE-FLOWERED BELLWORT (U. grandiflora) grows in like situations. Its greenish lemon-yellow flowers, an inch to an inch and a half long, appear from April to May, or when the female bumblebees, that fly before their lords, are the only insects large and strong enough to force an entrance. Mr. Trelease, who noted them on the flowers near Madison, Wisconsin, saw that one laden with pollen from another blossom came in contact with the three sticky branches of the style, protruding between the anthers, when she crawled between the anthers and sepals, as she must, to reach the nectar secreted at the base. But the linear anthers shedding their pollen longitudinally, there is a chance that the flower may fertilize itself should no bee arrive before a certain point is reached.

The SESSILE-LEAVED BELLWORT, or WILD OAT (U. sessifolia), as its name implies, has its thin, pale green leaves tapering at either end, seated on the stem, not surrounding it, or apparently strung on it. The smaller flower is cream colored. A sharply three-angled capsule about an inch long follows. Range from Minnesota and Arkansas to the Atlantic.

WILD YELLOW, MEADOW or FIELD LILY; CANADA LILY
 (Lilium Canadense) Lily family

Flowers - Yellow to orange-red, of a deeper shade within, and speckled with dark reddish-brown dots. One or several (rarely many) nodding on long peduncles from the summit. Perianth bell-shaped, of 6 spreading segments 2 to 3 in. long, their tips curved backward to the middle; 6 stamens, with reddish-brown linear anthers; 1 pistil, club-shaped; the stigma 3-lobed. Stem: 2 to 5 ft. tall, leafy, from a bulbous rootstock composed of numerous fleshy white scales. Leaves: Lance-shaped, to oblong; usually in whorls of fours to tens, or some alternate. Fruit: An erect, oblong, 3-celled capsule, the flat, horizontal seeds packed in 2 rows in each cavity. Preferred Habitat - Swamps, low meadows; moist fields. Flowering Season - June-July. Distribution - Nova Scotia to Georgia, westward beyond the Mississippi.

Not our gorgeous lilies that brighten the low-lying meadows in early summer with pendent, swaying bells; possibly not a true lily at all was chosen to illustrate the truth which those who listened to the Sermon on the Mount, and we, equally anxious, foolishly over-burdened folk of to-day, so little comprehend.

"Consider the lilies of the field, how they grow;
  they toil not, neither do they spin
And yet I say unto you,
That even Solomon in all his glory
  was not arrayed like one of these."

Opinions differ as to the lily of Scripture. Eastern peoples use the same word interchangeably for the tulip, anemone, ranunculus, iris, the water-lilies, and those of the field. The superb Scarlet Martagon Lily (L. chalcedonicum), grown in gardens here, is not uncommon wild in Palestine; but whoever has seen the large anemones there "carpeting every plain and luxuriantly pervading the land" is inclined to believe that Jesus, who always chose the most familiar objects in the daily life of His simple listeners to illustrate His teachings, rested His eyes on the slopes about Him glowing with anemones in all their matchless loveliness. What flower served Him then matters not at all. It is enough that scientists - now more plainly than ever before - see the universal application of the illustration the more deeply they study nature, and can include their "little brothers of the air" and the humblest flower at their feet when they say with Paul, "In God we live and move and have our being."

Tallest and most prolific of bloom among our native lilies, as it is the most variable in color, size, and form, the TURK'S CAP, or TURBAN LILY (L. superburn), sometimes nearly merges its identity into its Canadian sister's. Travelers by rail between New York and Boston know how gorgeous are the low meadows and marshes in July or August, when its clusters of deep yellow, orange, or flame-colored lilies tower above the surrounding vegetation. Like the color of most flowers, theirs intensifies in salt air. Commonly from three to seven lilies appear in a terminal group; but under skilful cultivation even forty will crown the stalk that reaches a height of nine feet where its home suits it perfectly; or maybe only a poor array of dingy yellowish caps top a shriveled stem when unfavorable conditions prevail. There certainly are times when its specific name seems extravagant.

Its range is from Maine to the Carolinas, westward to Minnesota and Tennessee. A well-conducted Turk's cap is not bell-shaped at maturity, like the Canada lily: it should open much farther, until the six points of its perianth curve so far backward beyond the middle as to expose the stamens for nearly their entire length. One of the purple-dotted divisions of the flower when spread out flat may measure anywhere from two and a half to four inches in length. Smooth, lance-shaped leaves, tapering at both ends, occur in whorls of threes to eights up the stem, or the upper ones may be alternate. Abundant food, hidden in a round, white-shingled storehouse under ground, nourishes the plant, and similarly its bulb-bearing kin, when emergency may require - a thrifty arrangement that serves them in good stead during prolonged drought and severe winters.

Why, one may ask, are some lilies radiantly colored and speckled; others, like the Easter lily, deep chaliced, white, spotless? Now, in all our lily kin nectar is secreted in a groove at the base of each of the six divisions of the flower, and upon its removal by that insect best adapted to come in contact with anthers and stigma as it flies from lily to lily depends all hope of perpetuating the lovely race. For countless ages it has been the flower's business to find what best pleased the visitors on whom so much depended. Some lilies decided to woo one class of insects; some, another. Those which literally set their caps for color-loving bees and butterflies whose long tongues could easily drain nectar deeply hidden from the mob for their special benefit, assumed gay hues, speckling the inner side of their spreading divisions, even providing lines as pathfinders to their nectaries in some cases, lest a visitor try to thrust in his tongue between the petal-like parts while standing on the outside, and so defeat their well-laid plan. It is almost pathetic to see how bright and spotted they are inside, that the visitor may not go astray. Thus we find the chief pollenizers of the Canada and the Turk's cap lilies to be specialized bees, the interesting upholsterers, or 1eaf-cutters, conspicuous among the throng. Nectar they want, of course; but the dark, rich pollen is needed also to mix with it for the food supply of a generation still unborn. Anyone who has smelled a lily knows how his nose looks afterward. The bees have no difficulty whatever in removing lily pollen and transferring it. So much for the colored lilies.

The long, white, trumpet-shape type of lily chooses for her lover the sphinx moth. For him she wears a spotless white robe - speckles would be superfluous - that he may see it shine in the dusk, when colored flowers melt into the prevailing blackness; for him she breathes forth a fragrance almost overwhelming at evening, to guide him to her neighborhood from afar; in consideration of his very long, slender tongue she hides her sweets so deep that none may rob him of it, taking the additional precaution to weld her six once separate parts together into a solid tube lest any pilferer thrust in his tongue from the side.

The common orange-tan DAY LILY (Hemerocallis fulva) and the commoner speckled, orange-red TIGER LILY (L. tigrinum) are not slow in seizing opportunities to escape from gardens into roadsides and fence corners.

YELLOW ADDER'S TONGUE; TROUT LILY; DOG-TOOTH "VIOLET"
  (Erythronium Americanum) Lily family

Flower - Solitary, pale russet yellow, rarely tinged with purple, slightly fragrant, 1 to 2 in. long, nodding from the summit of a footstalk 6 to 12 in. high, or about as tall as the leaves. Perianth bell-shaped, of 6 petal-like, distinct segments, spreading at tips, dark spotted within; 6 stamens; the club-shaped style with 3 short, stigmatic ridges. Leaves: 2, unequal, grayish green, mottled and streaked with brown or all green, oblong, 3 to 8 in. long, narrowing into clasping petioles. Preferred Habitat - Moist open woods and thickets, brooksides. Flowering Season - March-May Distribution - Nova Scotia to Florida, westward to the Mississippi.

Colonies of these dainty little lilies, that so often grow beside leaping brooks where and when the trout hide, justify at least one of their names; but they have nothing in common with the violet or a dog's tooth. Their faint fragrance rather suggests a tulip; and as for the bulb, which in some of the lily-kin has tooth-like scales, it is in this case a smooth, egg-shaped corm, producing little round offsets from its base. Much fault is also found with another name on the plea that the curiously mottled and delicately pencilled leaves bring to mind, not a snake's tongue, but its skin, as they surely do. Whoever sees the sharp purplish point of a young plant darting above ground in earliest spring, however, at once sees the fitting application of adder's tongue. But how few recognize their plant friends at all seasons of the year!

Every one must have noticed the abundance of low-growing spring flowers in deciduous woodlands, where, later in the year, after the leaves overhead cast a heavy shade, so few blossoms are to be found, because their light is seriously diminished. The thrifty adder's tongue, by laying up nourishment in its storeroom underground through the winter, is ready to send its leaves and flower upward to take advantage of the sunlight the still naked trees do not intercept, just as soon as the ground thaws. But the spring beauty, the rue-anemone, bloodroot, toothwort, and the first blue violet (palmata) among other early spring flowers, have not been slow to take advantage of the light either. Fierce competition, therefore, rages among them to secure visits from the comparatively few insects then flying - a competition so severe that the adder's tongue often has to wait until afternoon for the spring beauty to close before receiving a single caller. Hive-bees, and others only about half their size, of the Andrena and Halictus clans, the first to fly, the Bombylius frauds, and common yellow butterflies, come in numbers then. Guided by the speckles to the nectaries at the base of the flower, they must either cling to the stamens and style while they suck, or fall out. Thus cross-fertilization is commonly effected; but in the absence of insects the lily can fertilize itself. Crawling pilferers rarely think it worthwhile to slip and slide up the smooth footstalk and risk a tumble where it curves to allow the flower to nod - the reason why this habit of growth is so popular. The adder's tongue, which is extremely sensitive to the sunlight, will turn on its stalk to follow it, and expand in its warmth. At night it nearly closes.

A similar adder's tongue, bearing a white flower, purplish tinged on the outside, yellow at the base within to guide insects to the nectaries, is the WHITE ADDER'S TONGUE (E. albidum), rare in the Eastern States, but quite common westward as far as Texas and Minnesota.

YELLOW CLINTONIA
(Clintonia borealis) Lily-of-the-valley family

Flowers - Straw color or greenish yellow, less than 1 in. long, 3 to 6 nodding on slender pedicels from the summit of a leafless scape 6 to 15 in. tall. Perianth of 6 spreading divisions, the 6 stamens attached; style, 3-lobed. Leaves: Dark, glossy, large, oval to oblong, 2 to 5 (usually 3), sheathing at the base. Fruit. Oval blue berries on upright pedicels. Preferred Habitat - Moist, rich, cool woods and thickets. Flowering Season - May-June. Distribution - From the Carolinas and Wisconsin far northward.

To name canals, bridges, city thoroughfares, booming factory towns after DeWitt Clinton seems to many appropriate enough; but why a shy little woodland flower? As fitly might a wee white violet carry down the name of Theodore Roosevelt to posterity! "Gray should not have named the flower from the Governor of New York," complains Thoreau. "What is he to the lovers of flowers in Massachusetts? If named after a man, it must be a man of flowers."

So completely has Clinton, the practical man of affairs, obliterated Clinton, the naturalist, from the popular mind, that, were it not for this plant keeping his memory green, we should be in danger of forgetting the weary, overworked governor, fleeing from care to the woods and fields; pursuing in the open air the study which above all others delighted and refreshed him; revealing in every leisure moment a too-often forgotten side of his many-sided greatness.

INDIAN CUCUMBER-ROOT
(Medeola Virginiana) Lily-of-the-valley family

Flowers - Greenish yellow, on fine, curving footstalks, in a loose cluster above a circle of leaves. Perianth of 6 wide-spread divisions about 1/4 in. long; 6 reddish-brown stamens; 3 long reddish-brown styles, stigmatic on inner side. Stem: 1 to 2 1/2 ft. high, unbranched, cottony when young. Leaves: Of flowering plants, in 2 whorls; lower whorl of 5 to 9 large, thin, oblong, taper-pointed leaves above the middle of stem; upper whorl of 3 to 5 small, oval, pointed leaves 1 to 2 in. long, immediately under flowers. Flowerless plants with a whorl at summit. Fruit: Round, dark-purple berries. Preferred Habitat - Moist woods and thickets. Flowering Season - May-June. Distribution - Nova Scotia and Minnesota, southward nearly to the Gulf of Mexico.

Again we see the leaves of a plant coming to the aid of otherwise inconspicuous flowers to render them more attractive. By placing themselves in a circle just below these little spidery blossoms of weak and uncertain coloring, some of the Indian cucumber's leaves certainly make them at least noticeable, if not showy. It would be short-sighted philanthropy on the leaves' part to help the flowers win insect wooers at the expense of the plant's general health; therefore those in the upper whorl are fewer and much smaller than the leaves in the lower circle, and a sufficient length of stem separates them to allow the sunlight and rain to conjure with the chlorophyll in the group below. While there is a chance of nectar being pilfered from the flowers by ants, the stem is cottony and ensnares their feet. In September, when small clusters of dark-purple berries replace the flowers, and rich tints dye the leaves, the plant is truly beautiful - of course to invite migrating birds to disperse its seeds. It is said the Indians used to eat the horizontal, white, fleshy rootstock, which has a flavor like a cucumber's.

CARRION-FLOWER
(Smilax herbacea) Smilax family

Flowers - Carrion-scented, yellowish-green, 15 to 80 small, 6-parted ones clustered in an umbel on a long peduncle. Stem: Smooth, unarmed, climbing with the help of tendril-like appendages from the base of leafstalks. Leaves: Egg-shaped, heart-shaped, or rounded, pointed tipped, parallel-nerved, petioled. Fruit: Bluish-black berries. Preferred Habitat - Moist soil, thickets, woods, roadside fences. Flowering Season - April-June. Distribution - Northern Canada to the Gulf States, westward to Nebraska.

"It would be safe to say," says John Burroughs, "that there is a species of smilax with an unsavory name, that the bee does not visit, herbacea. The production of this plant is a curious freak of nature.... It would be a cruel joke to offer it to any person not acquainted with it, to smell. It is like the vent of a charnel-house." (Thoreau compared its odor to that of a dead rat in a wall!) "It is first cousin to the trilliums, among the prettiest of our native wild flowers," continues Burroughs, "and the same bad blood crops out in the purple trillium or birthroot."

Strange that so close an observer as Burroughs or Thoreau should not have credited the carrion-flower with being something more intelligent than a mere repellent freak! Like the purple trillium (q.v.), it has deliberately adapted itself to please its benefactors, the little green flesh flies so commonly seen about untidy butcher shops in summer. These, sharing with many beetles the unthankful task of removing putrid flesh and fowl from the earth, acting the part of scavengers for nature, are naturally attracted to carrion-scented flowers. Of these they have an ungrudged monopoly. But the purple trillium has an additional advantage in both smelling and looking like the same thing - a piece of raw meat past its prime. Bees and butterflies, with their highly developed aesthetic sense, ever delighting in beautiful colors, perfume, and nectar, naturally let such flowers as these alone - another object aimed at by them, for then the flies get all the pollen they can eat. Some they transfer, of course, from the larger staminate flowers to the smaller pistillate ones as they crawl over one umbel of the carrion-flower, then alight on another.

Presently fruit begins to set, and we can approach the luxuriant vine without offence to our noses. The beautiful glossy green foliage takes on resplendent tints in early autumn - again with interested motives, for are there not seeds within the little bluish-black berries, waiting for the birds to distribute them during their migration?

The vicious CATBRIER, GREENBRIER, or HORSEBRIER (S. rotundifolia), similar to the preceding, except that its four-angled stem is well armed with green prickles, its beautiful glossy, decorative leaves are more rounded, and its greenish flower umbels lack foul odor, scarcely needs description. Who has not encountered it in the roadside and woodland thickets, where it defiantly bars the way?

In the most inaccessible part of such a briery tangle, that rollicking polyglot, the yellow-breasted chat, loves to hide its nest. Indeed, many birds can say with Br'er Rabbit that they were "bred en bawn in a brier-patch." Throughout the eastern half of the United States and Upper Canada the catbrier displays its insignificant little blossoms from April to June for a miscellaneous lot of flies - insects which are content with the slightest floral attractions offered. The florist's staple vine popularly known as "SMILAX" (Myrslphyllum asparagoides), a native of the Cape of Good Hope, is not even remotely connected with true Smilaceae.

YELLOW STAR-GRASS
 (Hypoxis hirsuta; H. erecta of Gray) Amaryllis family

 Flowers - Bright yellow within, greenish and hairy outside, about 1/2 in. across, 6-parted; the perianth divisions spreading, narrowly oblong; a few flowers at the summit of a rough, hairy scape 2 to 6 in. high. Leaves: All from an egg-shaped corm; mostly longer than scapes, slender, grass-like, more or less hairy. Preferred Habitat -

Dry, open woods, prairies, grassy waste places, fields. Flowering Season - May-October. Distribution - From Maine far westward, and south to the Gulf of Mexico.

Usually only one of these little blossoms in a cluster on each plant opens at a time; but that one peers upward so brightly from among the grass it cannot well be overlooked. Sitting in a meadow sprinkled over with these yellow stars, we see coming to them many small bees - chiefly Halictus - to gather pollen for their unhatched babies' bread. Of course they do not carry all the pollen to their tunneled nurseries; some must often be rubbed off on the sticky pistil tip in the center of other stars. The stamens radiate, that self-fertilization need not take place except as a last extremity. Visitors failing, the little flower closes, bringing its pollen-laden anthers in contact with its own stigma.

BLACKBERRY LILY (Gemmingia Ciminensis; Pardanthus Chinensis of Gray) Iris family

Flowers - Deep orange color, speckled irregularly with crimson and purple within (Pardos = leopard; anthos = flower); borne in terminal, forked clusters. Perianth of 6 oblong, petal-like, spreading divisions; 6 stamens with linear anthers; style thickest above, with 3 branches. Stem: 1 1/2 to 4 ft. tall, leafy. Leaves: Like the iris; erect, folded blades, 8 to 10 in. long. Fruit: Resembling a blackberry; an erect mass of round, black, fleshy seeds, at first concealed in a fig-shaped capsule, whose 3 valves curve backward, and finally drop off. Preferred habitat - Roadsides and hills. Flowering Season - June-July. Distribution - Connecticut to Georgia, westward to Indiana and Missouri.

How many beautiful foreign flowers, commonly grown in our gardens here, might soon become naturalized Americans were we only generous enough to lift a few plants, scatter a few seeds over our fences into the fields and roadsides - to raise the bars of their prison, as it were, and let them free! Many have run away, to be sure. Once across the wide Atlantic, or wider Pacific, their passage paid (not sneaking in among the ballast like the more fortunate weeds), some are doomed to stay in prim, rigidly cultivated flower beds forever; others, only until a chance to bolt for freedom presents itself, and away they go. Lucky are they if every flower they produce is not picked before a single seed can be set.

This blackberry lily of gorgeous hue originally came from China. Escaping from gardens here and there, it was first reported as a wild flower at East Rock, Connecticut; other groups of vagabonds were met marching along the roadsides on Long Island; near Suffern, New York; then farther southward and westward, until it has already attained a very respectable range. Every plant has some good device for sending its offspring away from home to found new colonies, if man would but let it alone. Better still, give the eager travelers a lift!

LARGE YELLOW LADY'S SLIPPER; WHIPPOORWILL'S SHOE; YELLOW MOCCASIN
FLOWER
 (Cypripedium hirsutum; C. pubescens of Gray) Orchid family

Flower - Solitary, large, showy, borne at the top of a leafy stem to 2 ft. high. Sepals 3, 2 of them united, greenish or yellowish, striped with purple or dull red, very long, narrow; 2 petals, brown, narrower, twisting; the third an inflated sac, open at the top, 1 to 2 in. long, pale yellow, purple lined white hairs within; sterile stamen triangular; stigma thick. Leaves: Oval or elliptic, pointed, 3 to 5 in, long, parallel-nerved, sheathing. Preferred Habitat - Moist or boggy woods and thickets; hilly ground. Flowering Season - May-July. Distribution - Nova Scotia to Alabama, westward to Minnesota and Nebraska.

Swinging outward from a leaf-clasped stem, this orchid attracts us by its flaunted beauty and decorative form from tip to root, not less than the aesthetic little bees for which its adornment and mechanism are so marvelously adapted. Doubtless the heavy, oily odor is an additional attraction to them. Parallel purplish lines, converging toward the circular opening of the pale yellow, inflated pouch, guide the visitor into a spacious banquet-hall (labellum) such as the pink lady's slipper (q.v.) also entertains her guests in.

Fine hairs within secrete tiny drops of fluid at their tips - a secretion which hardens into a brittle crust, like a syrup's, when it dries. Darwin became especially interested in this flower through a delightful correspondence with Professor Asa Gray, who was the first to understand it, and he finally secured a specimen to experiment on.

"I first introduced some flies into the labellum through the large upper opening," Darwin wrote, "but they were either too large or too stupid, and did not crawl out properly. I then caught and placed within the labellum a very small bee which seemed of about the right size, namely Andrena parvula.... The bee vainly endeavored to crawl out again the same way it entered, but always fell backwards, owing to the margins being inflected. The labellum thus acts like one of those conical traps with the edges turned inwards, which are sold to catch beetles and cockroaches in London kitchens. It could not creep out through the slit between the folded edges of the basal part of the labellum, as the elongated, triangular, rudimentary stamen here closes the passage. Ultimately it forced its way out through one of the small orifices close to one of the anthers, and was found when caught to be smeared with the glutinous pollen. I then put the same bee into another labellum; and again it crawled out through one of the small orifices, always covered with pollen. I repeated the operation five times, always with the same result. I afterwards cut away the labellum, so as to examine the stigma, and found its whole surface covered with pollen. It should be noticed that an insect in making its escape, must first brush past the stigma and afterwards one of the anthers, so that it cannot leave pollen on the stigma, until being already smeared with pollen from one flower it enters another; and thus there will be a good chance of cross-fertilization between two distinct plants.... Thus the use of all parts of the flower, - namely, the inflected edges, or the polished inner sides of the labellum; the two orifices and their position close to the anthers and stigma, - the large size of the medial rudimentary stamen, - are rendered intelligible. An insect which enters the labellum is thus compelled to crawl out by one of the two narrow passages, on the sides of which the pollen-masses and stigma are placed."

These common orchids, which are not at all difficult to naturalize in a well-drained, shady spot in the garden, should be lifted with a good ball of earth and plenty of leaf-mould immediately after flowering. Here we can note little American Andrena bees unwittingly becoming the flower's slaves. Several species of exotic cypripediums are so common in the city florist's shops every one has an opportunity to study their marvelous structure.

The similar SMALL YELLOW LADY'S SLIPPER (C. parviflorum), a delicately fragrant orchid about half the size of its big sister, has a brighter yellow pouch, and occasionally its sepals and petals are purplish. As they usually grow in the same localities, and have the same blooming season, opportunities for comparison are not lacking. This fairer, sweeter, little orchid roams westward as far as the State of Washington.

YELLOW FRINGED ORCHIS
(Habenaria ciliaris) Orchid family

Flowers - Bright yellow or orange, borne in a showy, closely set, oblong spike, 3 to 6 in. long. The lip of each flower copiously fringed; the slender spur 1 to 1 1/2 in. long; similar to white fringed orchis (q.v.); and between the two, intermediate pale yellow hybrids may be found. Stem: Slender, leafy, 1 to 2 1/2 feet high. Leaves: Lance-shaped, clasping. Preferred Habitat - Moist meadows and sandy bogs. Flowering Season - July-August. Distribution - Vermont to Florida; Ontario to Texas.

Where this brilliant, beautiful orchid and its lovely white sister grow together in the bog - which cannot be through a very wide range, since one is common northward, where the other is rare, and vice versa - the yellow fringed orchis will be found blooming a few days later. In general structure the plants closely resemble each other. Their similar method of enforcing payment for a sip of nectar concealed in a tube so narrow and deep none but a sphinx moth or butterfly may drain it all (though large bumblebees occasionally get some too, from brimming nectaries) has been described (q.v.), to which the interested reader is referred. Both these orchids have their sticky discs projecting unusually far, as if raised on a pedicel - an arrangement which indicates that they "are to be stuck to the face or head of some nectar-sucking insect of appropriate size that visits the flowers," wrote Dr. Asa Gray over forty years ago. Various species of hawk moths, common in different parts of our area, of course have tongues of various lengths, and naturally every visitor does not receive his load of pollen on the same identical spot. At dusk, when sphinx moths begin their rounds, it will be noticed that the white and yellow flowers remain conspicuous long after blossoms of other colors have melted into the general darkness. Such flowers as cater to these moths, if they have fragrance, emit it then most strongly, as an additional attraction. Again, it will be noticed that few such flowers provide a strong projecting petal-platform for visitors to alight on; that would be superfluous, since sphinx moths suck while hovering over a tube, with their wings in exceedingly rapid motion, just like a hummingbird, for which the larger species are so often mistaken at twilight. This deep-hued orchid apparently attracts as many butterflies as sphinx moths, which show a predilection for the white species.

>From Ontario and the Mississippi eastward, and southward to the Gulf, the TUBERCLED or SMALL PALE GREEN ORCHIS (H. flava) lifts a spire of inconspicuous greenish-yellow flowers, more attractive to the eye of the structural botanist than to the aesthete. It blooms in moist places, as most orchids do, since water with which to manufacture nectar enough to fill their deep spurs is a prime necessity. Orchids have arrived at that pinnacle of achievement that it is impossible for them to fertilize themselves. More than that, some are absolutely sterile to their own pollen when it is applied to their stigmas artificially with insect aid, however, a single plant has produced over 1,000,700 seeds. No wonder, then, that, as a family, they have adopted the most marvelous blandishments and mechanism in the whole floral kingdom to secure the visits of that special insect to which each is adapted, and, having secured him, to compel him unwittingly to do their bidding. In the steaming tropical jungles, where vegetation is luxuriant to the point of suffocation, and where insect life swarms in mvriads undreamed of here, we can see the best of reasons for orchids mounting into trees and living on air to escape strangulation on the ground, and for donning larger and more gorgeous apparel to attract attention in the fierce competition for insect trade waged about them. Here, where the struggle for survival is incomparably easier, we have terrestrial orchids, small, and quietly clad, for the most part.

Having the gorgeous, exotic air plants of the hothouse in mind, this little tubercled orchis seems a very poor relation indeed. In June and July, about a week before the ragged orchis comes out, we may look for this small, fringeless sister. Its clasping leaves, which decrease in size as they ascend the stem (not to shut off the light and rain from the lower ones), are parallel-veined, elliptic, or, the higher ones, lance-shaped. A prominent tubercle, or palate, growing upward from the lip, almost conceals the entrance to the nectary. and makes a side approach necessary. Why? Usually an insect has free, straight access down the center of a flower's throat, but here he cannot have it. A slender tongue must be directed obliquely from above into the spur, and it will enter the discal groove as a thread enters the eye of a needle. By this arrangement the tongue must certainly come in contact with one of the sticky discs to which an elongated pollen gland is attached. The cement on the disc hardening even while the visitor sucks, the pollen gland is therefore drawn out, because firmly attached to his tongue. At first the pollen mass stands erect on the proboscis; but in the fraction of a moment which it takes a butterfly to flit to another blossom, it has bent forward automatically into the exact position required for it to come in contact with the sticky stigma of the next tubercled orchis entered, where it will be broken off. Now we understand the use of the palate. Butterfly collectors often take specimens with remnants of these pollen stumps stuck to their tongues. In his classical work "On the Fertilization of Orchids by Insects," Darwin tells of finding a mottled rustic butterfly whose proboscis was decorated with eleven pairs of pollen masses, taken from as many blossoms of the pyramidal orchis. Have these flowers no mercy on their long-suffering friends? A bee with some orchid pollen-stumps attached to its head was once sent to Mr. Frank Cheshire, the English expert who had just discovered some strange bee diseases. He was requested to name the malady that had caused so abnormal an outgrowth on the bee's forehead!

Often found growing in the same bog with the tubercled species is the RAGGED or FRINGED GREEN ORCHIS (H. lacera), so inconspicuous we often overlook it unawares. Examine one of the dingy, greenish-yellow flowers that are set along the stern in a spike to make all the show in the world possible, each with its three-parted, spreading lip finely and irregularly cut into thread-like fringe to hail the passing butterfly, and we shall see that it, too, has made ingenious provision against the draining of its spur by a visitor without proper pay for his entertainment. Even without the gay color that butterflies ever delight in, these flowers contain so much nectar in their spurs, neither butterflies nor large bumblebees are long in hunting them out. In swamps and wet woodland from Nova Scotia to Georgia, and westward to the Mississippi, the ragged orchis blooms in June or July.

LARGE YELLOW POND or WATER LILY; COW LILY; SPATTER-DOCK
(Nymphaea advena; Nupisar advena of Gray) Water-lily family

Flowers - Yellow or greenish outside, rarely purple tinged, round, depressed, 1 1/2 to 3 1/2 in. across. Sepals 6, unequal, concave, thick, fleshy; petals stamen-like, oblong, fleshy, short; stamens very numerous, in 5 to 7 rows; pistil compounded of many carpels, its stigmatic disc pale red or yellow, with 12 to 24 rays. Leaves: Floating, or some immersed, large, thick, sometimes a foot long, egg-shaped or oval, with a deep cleft at base, the lobes rounded. Preferred Habitat - Standing water, ponds, slow streams. Flowering Season - April-September. Distribution - Rocky Mountains eastward, south to the Gulf of Mexico, north to Nova Scotia.

Comparisons were ever odious. Because the yellow water lily has the misfortune to claim relationship with the sweet-scented white species (q.v.), must it never receive its just meed of praise? Hiawatha's canoe, let it be remembered,

"Floated on the river
Like a yellow leaf in autumn,
Like a yellow water-lily."

But even those who admire Longfellow's lines see no beauty in the golden flower-bowls floating among the large, lustrous, leathery leaves.

By assuming the functions of petals, the colored sepals advertise for insects. Beetles, which answer the first summons to a free lunch, crowd in as the sepals begin to spread. In the center the star-like disc, already sticky, is revealed, and on it any pollen they have carried with them from older flowers necessarily rubs off. At first, or while the stigma is freshly receptive to pollen, an insect cannot make his entrance except by crawling over this large, sticky plate. At this time, the anthers being closed, self-fertilization is impossible. A day or two later, after the pollen begins to ripen on countless anthers, the flower is so widely open that visitors have no cause to alight in the center; anyway, no harm could result if they did, cross-fertilization having been presumably accomplished. While beetles (especially Donacia) are ever abundant visitors, it is likely they do much more harm than good. So eagerly do they gnaw both petals and stamens, which look like loops of narrow yellow ribbon within the bowl of an older flower, that, although they must carry some pollen to younger flowers as they travel on, it is probable they destroy ten times more than their share. Flies transport pollen too. The smaller bees (Halictus and Andrena chiefly) find some nectar secreted on the outer faces of the stamen-like petals, which they mix with pollen to make their babies' bread.

The very beautiful native AMERICAN LOTUS (Nelumbo lutea), also known as WATER CHINKAPIN or WANKAPIN, found locally in Ontario, the Connecticut River, some lakes, slow streams, and ponds in New Jersey, southward to Florida, and westward to Michigan and Illinois, Indian Territory and Louisiana, displays its pale yellow flowers in July and August. They measure from four to ten inches across, and suggest a yellow form of the sweet-scented white water lily; but there are fewer petals, gradually passing into an indefinite number of stamens. The great round, ribbed leaves, smooth above, hairy beneath, may be raised high above the water, immersed or floating. Both leaf and flower stalks contain several large air canals. The flowers which are female when they expand far enough for a pollen-laden guest to crawl into the center, are afterward male, securing cross-fertilization by this means, just as the yellow pond lily does; only the small bees must content themselves here with pollen only - a diet that pleases the destructive beetles and the flies (Syrphidae) perfectly.

Japanese artists especially have taught us how much of the beauty of a Nelumbo we should lose if it ripened its decorative seed-vessel below the surface as the sweet-scented white water lily does. This flat-topped receptacle, held erect, has its little round nuts imbedded in pits in its surface, ready to be picked out by aquatic birds, and distributed by them in their wanderings. Both seeds and tubers are farinaceous and edible. In some places it is known the Indians introduced the plant for food. Professor Charles Goodyear has written an elaborate, plausible argument, illustrated, with many reproductions of sculpture, pottery, and mural painting in the civilized world of the ancients to prove that all decorative ornamental design has been evolved from the sacred Egyptian lotus (Nelumbo Nelumubo), still revered throughout the East (q.v.).

MARSH MARIGOLD; MEADOW-GOWAN; AMERICAN
COWSLIP
(Caltha palustris) Crowfoot family[1]

Flowers - Bright, shining yellow, 1 to 1 1/2 in. across, a few in terminal and axillary groups. No petals; usually 5 (often more) oval, petal-like sepals; stamens numerous; many pistils (carpels) without styles. Stem: Stout, smooth, hollow, branching, 1 to 2 ft. high. Leaves: Mostly from root, rounded, broad, and heart-shaped at base, or kidney-shaped, upper ones almost sessile, lower ones on fleshy petioles. Preferred Habitat - Springy ground, low meadows, swamps, river banks, ditches. Flowering Season - April-June. Distribution - Carolina to Iowa, the Rocky Mountains, and very far north.

Not a true marigold, and even less a cowslip, it is by these names that this flower, which looks most like a buttercup, will continue to be called, in spite of the protests of scientific classifiers. Doubtless the first of these folk-names refers to its use in church festivals during the Middle Ages as one of the blossoms devoted to the Virgin Mary.

"And winking Mary-buds begin
To ope their golden eyes,"

sing the musicians in "Cymbeline." Whoever has seen the watery Avon meadows in April, yellow and twinkling with marsh marigolds when "the lark at heaven's gate sings," appreciates why the commentators incline to identify Shakespeare's Mary-buds with the Caltha of these and our own marshes.

Not for poet's rhapsodies, but for the more welcome hum of small bees and flies intent on breakfasting do these flowers open in the morning sunshine. Nectar secreted on the sides of each of the many carpels invites a conscientious bee all around the center, on which she should alight to truly benefit her entertainer. Honey bees may be seen sucking only enough nectar to aid them in storing pollen; bumblebees feasting for their own benefit, not their descendants'; little mining bees and quantities of flies also, although not many species are represented among the visitors, owing to the flower's early blooming season. Always conspicuous among the throng are the brilliant Syrphidae flies - gorgeous little creatures which show a fondness for blossoms as gaily colored as their own lustrous bodies. Indeed, these are the principal pollinators.

Some country people who boil the young plants declare these "greens" are as good as spinach. What sacrilege to reduce crisp, glossy, beautiful leaves like these to a slimy mess in a pot! The tender buds, often used in white sauce as a substitute for capers, probably do not give it the same piquancy where piquancy is surely most needed - on boiled mutton, said to be Queen Victoria's favorite dish. Hawked about the streets in tight bunches, the marsh-marigold blossoms - with half their yellow sepals already dropped - and the fragrant, pearly-pink arbutus are the most familiar spring wild flowers seen in Eastern cities.

COMMON MEADOW BUTTERCUP; TALL CROW-FOOT; KINGCUPS; CUCKOO FLOWER; GOLDCUPS; BUTTER-FLOWERS; BLISTER-FLOWERS

(Ranunculus acris) Crowfoot family

Flowers - Bright, shining yellow, about 1 in. across, numerous, terminating long slender footstalks. Calyx of 5 spreading sepals; corolla of 5 petals; yellow stamens and carpels. Stem: Erect, branched above, hairy (sometimes nearly smooth), 2 to 3 feet tall, from fibrous roots. Leaves: In a tuft from the base, long petioled, of 3 to 7 divisions cleft into numerous lobes; stem leaves nearly sessile, distant, 3-parted. Preferred Habitat - Meadows, fields, roadsides, grassy places. Flowering Season - May-September. Distribution - Naturalized from Europe in Canada and the United States; most common North.

What youngster has not held these shining golden flowers under his chin to test his fondness for butter? Dandelions and marsh-marigolds may reflect their color in his clear skin too, but the buttercup is every child's favorite. When

"Cuckoo-buds of yellow hue
Do paint the meadows with delight,"

daisies, pink clover, and waving timothy bear them company here; not the "daisies pied," violets, and lady-smocks of Shakespeare's England. How incomparably beautiful are our own meadows in June! But the glitter of the buttercup, which is as nothing to the glitter of a gold dollar in the eyes of a practical farmer, fills him with wrath when this immigrant takes possession of his pastures. Cattle will not eat the acrid, caustic plant - a sufficient reason for most members of the Ranunculaceae to stoop to the low trick of secreting poisonous or bitter juices. Self-preservation leads a cousin, the garden monk's hood, even to murderous practices. Since children will put everything within reach into their mouths, they should be warned against biting the buttercup's stem and leaves, that are capable of raising blisters. "Beggars use the juice to produce sores upon their skin," says Mrs. Creevy. A designer might employ these exquisitely formed leaves far more profitably.

This and the bulbous buttercup, having so much else in common, have also the same visitors. "It is a remarkable fact," says Sir John Lubbock, "as Aristotle long ago mentioned, that in most cases bees confine themselves in each journey to a single species of plant; though in the case of some very nearly allied forms this is not so; for instance, it is stated on good authority (Muller) that Ranunculus acris, R. repens, and R. bulbosus are not distinguished by the bees, or at least are visited indifferently by them, as is also the case with two of the species of clover." From what we already know of the brilliant Syrphidae flies' fondness for equally brilliant colors, it is not surprising to find great numbers of them about the buttercups, with bees, wasps, and beetles - upwards of sixty species. Modern scientists believe that the habit of feeding on flowers has called out the color-sense of insects and the taste for bright colors, and that sexual selection has been guided by this taste. The most unscientific among us soon finds evidence on every hand that flowers and insects have developed together through mutual dependence.

By having its nourishment thriftily stored up underground all winter, the BULBOUS BUTTERCUP (R. bulbosus) is able to steal a march on its fibrous-rooted sister that must accumulate hers all spring; consequently it is first to flower, coming in early May, and lasting through June. It is a low and generally more hairy plant, but closely resembling the tall buttercup in most respects, and, like it, a naturalized European immigrant now thoroughly at home in fields and roadsides in most sections of the United States and Canada.

Much less common is the CREEPING BUTTERCUP (R. repens), which spreads by runners until it forms large patches in fields and roadsides, chiefly in the Eastern States. Its leaves, which are sometimes blotched, are divided into three parts, the terminal one, often all three, stalked. May-July.

First to bloom in the vicinity of New York (from March to May) is the HISPID BUTTERCUP (R. hispidus), densely hairy when young. The leaves, which are pinnately divided into from three to five leaflets, cleft or lobed, chiefly arise on long petioles from a cluster of thickened fibrous roots. The flower may be only half an inch or an inch and a half across. It is found in dry woods and thickets throughout the eastern half of the United States; whereas the much smaller flowered BRISTLY BUTTERCUP (R. Pennsylvanicus) shows a preference for low-lying meadows and wet, open ground through a wider, more westerly range. Its stout, hollow, leafy stem, beset with stiff hairs, discourages the tongues of grazing animals. June-August.

Commonest of the early buttercups is the TUFTED BUTTERCUP (R. fascicularis), a little plant seldom a foot high, found in the woods and on rocky hillsides from Texas and Manitoba, east to the Atlantic, flowering in April or May. The long-stalked leaves are divided into from three to five parts; the bright yellow flowers, with rather narrow, distant petals, measure about an inch across. They open sparingly, usually only one or two at a time on each plant, to favor pollination from another one.

Scattered patches of the SWAMP or MARSH BUTTERCUP (P. septentrionalis) brighten low, rich meadows also with their-large satiny yellow flowers, whose place in the botany even the untrained eye knows at sight. The smooth, spreading plant sometimes takes root at the joints of its branches and sends forth runners, but the stems mostly ascend. The large lower mottled leaves are raised well out of the wet, or above the grass, on long petioles. They have three divisions, each lobed and cleft. From Georgia and Kentucky far northward this buttercup blooms from April to July, opening only a few flowers at a time-a method which may make it less showy, but more certain to secure cross-pollination between distinct plants.

The YELLOW WATER BUTTERCUP or CROWFOOT (R. deiphinifolius; R. multifidus of Gray) found blooming in ponds through the summer months, certainly justifies the family name derived from rana = a frog. Many other members grow in marshes, it is true, but this ranunculus lives after the manner of its namesake, sometimes immersed, sometimes stranded on the muddy shore. Two types of leaves occur on the same stem. Their waving filaments, which make the immersed leaves look fringy, take every advantage of what little carbonic acid gas is dissolved under the surface. Moreover, they are better adapted to withstand the water's pressure and possible currents than solid blades would be. The floating leaves which loll upon the surface to take advantage of the air and sunlight, expand three, four, or five divisions, variously lobed. On this plant we see one set of leaves perfectly adapted to immersion, and another set to aerial existence. The stem, which may measure several feet in length, roots at the joints when it can. Range from the Mississippi and Ontario eastward to the Atlantic Ocean.

The WHITE WATER-CROWFOOT (Batrachium trichophyllum; Ranunculus aquatilis of Gray) has its fine thread-like leaves entirely submerged; but the flowers, like a whale, as the old conundrum put it, come to the surface to blow. The latter are small, white, or only yellow at the base, where each petal bears a spot or little pit that serves as a pathfinder to the flies. When the water rises unusually high, the blossoms never open, but remain submerged, and fertilize themselves. Seen underwater, the delicate leaves, which are little more than forked hairs, spread abroad in dainty patterns; lifted cut of the water these flaccid filaments utterly collapse. In ponds and shallow, slow streams, this common plant flowers from June to September almost throughout the Union, the British Possessions north of us, and in Europe and Asia.

The WATER PLANTAIN SPEARWORT (K. obtusiusculus; R. a/isrnaefoiius of Gray) flecks the marshes from June to August with its small golden flowers, which the merest novice knows must be kin to the buttercup. The smooth, hollow stem, especially thick at the base, likes to root from the lower joints. A peculiarity of the lance-shaped or oblong lance-shaped leaves is that the lower ones have petioles so broad where they clasp the stem that they appear to be long blades suddenly contracted just above their base.

BARBERRY; PEPPERIDGE-BUSH
(Berberis vulgaris) Barberry family

Flowers - Yellow, small, odor disagreeable, 6-parted, borne in drooping, many-flowered racemes from the leaf axils along arching twigs. Stem: A much branched, smooth, gray shrub, to 8 ft. tall, armed with sharp spines. Leaves: From the 3-pronged spines (thorns); oval or obovate, bristly edged. Fruit: Oblong, scarlet, acid berries. Preferred Habitat - Thickets; roadsides; dry or gravelly soil. Flowering Season - May-June. Distribution - Naturalized in New England and Middle States; less common in Canada and the West. Europe and Asia.

When the twigs of barberry bushes arch with the weight of clusters of beautiful bright berries in September, everyone must take notice of a shrub so decorative, which receives scant attention from us, however, when its insignificant little flowers are out. Yet these blossoms, small as they are, are up to a marvelous trick, quite as remarkable as the laurel's (q.v.) or the calopogon's (q.v.), to compel insects to do their bidding. Three of the six sepals, by their size and color, attend to the advertising, playing the part of a corolla; and partly by curving inward at the tip, partly by the drooping posture of the flower, help protect the stamens, pistil, and nectar glands within from rain. Did the flowers hang vertically, not obliquely, such curvature of the tips of sepals and petals would be unnecessary. Six stamens surround a pistil, but each of their six anthers, which are in reality little pollen boxes opening by trap-doors on either side, is tucked under the curving tip of a petal at whose base lie two orange-colored nectar glands. A small bee or fly enters the flower: what happens? To reach the nectar, he must probe between the bases of two exceedingly irritable stamens. The merest touch of a visitor's tongue against them releases two anthers, just as the nibbling mouse all unsuspectingly releases the wire from the hook of the wooden trap he is caught in. As the two stamens spring upward on being released, pollen instantly flies out of the trap-doors of the anther boxes on the bee, which suffers no greater penalty than being obliged to carry it to the stigma of another flower. So short are the stamens, it is improbable that a flower's pollen ever reaches its own stigma except through the occasional confused fumbling of a visitor. Usually he is so startled by the sudden shower of pollen that he flies away instantly.

In the barberry bushes, as in the gorse, when grown in dry, gravelly situations, we see many leaves and twigs modified into thorns to diminish the loss of water through evaporation by exposing too much leaf surface to the sun and air. That such spines protect the plants which bear them from the ravages of grazing cattle is, of course, an additional motive for their presence. Under cultivation, in well-watered garden soil - and how many charming varieties of barberries are cultivated - the thorny shrub loses much of its armor, putting forth many more leaves, in rosettes, along more numerous twigs, instead. Even the prickly-pear cactus might become mild as a lamb were it to forswear sandy deserts and live in marshes instead. Country people sometimes rob the birds of the acid berries to make preserves. The wood furnishes a yellow dye.

Curiously enough it is the EUROPEAN BARBERRY that is the common species here. The AMERICAN BARBERRY (B. Canadensis), a lower shrub, with dark reddish-brown twigs; its leaves more distantly toothed; its flowers, and consequently its berries, in smaller clusters, keeps almost exclusively to the woods in the Alleghany region and in the southwest, in spite of its specific name.

SPICE-BUSH; BENJAMIN-BUSH; WILD ALLSPICE; FEVER-BUSH
(Benzoin Benzoin; Lindera Benzoin of Gray) Laurel family

Flowers - Before the leaves, lemon yellow, fragrant, small, in clusters close to the slender, brittle twigs. Six petal-like sepals; sterile flowers with 9 stamens in 3 series; fertile flowers with a round ovary encircled by abortive stamens. Stem: A smooth shrub 4 to 20 ft. tall. Leaves: Alternate, entire, oval or elliptic, 2 to 5 in, long. Fruit: Oblong, red, berry-like drupes. Preferred Habitat - Moist woodlands, thickets, beside streams. Flowering Season - March-May. Distribution - Central New England, Ontario, and Michigan, southward to Carolina and Kansas.

Even before the scaly catkins on the alders become yellow, or the silvery velvet pussy willows expand to welcome the earliest bees that fly, this leafless bush breathes a faint spicy fragrance in the bleak gray woods. Its only rivals among the shrubbery, the serviceberry and its twin sister the shad-bush, have scarcely had the temerity to burst into bloom when the little clusters of lemon-yellow flowers, cuddled close to the naked branches, give us our first delightful spring surprise. All the favor they ask of the few insects then flying is that they shall transfer the pollen from the sterile to the fertile flowers as a recompense for the early feast spread. Inasmuch as no single blossom contains both stamens and pistil, little wonder the flowers should woo with color and fragrance the guests on whose ministrations the continuance of the species absolutely depends. Later, when the leaves appear, we may know as soon as we crush them in the hand that the aromatic sassafras is next of kin. But ages before Linnaeus published "Species Plantarum" butterflies had discovered floral relationships.

Sharp eyes may have noticed how often the leaves on both the spice-bush and the sassafras tree are curled. Have you ever drawn apart the leaf edges and been startled by the large, fat green caterpillar, speckled with blue, whose two great black "eyes" stare up at you as he reposes in his comfortable nest - a cradle which also combines the advantages of a restaurant? This is the caterpillar of the common spice-bush swallow-tail butterfly (Papilio troilus), an exquisite, dark, velvety creature with pale greenish-blue markings on its hind wings. (See Dr. Holland's "Butterfly Book," Plate XLI.) The yellow stage of this caterpillar (which William Hamilton Gibson calls the "spice-bush bugaboo") indicates, he says, that "its period of transformation is close at hand. Selecting a suitable situation, it spins a tiny tuft of silk, into which it entangles its hindmost pair of feet, after which it forms a V-shaped loop about the front portion of its body, and hangs thus suspended, soon changing to a chrysalis of a pale wood color. These chrysalides commonly survive the winter, and in the following June the beautiful 'blue swallow-tail' will emerge, and may be seen suggestively fluttering and poising about the spice and sassafras bushes." After the eggs she lays on them hatch, the caterpillars live upon the leaves. Mrs. Starr Dana says the leaves were used as a substitute for tea during the Rebellion; and the powdered berries for allspice by housekeepers in Revolutionary days.

GREATER CELANDINE; SWALLOW-WORT
(Chelidonium majus) Poppy family

Flowers - Lustreless yellow, about 1/2 in. across, on slender pedicels, in a small umbel-like cluster. Sepals 2, soon falling; 4 petals, many yellow stamens, pistil prominent. Stem: Weak, to 2 ft. high, branching, slightly hairy, containing bright orange acrid juice. Leaves: Thin, 4 to 8 in. long, deeply cleft into 5 (usually) irregular oval lobes, the terminal one largest. Fruit: Smooth, slender, erect pods, 1 to 2 in, long, tipped with the persistent style. Preferred Habitat - Dry waste land, fields, roadsides, gardens, near dwellings. Flowering Season - April-September. Distribution - Naturalized from Europe in Eastern United States.

Not this weak invader of our roadsides, whose four yellow petals suggest one of the cross-bearing mustard tribe, but the pert little LESSER CELANDINE, PILEWORT, or FIGWORT BUTTERCUP (Ficaria Ficaria), one of the Crowfoot family, whose larger solitary satiny yellow flowers so commonly star European pastures, was Wordsworth's special delight - a tiny, turf-loving plant, about which much poetical association clusters. Having stolen passage across the Atlantic, it is now making itself at home about College Point, Long Island; on Staten Island; near Philadelphia, and maybe elsewhere. Doubtless it will one day overrun our fields, as so many other European immigrants have done.

The generic Greek name of the greater celandine, meaning a swallow, was given it because it begins to bloom when the first returning swallows are seen skimming over the water and freshly ploughed fields in a perfect ecstasy of flight, and continues in flower among its erect seed capsules until the first cool days of autumn kill the gnats and small winged insects not driven to cover. Then the swallows, dependent on such fare, must go to warmer climes where plenty still fly. Quaint old Gerarde claims that the swallow-wort was so called because "with this herbe the dams restore eye-sight to their young ones when their eye be put out" by swallows. Coles asserts "the swallow cureth her dim eyes with celandine."

There can be little satisfaction in picking a weed which droops immediately, poppy fashion, and whose saffron juice stains whatever it touches. A drop of this acrid fluid on the tip of the tongue is not soon forgotten. The luminous experiments of Darwin, Lubbock, Wallace, Muller, and Sprengel, among others, have proved that color in flowers exists for the purpose of attracting insects. But how about colored juices in the blood-roots' and poppies' stems, for example; the bright stalk of the pokeweed, the orange-yellow root of the carrot, the exquisite tints of autumn leaves, fungi, and seaweed? Besides the green color (chlorophyll), the most necessary of all ingredients to a plant are the lipochromes, which vary from yellow to red. These are most conspicuous when they displace the chlorophyll in autumn foliage. Then there are the anthocyans, ranging from magenta to blue and violet. These vary according to the amount of acid or alkali in the sap. Try the effect of immersing a blue morning glory in an acid solution, or a deep pink one in an alkaline solution. One theory to account for the presence of color is that it exists to screen the plant's protoplasm from light; that it has a physiological function with which insects have nothing whatever to do; and that by its presence the temperature is raised and the plant is protected from cold. Every one who has handled the colorless Indian pipe knows how cold and clammy it is.

The YELLOW or CELANDINE POPPY (Stylophorum diphyllum), with shining yellow flowers double the size of the greater celandine's, and similar pinnatifid leaves springing chiefly from the base, blooms even in March and through the spring in the Middle States and westward to Wisconsin and Missouri. Usually only one of the few terminal blossoms opens at a time, but in low, open woodlands it gleams like a miniature sun. Alas! that the glorious CALIFORNIA POPPY, so commonly grown in Eastern gardens (Eschscholtzia Californica), should confine itself to a limited range on the Pacific Coast. We have no true native poppies (Papaver) in America; such as are rarely to be seen in a wild state, have only locally escaped from cultivation.

GOLDEN CORYDALIS
(Capnoides aureum; Corydalis aurea of Gray) Poppy family

Flowers - Bright yellow, about 1/2 in. long, with a spur half the length of the tubular corolla; irregular, lipped; each upheld by a little bract, mostly at a horizontal; borne in a terminal, short raceme. Stem: Smooth, 6 to 14 in. high, branching. Leaves: Finely dissected, decom pound, petioled. Fruit: Sickle-shaped, drooping pods, wavy lumped, and tipped with the style. Preferred Habitat - Woods, rocky banks. Flowering Season - March-May. Distribution - Minnesota to Nova Scotia and Pennsylvania.

A dainty little plant, next of kin to the pink corydalis (q.v.).

BLACK MUSTARD
(Brassica nigra) Mustard family

Flowers - Bright yellow, fading pale, 1/4 to 1/2 in. across, 4-parted, in elongated racemes; quickly followed by narrow upright

4-sided pods about 1/2 in. long appressed against the stem. Stem: Erect, 2 to 7 ft. tall, branching. Leaves: Variously lobed and divided, finely toothed, the terminal lobe larger than the 2 to 4 side ones. Preferred Habitat - Roadsides, fields, neglected gardens. Flowering Season - June-November. Distribution - Common throughout our area; naturalized from Europe and Asia.

"The kingdom of heaven is like unto a grain of mustard seed, which a man took and sowed in his field which indeed is less than all seeds but when it is grown, it is greater than the herbs, and becometh a tree, so that the birds of the air come and lodge in the branches thereof."

Commentators differ as to which is the mustard of the parable - this common black mustard, or a rarer shrub-like tree (Salvadora Persica), with an equivalent Arabic name, a pungent odor, and a very small seed. Inasmuch as the mustard which is systematically planted for fodder by Old World farmers grows with the greatest luxuriance in Palestine, and the comparison between the size of its seed and the plant's great height was already proverbial in the East when Jesus used it, evidence strongly favors this wayside weed. Indeed, the late Dr. Royle, who endeavored to prove that it was the shrub that was referred to, finally found that it does not grow in Galilee.

Now, there are two species which furnish the most powerfully pungent condiment known to commerce; but the tiny dark brown seeds of the black mustard are sharper than the serpent's tooth, whereas the pale brown seeds of the WHITE MUSTARD, often mixed with them, are far more mild. The latter (Sinapis alba) is a similar, but more hairy, plant, with slightly larger yellow flowers. Its pods are constricted like a necklace between the seeds.

The coarse HEDGE MUSTARD (Sisymbrium officinale), with rigid, spreading branches, and spikes of tiny pale yellow flowers, quickly followed by awl-shaped pods that are closely appressed to the stem, abounds in waste places throughout our area. It blooms from May to November, like the next species.

Another common and most troublesome weed from Europe is the FIELD or CORN MUSTARD, CHARLOCK or FIELD KALE (Brassica arvensis; Sinapis arvensis of Gray) found in grain fields, gardens, rich waste lands, and rubbish heaps. The alternate leaves, which stand boldly out from the stem, are oval, coarsely saw-toothed, or the lower ones more irregular, and lobed at their bases, all rough to the touch, and conspicuously veined. The four-parted yellow flowers, measuring half an inch or more across, have six stamens (like the other members of this cross-bearing family), containing nectar at their bases. Two of them are shorter than the other four. Honeybees, ever abundant, the brilliant Syrphidae flies which love yellow, and other small visitors after pollen and nectar, to obtain the latter insert their tongues between the stamens, and usually cross-fertilize the flowers. In stormy weather, when few insects fly, the anthers finally turn their pollen-covered tips upward; then, by a curvature of the tip of the stamens, they are brought in contact with the flower's own stigma; for it is obviously better that even self-fertilized seed should be set than none at all. (See Ladies'-smock.) "The birds of the air" may not lodge in the charlock's few and feeble branches; nevertheless they come seeking the mild seeds in the strongly nerved, smooth pods that spread in a loose raceme. Domestic pigeons eat the seeds greedily.

The highly intelligent honey-bee, which usually confines itself to one species of plant on its flights, apparently does not know the difference between the field mustard and the WILD RADISH, or JOINTED or WHITE CHARLOCK (Raphanus Raphanistrum); or, knowing it, does not care to make distinctions, for it may be seen visiting these similar flowers indiscriminately. At first the blossoms of the radish are yellow, but they quickly fade to white, and their purplish veins become more conspicuous. Rarely the flowers are all purplish. The entire plant is rough to the touch; the leaves, similar to those of the garden radish, are deeply cleft (lyrate-pinnatifid); the seed pods, which soon follow the flowers up the spike, are nearly cylindric when fresh, but become constricted between the seeds, as they dry, until each little pod looks like a section of a bead necklace.

The GARDEN RADISH of the market (R. sativus), occasionally escaped from cultivation, although credited to China, is entirely unknown in its native state. "It has long been held in high esteem," wrote Peter Henderson, "and before the Christian era a volume was written on this plant alone. The ancient Greeks, in offering their oblations to Apollo, presented turnips in lead, beets in silver, and radishes in vessels of beaten gold." Pliny describes a radish eaten in Rome as being so transparent one might see through the root. It was not until the sixteenth century that the plant was introduced into England. Gerarde mentions cultivating four varieties for Queen Elizabeth in Lord Burleigh's garden.

The YELLOW ROCKET, HERB OF ST. BARBARA, YELLOW BITTER-CRESS, WINTER- or ROCKET-CRESS (Barbarca Barbarea; B. vulgaris of Gray) sends up spikes of little flowers like a yellow sweet alyssum as early as April, and continues in bloom through June. Smooth pods about one inch long quickly follow. The thickish, shining, tufted leaves, very like the familiar WATER-CRESS (Roripa Nasturtium), were formerly even more commonly eaten as a salad. In rich but dry soil the plant flourishes from Virginia far northward, locally in the interior of the United States and on the Pacific Coast.

WITCH-HAZEL
(Hamamelis Virginiana) Witch-hazel family

Flowers - Yellow, fringy, clustered in the axils of branches. Calyx 4-parted; 4 very narrow curving petals about 34 in. long; 4 short stamens, also 4 that are scale-like; 2 styles. Stem: A tall, crooked shrub. Leaves: Broadly oval, thick, wavy-toothed, mostly fallen at flowering time. Fruit: Woody capsules maturing the next season and remaining with flowers of the succeeding year (Hama = together with; mela = fruit). Preferred Habitat - Moist woods or thickets near streams. Flowering Season - August-December. Distribution - Nova Scotia and Minnesota, southward to the Gulf States.

To find a stray. apple blossom among the fruit in autumn, or an occasional violet deceived by caressing Indian Summer into thinking another spring has come, surprises no one; but when the witch-hazel bursts into bloom for the first time in November, as if it were April, its leafless twigs conspicuous in the gray woods with their clusters of spidery pale yellow flowers, we cannot but wonder with Edward Rowland Sill:

"Has time grown sleepy at his post
And let the exiled Summer back?

Or is it her regretful ghost,
Or witchcraft of the almanac?"

Not to the blue gentian but to the witch-hazel should Bryant have addressed at least the first stanza of his familiar lines (See Fringed Gentian). The shrub doubtless gives the small bees and flies their last feast of the season in consideration of their services in transferring pollen from the staminate to the fertile flowers. Very slowly through the succeeding year the seeds within the woody capsules mature until, by the following autumn, when fresh flowers appear, they are ready to bombard the neighborhood after the violets' method, in the hope of landing in moist yielding soil far from the parent shrub to found a new colony. Just as a watermelon seed shoots from between the thumb and forefinger pinching it, so the large, bony, shining black, white-tipped witch-hazel seeds are discharged through the elastic rupture of their capsule whose walls pinch them out. To be suddenly hit in the face by such a missile brings no smile while the sting lasts. Witch-hazel twigs ripening indoors transform a peaceful living room into a defenseless target for light artillery practice.

Nowhere more than in the naming of wild flowers can we trace the homesickness of the early English colonists in America. Any plant even remotely resembling one they had known at home was given the dear familiar name. Now our witch-hazel, named for an English hazel tree of elm lineage, has similar leaves it is true, but likeness stops there; nevertheless, all the folklore clustered about that mystic tree has been imported here with the title. By the help of the hazel's divining-rod the location of hidden springs of water, precious ore, treasure, and thieves may be revealed, according to old superstition. Cornish miners, who live in a land so plentifully stored with tin and copper lodes they can have had little difficulty in locating seams of ore with or without a hazel rod, scarcely ever sink a shaft except by its direction.

The literature of Europe is filled with allusions to it. Swift wrote:

"They tell us something strange and odd
 About a certain magic rod
 That, bending down its top divines

Where'er the soil has hidden mines
Where there are none, it stands erect
Scorning to show the least respect."

A good story is told on Linnaeus in Baring-Gould's "Curious Myths of the Middle Ages": "When the great botanist was on one of his voyages, hearing his secretary highly extol the virtues of his divining-wand, he was willing to convince him of its insufficiency, and for that purpose concealed a purse of one hundred ducats under a ranunculus, which grew by itself in a meadow, and bid the secretary find it if he could. The wand discovered nothing, and Linnaeus's mark was soon trampled down by the company present, so that when he went to finish the experiment by fetching the gold himself, he was utterly at a loss where to find it. The man with the wand assisted him, and informed him that it could not lie in the way they were going, but quite the contrary so they pursued the direction of the wand, and actually dug out the gold. Linnaeus said that another such experiment would be sufficient to make a proselyte of him."

Many a well has been dug even in this land of liberty where our witch-hazel indicated; but here its kindly magic is directed chiefly through the soothing extract distilled from its juices.

FIVE-FINGER; COMMON CINQUEFOIL
(Potentilla Canadensis) Rose family

Flowers - Yellow, 1/4 to 1/2 in. across, growing singly on long peduncles from the leaf axils. Five petals longer than the 5 acute calyx lobes with 5 linear bracts between them; about 20 stamens; pistils numerous, forming a head. Stem: Spreading over ground by slender runners or ascending. Leaves: 5-fingered, the digitate, saw-edged leaflets (rarely 3 or 4) spreading from a common point, petioled; some in a tuft at base. Preferred Habitat - Dry fields, roadsides, hills, banks. Flowering Season - April-August. Distribution - Quebec to Georgia, and westward beyond the Mississippi.

Everyone crossing dry fields in the eastern United States and Canada at least must have trod on a carpet of cinquefoil (cinque = five, feuilles = leaves), and have noticed the bright little blossoms among the pretty foliage, possibly mistaking the plant for its cousin, the trefoliate barren strawberry (q.v.). Both have flowers like miniature wild yellow roses. During the Middle Ages, when misdirected zeal credited almost any plant with healing virtues for every ill that flesh is heir to, the cinquefoils were considered most potent remedies, hence their generic name.

The SHRUBBY CINQUEFOIL, or PRAIRIE WEED (P. fructicosa), becomes fairly troublesome in certain parts of its range, which extends from Greenland to Alaska, and southward to New Jersey, Arizona, and California; as well as over northern Europe and Asia. It is a bushy, much branched, and leafy shrub, six inches to four feet high), with bright yellow, five-parted flowers an inch across, more or less, either solitary or in cymes at the tips of the branches. They appear from June to September. The honeybee, alighting in the center of a blossom and turning around, passes its tongue over the entire nectar-bearing ring at the base of the stamens, then proceeding to another flower to do likewise, effects cross-fertilization regularly. On a sunny day the bright blossoms attract many visitors of the lower grade out after nectar and pollen, the beetles often devouring the anthers in their greed. The leaves on this cinquefoil are usually compounded of one terminal and four side leaflets that are narrowly oblong, an inch or less in length, and silky hairy. Sometimes there may be seven leaflets pinnately, not digitately, arranged. Although the shrubby cinquefoil prefers swamps and moist, rocky places to dwell in, it wisely adapts itself, as globe-trotters should, to whatever conditions it meets.

SILVERY or HOARY CINQUEFOIL (P. argentea), found in dry soil, blooming from May to September from Canada to Delaware, Indiana, Kansas, and Dakota, also in Europe and Asia, has yellow flowers only about a quarter of an inch across, but foliage of special beauty. From the tufted, branching, ascending stems, four to twelve inches long, the finely cleft, five-foliate leaves are spread on foot stems that diminish in size as they ascend, not to let the upper leaves shut off the light from the lower ones. These leaves are smooth and green above, silvery on the under side, with fine white hairs, adapted for protection from excessive sunlight and too rapid transpiration of precious moisture. They entirely conceal the sensitive epidermis from which they grow.

YELLOW AVENS; FIELD AVENS
(Geum strictum) Rose family

Flowers - Golden yellow, otherwise much resembling the lower growing white avens (q.v.). Preferred Habitat - Low ground, moist meadows, swamps. Flowering Season - June-August. Distribution - Pennsylvania, Missouri, and Arizona, far northward.

After the marsh marigolds have withdrawn their brightness from low-lying meadows, blossoms of yellow avens twinkle in their stead. In autumn the jointed, barbed styles, protruding from the seed clusters, steal a ride by the same successful method of travel to new colonizing ground adopted by burdocks, goose-grass, tick-trefoils (q.v.), agrimony, and a score of other "tramps of the vegetable world."

TALL or HAIRY AGRIMONY
(Agrimonia hirsuta; Eupatoria of Gray) Rose family

Flowers - Yellow, small, 5-parted, in narrow, spike-like racemes. Stem: Usually 3 to 4 ft. tall, sometimes less or more clothed, with long, soft hairs. Leaves: Large, thin, bright green, compounded of (mostly) 7 principal oblong, coarsely saw-edged leaflets, with pairs of tiny leaflets between. Preferred Habitat - Woods, thickets, edges of fields. Flowering Season - June-August. Distribution - North Carolina, westward to California, and far north.

Quite a different species, not found in this country, is the common European Agrimony - A. Eupatoria of Linnaeus - which figures so prominently in the writings of medieval herbalists as a cure-all. Slender spires of green fruit below and yellow flowers above curve and bend at the borders of woodlands here apparently for no better reason than to enjoy life. Very few insects visit them, owing to the absence of nectar - certainly not the highly specialized and intelligent "Humble-Bee," to whom Emerson addressed the lines:

"Succory to match the sky,
Columbine with horn of honey,
Scented fern and agrimony,
Clover, catch-fly, adder's-tongue,
And brier-roses, dwelt among."

It is true the bumblebee may dwell among almost any flowers, but he has decided preferences for such showy ones as have adapted themselves to please his love of certain colors (not yellow), or have secreted nectar so deeply hidden from the mob that his long tongue may find plenty preserved when he calls. Occasional visitors alighting on the agrimony for pollen may distribute some, but the little blossoms chiefly fertilize themselves. When crushed they give forth a faint, pleasant odor. Pretty, nodding seed urns, encircled with a rim of hooks, grapple the clothing of man or beast passing their way, in the hope of dropping off in a suitable place to found another colony.

SENSITIVE PEA; WILD or SMALL-FLOWERED SENSITIVE PLANT
(Cassia nictitans) Senna family

Flowers - Yellow, regular, 5-parted, about 1/4 in. across; 2 or 3 together in the axils. Stem: Weak, 6 to 15 in. tall, branching, leafy. Leaves: Alternate, sensitive, compounded of 12 to 44 small, narrowly oblong leaflets; a cup-shaped gland below lowest pair; stipules persistent. Fruit: A pod, an inch long or more, containing numerous seeds. Preferred Habitat - Dry fields, sandy wasteland, roadsides. Flowering Season - July-October. Distribution - New England westward to Indiana, south to Georgia and Texas.

How many of us ever pause to test the sensitiveness of this exquisite foliage that borders the roadsides, and in appearance is almost identical with the South American sensitive plant's, so commonly cultivated in hothouses here? Failing to see its fine little leaflets fold together instantly when brushed with the hand, as they do in the tropical species (Mimosa pudica), many pass on, concluding its title a misnomer. By simply touching the leaves, however roughly, only a tardy and slight movement follows. A sharp blow produces quicker effect, while if the whole plant be shaken by forcibly snapping the stem with the finger, all the leaves will be strongly affected; their sensitiveness being apparently more aroused by vibration through jarring than by contact with foreign bodies. The leaves, which ordinarily spread out flat, partly close in bright sunshine and "go to sleep" at night, not to expose their sensitive upper surfaces to fierce heat in the first case, and to cold by radiation in the second. "Lifeless things may be moved or acted on," says Asa Gray; "living beings move and act - plants less conspicuously, but no less really than animals. In sharing the mysterious gift of life they share some of its simpler powers."

The PARTRIDGE PEA or LARGE-FLOWERED SENSITIVE PLANT (C. Chamaecrista) likewise goes to sleep; the ten to fifteen pairs of leaflets which, with a terminal one, make up each pinnate leaf, slowly turning their outer edges uppermost after sunset, and overlapping as they flatten themselves against their common stem until the entire aspect of the plant is changed. By day the expanded foliage is feathery, fine, acacia-like; at night the bushy, branching, spreading plant, that measures only a foot or two high, appears to produce nothing but pods. These leaves respond slowly to vibration, just as the sensitive pea's do. In spite of their names, neither produces the butterfly-shaped (papilionaceous) blossom of true peas. The partridge pea bears from two to four showy flowers together, each measuring an inch or more across, on a slender pedicel from the axils. It fully expands only four of its five bright yellow petals; they are somewhat unequal in size, the upper ones, with touches of red at the base, as pathfinders, not, however, as nectar-guides, since no sweets are secreted here. Curiously enough, both right and left hand flowers are found upon the same plant; that is to say, the sickle-shaped pistil turns either to the right or the left. One lateral petal, instead of being flexible and spread like the rest, stands so stiffly erect and incurved that it commonly breaks on being bent back. Why? The pistil, it will be noticed, points away from the ten long black anthers. Obviously, then, the flower cannot fertilize itself.

Its benefactors are bumblebee females and workers out after pollen. Cup-shaped nectaries ("extra nuptial") are situated on the upper side and near the base of the leaf stalks on these cassia plants, where they can have no direct influence on the fertilization of the blossoms. Apparently, they are free lunch-counters, kept open out of pure charity. Landing upon the long black anthers with pores in their tips to let out the pollen, the bumblebees "seize them between their mandibles, says Professor Robertson, "and stroke them downward with a sort of milking motion. The pollen…falls either directly upon the bee or upon the erect lateral petal which is pressed close against the bee's side. In this way the side of the bee which is next to the incurved petal receives the most pollen…. A bee visiting a left-hand flower receives pollen upon the right side, and then flying to a right-hand flower, strikes the same side against the stigma." When we find circular holes in these petals we may know the leaf-cutter or upholsterer bee (Megachile brevis) has been at work collecting roofs for her nurseries (see Hairy Ruellia). The partridge pea, which has a more westerly range than the sensitive pea's, extends it southward even to Bolivia. Game birds, migrants and rovers, which feed upon the seeds, have of course helped in their wider distribution. The plant blooms from July to September.

WILD or AMERICAN SENNA
(Cassia Marylandica) Senna family

Flowers - Yellow, about 3/4 in. broad, numerous, in short axillary clusters on the upper part of plant. Calyx of 5 oblong lobes; 5 petals, 3 forming an upper lip, 2 a lower one; 10 stamens of 3 different kinds; 1 pistil. Stem: 3 to 8 ft. high, little branched. Leaves: Alternate, pinnately compounded of 6 to 10 pairs of oblong leaflets. Fruit: A narrow, flat curving pod, 3 to 4 in. long. Preferred Habitat - Alluvial or moist, rich soil, swamps, roadsides. Flowering Season - July-August. Distribution - New England, westward to Nebraska, south to the Gulf States.

Whoever has seen certain Long Island roadsides bordered with wild senna, the brilliant flower clusters contrasted with the deep green of the beautiful foliage, knows that no effect produced by art along the drives of public park or private garden can match these country lanes in simple charm. Bumblebees, buzzing about the blossoms, may be observed "milking" the anthers just as they do those of the partridge pea. No red spots on any of these petals guide the visitors, as in the previous species, however; for do not the three small, dark stamens, which are reduced to mere scales, answer every purpose as pathfinders here? The stigma, turned sometimes to the right, sometimes to the left, strikes the bee on the side; the senna being what Delpino, the Italian botanist, calls a pleurotribe flower.

While leaves of certain African and East Indian species of senna are most valued for their medicinal properties, those of this plant are largely collected in the Middle and Southern States as a substitute. Caterpillars of several sulphur butterflies, which live exclusively on cassia foliage, appear to feel no evil effects from overdoses.

WILD INDIGO; YELLOW or INDIGO BROOM; HORSEFLY-WEED
(Baptisia tinctoria) Pea family

Flowers - Bright yellow, papilionaceous, about 1/2 in. long, on short pedicels, in numerous but few flowered terminal racemes. Calyx light green, 4 or 5-toothed; corolla of 5 oblong petals, the standard erect, the keel enclosing 10 incurved stamens and pistil. Stem: Smooth, branched, 2 to 4 ft. high. Leaves: Compounded of 3 ovate leaflets. Fruit: A many-seeded round or egg-shaped pod tipped with the awl-shaped style. Preferred Habitat - Dry, sandy soil. Flowering Season - June-September. Distribution - Maine and Minnesota to the Gulf States.

Dark grayish green, clover-like leaves, and small, bright yellow flowers growing in loose clusters at the ends of the branches of a bushy little plant, are so commonly met with they need little description. A relative, the true indigo-bearer, a native of Asia, once commonly grown in the Southern States when slavery made competition with Oriental labor possible, has locally escaped and become naturalized. But the false species, although, as Dr. Gray says, it yields "a poor sort of indigo," yields a most valuable medicine

363

employed by the homeopathists in malarial fevers. The plant turns black in drying. As in the case of other papilionaceous blossoms, bees are the visitors best adapted to fertilize the flowers. When we see the little, sleepy, dusky-winged butterfly (Thanaos brizo) around the plant we may know she is there only to lay eggs, that the larvae and caterpillars may find their favorite food at hand on waking into life.

RATTLE-BOX
(Crotalaria sagittalis) Pea family

Flowers - Yellow, 1/2 in. long or less, usually only 2 or 3 on a long peduncle. Calyx 5-toothed, slightly 2-lipped; corolla papilionaceous. Stem: 3 to 10 in. high, weak, hairy. Leaves: Alternate, simple, oval to lance-shaped; stipules arrow-shaped above and running along stem. Fruit: An inflated oblong pod 1 in, long, blackish, seedy. Preferred Habitat - Dry, sandy, open situations. Flowering Season - June-September. Distribution - New England and Minnesota to the Gulf of Mexico.

These insignificant little yellow flowers attract scant notice from human observers accustomed to associate their generic name with some particularly beautiful relatives from the West Indies grown in hothouses here. But did not small bees alight on the keel and depress it, as in the lupine, next of kin (q.v.) there might be no seeds to rattle in the dark inflated pods that so delight children. (Krotalon = a castanet.)

YELLOW SWEET CLOVER; YELLOW MELILOT
(Melilotus officinalis) Pea family

Resembling the white sweet clover, except in color. (q.v.)

YELLOW or HOP CLOVER
(Trifotium agrarium) Pea family

Flowers - Yellow, scale-like, overlapping in a densely many-flowered oblong head about 1/2 in. long, becoming brown with age. Stem: Ascending, branched, 6 to 18 in. high. Leaves: 3-foliate, very finely toothed. Preferred Habitat - Waste places, fields, roadsides. Flowering Season - May-September. Distribution - Virginia to Iowa, and far northward.

What did the sulphur butterflies provide as food for their caterpillar babies before the commonest clovers came over from the Old World to possess the soil? Wherever a trifolium grows, there one is sure to see

"gallow-yellow butterflies,
Like blooms of lorn primroses blowing loose,
    when autumn winds arise."

The BLACKSEED HOP CLOVER, BLACK or HOP MEDIC (Medicago lupulina), with even smaller, bright yellow oblong heads which turn black when ripe, lies on the ground, its branches spreading where they leave the root. A native of Europe and Asia, it is now distributed as a common weed throughout our area, for there is scarcely a month in the year when it does not bloom and set seed. It is still another of the many plants known as the shamrock.

YELLOW WOOD-SORREL; LADY'S SORREL
(Oxalis stricta) Wood-sorrel family

Flowers - Golden, fragrant, in long peduncled, small, terminal groups. Calyx of 5 sepals; corolla of 5 petals, usually reddish at base; stamens, 10; 1 pistil with 5 styles; followed by slender pods. Stem: Pale, erect, 3 to 12 in. high, the sap sour. Leaves: Palmately compound, of 3 heart-shaped, clover-like leaflets on long petioles. Preferred Habitat - Open woodlands, waste or cultivated soil, roadsides. Flowering Season - April-October. Distribution - Nova Scotia and Dakota westward to the Gulf of Mexico.

An extremely common little weed, whose peculiarly sensitive leaves children delight to set in motion by rubbing, or to chew for the sour juice. Concerning the night "sleep" of wood-sorrel leaves and the two kinds of flowers these plants bear, see the white and violet wood-sorrels.

WILD or SLENDER YELLOW FLAX
 (Linum Virginianum) Flax family

Flowers - Yellow, about 1/3 in. across, each from a leaf axil, scattered along the slender branches. Sepals, 5; 5 petals, 5 stamens. Stem: 1 to 2 ft. high, branching, leafy. Leaves. Alternate, seated on the stem; small, oblong, or lance-shaped, 1 nerved.
Preferred Habitat - Dry woodlands and borders; shady places.
Flowering Season - June-August.
Distribution - New England to Georgia.

Certainly in the Atlantic States this is the commonest of its slender, dainty tribe; but in bogs and swamps farther southward and westward to Texas the RIDGED YELLOW FLAX (L. striatum), with leaves arranged opposite each other up to the branches and an angled stem so sticky it "adheres to paper in which it is dried," takes its place.

"Blue were her eyes as the fairy flax,"

wrote Longfellow, as if blue flax were a familiar sight on this side of the Atlantic. The charming little European plant (L. usitatissimum), which has furnished the fiber for linen and the oily seeds for poultices from time immemorial, is only a fugitive from cultivation here. Unhappily, it is rarely met with along the roadsides and railways as it struggles to gain a foothold in our waste places. Possibly Longfellow had in mind the blue toad flax (q.v.).

JEWEL-WEED; SPOTTED TOUCH-ME-NOT: SILVER CAP; WILD BALSAM: LADY'S

EARDROPS; SNAP WEED; WILD LADY'S SLIPPER
(Impatiens biflora; I. fulva of Gray) Jewel-weed family

Flowers - Orange yellow, spotted with reddish-brown, irregular, 1 in. long or less, horizontal, 2 to 4 pendent by slender footstalks on a long peduncle from leaf axils. Sepals, 3, colored; 1 large, sac-shaped, contracted into a slender incurved spur and 2-toothed at apex; 2 other sepals small. Petals, 3; 2 of them 2-cleft into dissimilar lobes; 5 short stamens, 1 pistil. Stem: 2 to 5 ft. high, smooth, branched, colored, succulent. Leaves: Alternate, thin, pale beneath, ovate, coarsely toothed, petioled. Fruit: An oblong capsule, its 5 valves opening elastically to expel the seeds. Preferred Habitat - Beside streams, ponds, ditches; moist ground. Flowering Season - July-October. Distribution - Nova Scotia to Oregon, south to Missouri and Florida.

These exquisite, bright flowers, hanging at a horizontal, like jewels from a lady's ear, may be responsible for the plant's folk name; but whoever is abroad early on a dewy morning, or after a shower, and finds notched edges of the drooping leaves hung with scintillating gems, dancing, sparkling in the sunshine, sees still another reason for naming this the jewel-weed. In a brook, pond, spring, or wayside trough, which can never be far from its haunts, dip a spray of the plant to transform the leaves into glistening silver. They shed water much as the nasturtium's do.

When the tiny ruby-throated hummingbird flashes northward out of the tropics to spend the summer, where can he hope to find nectar so deeply secreted that not even the long-tongued bumblebee may rob him of it all? Beyond the bird's bill his tongue can be run out and around curves no other creature can reach. Now the early blooming columbine, its slender cornucopias brimming with sweets, welcomes the messenger whose needle-like bill will carry pollen from flower to flower; presently the coral honeysuckle and the scarlet painted-cup attract him by wearing his favorite color; next the jewel-weed hangs horns of plenty to lure his eye; and the trumpet vine and cardinal flower continue to feed him successively in Nature's garden; albeit cannas, nasturtiums, salvia, gladioli, and such deep, irregular showy flowers in men's flower beds sometimes lure him away. These are bird flowers dependent in the main on the ruby-throat, which is not to say that insects never enter them, for they do; only they are not the visitors catered to. Watch the big, velvety bumblebee approach a roomy jewel-weed blossom and nearly disappear within. The large bunch of united stamens, suspended directly over the entrance, bears copious white pollen. So much comes off on his back that after visiting a flower or two he becomes annoyed; clings to a leaf with his fore legs while he thoroughly brushes his back and wings with his middle and hind pairs, and then collects the sticky grains into a wad on his feet which he presently kicks off with disgust to the ground. Examine a jewel-weed blossom to see that the clumsy bumblebee's pollen-laden back is not so likely to come in contact with the short five-parted stigma concealed beneath the stamens, as a hummingbird's slender bill that is thrust obliquely into the spur while he hovers above.

But, as if the plant had not sufficient confidence in its visitors to rely exclusively on them for help in continuing the lovely species, it bears also cleistogamous blossoms that never open - economical products without petals, which ripen abundant self-fertilized seed (see white wood sorrel). It is calculated that each jewel-weed blossom produces about two hundred and fifty pollen grains; yet each is by no means able to produce seed in spite of its prodigality. Nevertheless, enough cross-fertilized seed is set to save the species from the degeneracy that follows close inbreeding among plants as well as animals. In England, where this jewel-weed is rapidly becoming naturalized, Darwin recorded there are twenty plants producing cleistogamous flowers to one having showy blossoms which, even when produced, seldom set seed. What more likely, since hummingbirds are confined to the New World? Therefore why should the plant waste its energy on a product useless in England? It can never attain perfection there until hummingbirds are imported, as bumblebees had to be into Australia before the farmers could harvest seed from their clover fields (see red clover).

Familiar as we may be with the nervous little seedpods of the touch-me-not, which children ever love to pop and see the seeds fly, as they do from balsam pods in grandmother's garden, they still startle with the suddenness of their volley. Touch the delicate hair-trigger at the end of a capsule, and the lightning response of the flying seeds makes one jump. They sometimes land four feet away. At this rate of progress a year, and with the other odds against which all plants have to contend, how many generations must it take to fringe even one mill pond with jewel-weed; yet this is rapid transit indeed compared with many of Nature's processes. The plant is a conspicuous sufferer from the dodder (q.v.).

The PALE TOUCH-ME-NOT (I. aurea; I. pallida of Gray) most abundant northward, a larger, stouter species found in similar situations, but with paler yellow flowers only sparingly dotted if at all, has its broader sac-shaped sepal abruptly contracted into a short, notched, but not incurved spur. It shares its sister's popular names.

VELVET LEAF; INDIAN MALLOW; AMERICAN JUTE
(Abutilon Abulilon; A. Avicennae of Gray) Mallow family

Flowers - Deep yellow, 1/2 to 3/4 in. broad, 5-parted, regular, solitary on stout peduncles from the leaf axils. Stem: 3 to 6 ft. high, velvety, branched. Leaves: Soft velvety, heart-shaped, the lobes rounded, long petioled. Fruit: In a head about 1 in. across, 12 to 15 erect hairy carpels, with spreading sharp beaks. Preferred Habitat - Escaped from cultivation to waste sandy loam, fields, roadsides. Flowering Season - August-October. Distribution - Common or frequent, except at the extreme North.

There was a time, not many years ago, when this now common and often troublesome weed was imported from India and tenderly cultivated in flower gardens. In the Orient it and allied species are grown for their fiber, which is utilized for cordage and cloth; but the equally valuable plant now running wild here has yet to furnish American men with a profitable industry. Although the blossom is next of kin to the veiny Chinese bell-flower, or striped abutilon, so common in greenhouses, its appearance is quite different.

ST. ANDREW'S CROSS
(Ascyrum hypericoides; A. Crux-Andreae of Gray) St.
John's-wort family

Flowers - Yellow, 1/2 to 3/4 in. across, terminal and from the leaf axils. Calyx of 4 sepals in 2 pairs; 4 narrow, oblong petals; stamens numerous; 2 styles. Stem: Much branched and spreading from base, 5 to 10 in. high, leafy. Leaves: Opposite, oblong, small, seated on stem. Preferred Habitat - Dry, sandy soil; pine barrens. Flowering Season - July-August. Distribution - Nantucket Island (Mass.), westward to Illinois, south to Florida and Texas.

Because the four pale yellow petals of this flower approach each other in pairs, suggesting a cross with equals arms, the plant was given its name by Linnaeus in 1753. ST. PETER'S-WORT (A. stans), a similar plant, found in the same localities, in bloom at the same time, has larger flowers in small clusters at the tips only of its upright branches.

COMMON ST. JOHN'S-WORT
(Hypericum perforatum) St. John's-wort family

Flowers - Bright yellow, 1 in. across or less, several or many in terminal clusters. Calyx of 5 lance-shaped sepals; 5 petals dotted with black; numerous stamens in 3 sets 3 styles. Stem: to 2 ft. high, erect, much branched. Leaves: Small, opposite, oblong, more or less black-dotted. Preferred Habitat - Fields, waste lands, roadsides. Flowering Season - June-September. Distribution - Throughout our area, except the extreme North; Europe, and Asia.

"Gathered upon a Friday, in the hour of Jupiter when he comes to his operation, so gathered, or borne, or hung upon the neck, it mightily helps to drive away all phantastical spirits." These are the blossoms which have been hung in the windows of European peasants for ages on St. John's eve, to avert the evil eye and the spells of the spirits of darkness. "Devil chaser" its Italian name signifies. To cure demoniacs, to ward off destruction by lightning, to reveal the presence of witches, and to expose their nefarious practices, are some of the virtues ascribed to this plant, which superstitious farmers have spared from the scythe and encouraged to grow near their houses until it has become, even in this land of liberty, a troublesome weed at times. "The flower gets its name," says F. Schuyler Mathews, "from the superstition that on St. John's day, the 24th of June, the dew which fell on the plant the evening before was efficacious in preserving the eyes from disease. So the plant was collected, dipped in oil, and thus transformed into a balm for every wound." Here it is a naturalized, not a native, immigrant. A blooming plant, usually with many sterile shoots about its base, has an unkempt, untidy look; the seed capsules and the brown petals of withered flowers remaining among the bright yellow buds through a long season. No nectar is secreted by the St. John's-worts, therefore only pollen collectors visit them regularly, and occasionally cross-fertilize the blossoms, which are best adapted, however, to pollinate themselves.

The SHRUBBY ST. JOHN'S-WORT (H. prolificum) bears yellow blossoms, about half an inch across, which are provided with stamens so numerous, the many flowered terminal clusters have a soft, feathery effect. In the axils of the oblong, opposite leaves are tufts of smaller ones, the stout stems being often concealed under a wealth of foliage. Sandy or rocky places from New Jersey southward best suit this low, dense, diffusely branched shrub which blooms prolifically from July to September.

Farther north, and westward to Iowa, the GREAT or GIANT ST. JOHN'S-WORT (H. Ascyron) brightens the banks of streams at midsummer with large blossoms, each on a long footstalk in a few-flowered cluster.

LONG-BRANCHED FROST-WEED; FROST-FLOWER; FROST-WORT; CANADIAN
ROCK-ROSE
(Helianthemum Canadense) Rock-rose family

Flowers - Solitary, or rarely 2; about 1 in. across, 5-parted, with showy yellow petals; the 5 unequal sepals hairy. Also abundant small flowers lacking petals, produced from the axils later. Stem: Erect, 3 in. to 2 ft. high; at first simple, later with elongated branches. Leaves: Alternate, oblong, almost seated on stem. Preferred Habitat - Dry fields, sandy or rocky soil. Flowering Season - Petal-bearing flowers, May-July. Distribution - New England to the Carolinas, westward to Wisconsin and Kentucky.

Only for a day, and that must be a bright sunny one, does the solitary frost-flower expand its delicate yellow petals. On the next, after pollen has been brought to it by insect messengers and its own carried away, the now useless petal advertisements fall, and the numerous stamens, inserted upon the receptacle with them, also drop off, leaving the club-shaped pistil to develop with the ovary into a rounded, ovoid, three-valved capsule. Notice how flat the stamens lie upon the petals to keep safely out of reach of the stigma. Another flower, exactly like the first, now expands, and the bloom continues for weeks. Why does only one blossom open at a time? Because the whole aim of the showy flowers is to set cross-fertilized seed, and when only one at a time appears, pollination not only between distinct blossoms but between distinct plants insures the healthiest, most vigorous offspring - a wise precaution against degeneracy, in view of the quantities of self-fertilized seed that will be set late in summer by the tiny apetalous flowers that never open (see white wood sorrel). Surely two kinds of blossoms should be enough for any species; but why call this the frost-flower when its bloom is ended by autumn? Only the witch-hazel may be said to flower for the first time after frost. When the stubble in the dry fields is white some cold November morning, comparatively few notice the ice crystals, like specks of glistening quartz, at the base of the stems of this plant. The similar HOARY FROST-WEED (H. majus), whose showy flowers appear in clusters at the hoary stein's summit, in June and July, also bears them. Often this ice formation assumes exquisite feathery, whimsical forms, bursting the bark asunder where an astonishing quantity of sap gushes forth and freezes. Indeed, so much sap sometimes goes to the making of this crystal flower, that it would seem as if an extra reservoir in the soil must pump some up to supply it with its large fantastic corolla.

BEACH or FALSE HEATHER; POVERTY GRASS
(Hudsonia tomentosa) Rock-rose family

Flowers - Bright yellow, small, about 1/4 in. across, numerous, closely ascending the upper part of the heath-like branches. Sepals 5, unequal; 5 petals; stamens, 9 to 18. Stem: 4 to 8 in. tall, tufted, densely branched and matted, hoary hairy, pale. Leaves: Overlapping like scales, very small. Preferred Habitat - Sands of the seashore, pine barrens, beaches of rivers and lakes. Flowering Season - May-July. Distribution - New Brunswick to Maryland, west to Lake of the Woods.

Like the showy flowers of the frost-weed, these minute ones open in the sunshine only, and then but for a single day. Nevertheless, the hoary, heath-like little shrub, by growing in large colonies and keeping up a succession of bright bloom, tinges the sand dunes back of the beach with charming color that artists delight to paint in the foreground of their marine pictures.

YELLOW VIOLETS
(Viola) Violet family

Fine hairs on the erect, leafy, usually single stem of the DOWNY YELLOW VIOLET (V. pubescens), whose dark veined, bright yellow petals gleam in dry woods in April and May, easily distinguish it from the SMOOTH YELLOW VIOLET (V. scabriuscula), formerly considered a mere variety in spite of its being an earlier bloomer, a lover of moisture, and well equipped with basal leaves at flowering time, which the downy species is not. Moreover, it bears a paler blossom, more coarsely dentate leaves, often decidedly taper-pointed, and usually several stems together.

Our other common yellow species, the ROUND-LEAVED VIOLET (V. rotundifolia), lifts smaller, pale, brown-veined, and bearded blossoms above a tuffet of broad, shining leaves close to the ground.

The veins on the petals serve as pathfinders to the nectary for the bee, and the beard as footholds, while she probes the inverted blossoms. Such violets as have their side petals bearded are most frequently visited by small greenish mason bees (Osmia), with collecting brushes on their abdomen that receive the pollen as it falls. Abundant cleistogamous flowers (see blue violets and white wood sorrel) are borne on the runners late in the season. Bryant, whose botanical lore did not always keep step with his Muse, wrote of the yellow violet as the first spring flower, because he found it "by the snowbank's edges cold," one April day, when the hepaticas about his home at Roslyn, Long Island, had doubtless been in bloom a month.

"Of all her train the hands of Spring
First plant thee in the watery mould,"

he wrote, regardless of the fact that the round-leaved violet's preferences are for dry, wooded, or rocky hillsides. Muller believed that all violets were originally yellow, not white, after they evoluted from the green stage.

EASTERN CACTUS; PRICKLY PEAR; INDIAN FIG
(Opuntia Opuntia; 0. vulgaris of Gray) Cactus family

Flowers -Yellow, sometimes reddish at center, 2 to 3 in. across, solitary, mostly seated at the side of joints. Calyx tube not prolonged beyond ovary, its numerous lobes spreading. Petals numerous; stamens very numerous; ovary cylindric; the style longer than stamens, and with several stigmas. Stem: Prostrate or ascending, fleshy, juicy, branching, the thick, flattened joints oblong or rounded, 2 to 5 in. long. Leaves: Tiny, awl-shaped, dotting the joints, but usually falling early; tufts of yellowish bristles at their base. Plant unarmed, or with few solitary stout spines. Fruit: Pear-shaped, pulpy, red, nearly smooth, 1 in. long or over, edible. Preferred Habitat - Sandy or dry or rocky places. Flowering Season - June-August. Distribution - Massachusetts to Florida.

Upwards of one hundred and fifty species of Opuntia, which elect to grow in parching sands, beneath a scorching sun, often prostrate on baking hot rocks, on glaring plains, beaches, and deserts, from Massachusetts to Peru - for all are natives of the New World - show so marvelous an adaptation to environment in each instance that no group of plants is more interesting to the botanist, more decorative in form and color from an artistic standpoint, more distinctively characteristic. Plants choosing such habitats as they have adopted, usually in tropical or semi-tropical regions, had to resort to various expedients to save loss of water through transpiration and evaporation. Now, as leaves are the natural outlets for moisture thrown off by any plant, manifestly the first thing to do was either to reduce the number of branches and leaves, or to modify them into sharp spines (not surface prickles like the rose's); to cultivate a low habit of growth, not to expose unnecessary surface to sun and air; to thicken the skin until little moisture could evaporate through the leathery coat; and, finally, to utilize the material thus saved in developing stems so large, fleshy, and juicy that they should become wells in a desert, with powers of sustenance great enough to support the plant through its fiery trials. A common expedient of plants in dry situations, even at the north, is to modify their leaves into spines, as the gorse and the barberry, for example, have done. That such an armor also serves to protect them against the ravages of grazing animals is an additional advantage, of course; but not their sole motive in wearing it. Popular to destruction would the cool juices of the cacti be in thirsty lands, if only they might be obtained without painful and often poisonous scratches. Given moist soil and greater humidity of atmosphere to grow in, spiny plants at once show a tendency to grow taller, to branch and become leafy. A covering of hairs which reflect the light, thus diminishing the amount that might reach the juicy interior area, has likewise been employed by many cacti, among other denizens of dry soil.

In this common prickly pear cactus of the Atlantic seaboard, where the air is laden with moisture from the ocean, few or no spines are produced; and dotted over the surface of its branching, fleshy, flattened joints we find tiny, awl-shaped leaves, whereas foliage is entirely wanting in the densely prickly, rounded, solid, unbranched, hairy cacti of the southwestern deserts, and the arid plains of Mexico.

In sunshine the beautiful yellow blossom of our prickly pear expands to welcome the bees, folding up its petals again for several successive nights. William Hamilton Gibson says it "encloses its buzzing visitor in a golden bower, from which he must emerge at the roof as dusty as a miller," only to enter another blossom and leave some pollen on its numerous stigmas.

But the cochineal, not the bee, is forever associated with cacti in the popular mind. Indeed, several species are extensively grown on plantations, known as Nopaleries, which furnish food to countless trillions of these tiny insects. Like its relative the aphis of rose bushes (see wild roses), the cochineal fastens itself to a cactus plant by its sucking tube, to live on the juices. The males are winged, and only the female, which yields the valuable dye, sticks tight to the plant. Three crops of insects a year are harvested on a Mexican plantation. After three months' sucking, the females are brushed off, dried in ovens, and sold for about two thousand dollars a ton. The annual yield of Mexico amounting to many thousands of tons, it is no wonder the cactus plant, which furnishes so valuable an industry, should appear on the coat-of-arms of the Mexican republic. Some cacti are planted for hedges, the fruit of others furnishes a refreshing drink in tropical climates, the juices are used as a water color, and to dye candies - in short, this genus Opuntia and allied clans have great commercial value.

The WESTERN PRICKLY PEAR (0. humifusa; O. Rafnesquii of Gray) - a variable species ranging from Minnesota to Texas, is similar to the preceding, but bears a larger flower, and longer, more rounded, deeper green joints, beset with not numerous spines, scattered chiefly near their margins. A few deflexed spines in a cluster leave the surface where a tiny awl-shaped leaf and a tuft of reddish brown hairs are likewise usually found.

EVENING-PRIMROSE; NIGHT WILLOW-HERB (Onagra biennis; Qenothera biennis of Gray) Evening-primrose family

Flowers - Yellow, fragrant, opening at evening, 1 to 2 in. across, borne in terminal leafy-bracted spikes. Calyx tube slender, elongated, gradually enlarged at throat, the 4-pointed lobes bent backward; corolla of 4 spreading petals; 8 stamens; 1 pistil; the stigma 4-cleft. Stem: Erect, wand-like, or branched, to 1 to 5 ft. tall, rarely higher, leafy. Leaves: Alternate, lance-shaped, mostly seated on stem, entire, or obscurely toothed. Preferred Habitat - Roadsides, dry fields, thickets, fence-corners. Flowering Season - June-October. Distribution - Labrador to the Gulf of Mexico, west to the Rocky Mountains.

Like a ballroom beauty, the evening primrose has a jaded, bedraggled appearance by day when we meet it by the dusty roadside, its erect buds, fading flowers from last night's revelry, wilted ones of previous dissipations, and hairy oblong capsules, all crowded together among the willow-like leaves at the top of the rank growing plant. But at sunset a bud begins to expand its delicate petals slowly, timidly - not suddenly and with a pop, as the evening primrose of the garden does.

Now, its fragrance, that has been only faintly perceptible during the day, becomes increasingly powerful. Why these blandishments at such an hour? Because at dusk, when sphinx moths, large and small, begin to fly (see Jamestown weed), the primrose's special benefactors are abroad. All these moths, whose length of tongue has kept pace with the development of the tubes of certain white and yellow flowers dependent on their ministrations, find such glowing like miniature moons for their special benefit, when blossoms of other hues have melted into the deepening darkness. If such have fragrance, they prepare to shed it now. Nectar is secreted in tubes so deep and slender that none but the moths' long tongues can drain the last drop. An exquisite, little, rose-pink twilight flyer, his wings bordered with yellow, flutters in ecstasy above the evening primrose's freshly opened flowers, transferring in his rapid flight some of their abundant, sticky pollen that hangs like a necklace from the outstretched filaments. By day one may occasionally find a little fellow asleep in a wilted blossom, which serves him as a tent, under whose flaps the brightest bird eye rarely detects a dinner. After a single night's dissipation the corolla wilts, hangs a while, then drops from the maturing capsule as if severed with a sharp knife. Few flowers, sometimes only one opens on a spike on a given evening - a plan to increase the chances of cross-fertilization between distinct plants; but there is a very long succession of bloom. If a flower has not been pollenized during the night it remains open a while in the morning. Bumblebees now hurry in, and an occasional humming-bird takes a sip of nectar. Toward the end of summer, when so much seed has been set that the flower can afford to be generous, it distinctly changes its habit and keeps open house all day.

During our winter walks we shall see close against the ground the rosettes of year-old evening primrose plants - exquisitely symmetrical, complex stars from whose center the flower stalks of another summer will arise.

Floriform sunshine bursts forth from roadsides, fields, and prairies when the COMMON SUNDROPS (Kneiffia fructicosa; formerly Qenothera fructicosa) - is in flower. It is first cousin to the similar evening primrose of taller, ranker growth. Often only one blossom on a stalk expands at a time, to increase the chances of cross-fertilization between distinct plants; but where colonies grow it is a conspicuous acquaintance, for its large, bright yellow corollas remain open all day. Bumblebees with their long tongues, and some butterflies, drain the deeply hidden nectar; smaller visitors get some only when it wells up high in the tube. As the stigma surpasses the anthers, self-fertilization is impossible unless an insect blunders by alighting elsewhere than on the lower side, where the stigma is purposely turned to be rubbed against his pollen-laden ventral surface when he settles on a blossom. Unable to reach the nectar, mining and leaf-cutter bees, wasps, flower flies, and beetles visit it for the abundant pollen; and the common little white cabbage butterfly (Pieris protodice) sucks here constantly. The capsules of the sundrops are somewhat club-shaped and four-winged, angled above, with four intervening ribs between. Range from Nova Scotia to Georgia, west beyond the Mississippi.

A similar, but smaller, diurnal species (K. pumilla), likewise found blooming in dry soil from June to August, has a more westerly range North and South.

WILD OR FIELD PARSNIP; MADNEP; TANK
(Pastinaca sativa) Carrot family

Flowers - Dull or greenish yellow, small, without involucre or involucels; borne in 7 to 15 rayed umbels, 2 to 6 in. across. Stem: 2 to 5 ft. tall, stout, smooth, branching, grooved, from a long, conic, fleshy, strong-scented root. Leaves: Compounded (pinnately), of several pairs of oval, lobed, or cut, sharply toothed leaflets; the petioled lower leaves often 1 1/2 ft. long. Preferred Habitat - Waste places, roadsides, fields. Flowering Season - June-September. Distribution - Common throughout nearly all parts of the United States and Canada. Europe.

Men are not the only creatures who feed upon such of the umbel-bearing plants as are innocent - parsnips, celery, parsley, carrots, caraway, and fennel, among others; and even those which contain properties that are poisonous to highly organized men and beasts, afford harmless food for insects. Pliny says that parsnips, which were cultivated beyond the Rhine in the days of Tiberius, were brought to Rome annually to please the emperor's exacting palate; yet this same plant, which has overrun two continents, in its wild state (when its leaves are a paler yellowish green than under cultivation) often proves poisonous. A strongly acrid juice in the very tough stem causes intelligent cattle to let it alone - precisely the object desired. But caterpillars of certain swallow-tail butterflies, particularly of the common eastern swallow-tail (Papilio asterias), may be taken on it - the same greenish, black-banded, and yellow-dotted fat "worm" found on parsnips, fennel, and parsley in the kitchen garden. Insects understood plant relationships ages before Linnaeus defined them. When we see this dark, velvety butterfly, marked with yellow, hovering above the wild parsnip, we may know she is there only to lay eggs that her larvae may eat their way to maturity on this favorite food store. After the flat, oval, shining seeds with their conspicuous oil tubes are set in the spreading umbels, the strong, vigorous plant loses nothing of its decorative charm.

>From April to June the lower-growing EARLY or GOLDEN MEADOW PARSNIP (Zizia aurea) spreads its clearer yellow um-

bels above moist fields, meadows, and swamps from New Brunswick and Dakota to the Gulf of Mexico. Its leaves are twice or thrice compounded of oblong, pointed, saw-edged, but not lobed leaflets.

The HAIRY-JOINTED MEADOW PARSNIP (Thaspium barbinode), another early bloomer, with pale-yellow flowers, most common in the Mississippi basin, may always be distinguished by the little tufts of hair at the joints of the stem, the compound leaves, and often on the rays of the umbels.

A yellow variety of the PURPLE MEADOW PARSNIP, which is popularly known as GOLDEN ALEXANDERS (T. trifoliatum var. aureum), confines itself chiefly to woodlands. The leaves are compounded of three leaflets, longer and more lance-shaped in outline than those of other yellow species.

FOUR-LEAVED or WHORLED LOOSESTRIFE; CROSSWORT
 (Lysimachia quadrifolia) Primrose family

Flowers - Yellow, streaked with dark red, 1/2 in. across or less; each on a thread-like, spreading footstem from a leaf axil. Calyx, 5 to 7 parted; corolla of 5 to 7 spreading lobes, and as many stamens inserted on the throat; 1 pistil. Stem: Slender, erect, to 3 ft. tall, leafy. Leaves: In whorls of 4 (rarely in 3's to 7's), lance-shaped or oblong, entire, black dotted. Preferred Habitat - Open woodland, thickets, roadsides, moist, sandy soil. Flowering Season - June-August. Distribution - Georgia and Illinois, north to New Brunswick.

Medieval herbalists usually recorded anything that "Plinie saieth" with profoundest respect; not always so, quaint old Parkinson. Speaking of the common (vulgaris), wild loosestrife of Europe, a rather stout, downy species with terminal clusters of good-sized, yellow flowers, that was once cultivated in our Eastern States, and has sparingly escaped from gardens, he thus refers to the reputation given it by the Roman naturalist: "It is believed to take away strife, or debate between ye beasts, not onely those that are yoked together, but even those that are wild also, by making them tame and quiet...if it be either put about their yokes or their necks," significantly adding, "which how true, I leave to them shall try and find it soe." Our slender, symmetrical, common loosestrife, with its whorls of leaves and little star-shaped blossoms on thread-like pedicels at regular intervals up the stem, is not even distantly related to the wonderful purple loosestrife (q.v.).

Another common, lower-growing species, the BULB-BEARING LOOSESTRIFE (L. terrestris; L. stricta of Gray) - blooming from July to September, lifts a terminal, elongated raceme of even smaller, slender-pedicelled, yellow flowers streaked or dotted with reddish; and in the axils of its abundant, opposite, lance-shaped, black-dotted leaves, long bulblets, that are in reality suppressed branches, are usually borne after the flowering season. Occasionally no flowers are produced, only these strange bulblets. In this state Linnaeus mistook the plant for a terrestrial mistletoe. This species shows a decided preference for swamps, moist thickets, and ditches throughout a range which extends from Manitoba and Arkansas to the Atlantic Ocean.

MONEYWORT, or CREEPING LOOSESTRIFE (L. Nummularia), a native of Great Britain, which has long been a favorite vine in American hanging baskets and urns, when kept in moist soil, suspended from a veranda, will produce prolific shoots two or three feet in length, hanging down on all sides. Pairs of yellow, dark-spotted, five-lobed flowers grow from the axils of the opposite leaves from June to August. One often finds it running wild in moist soil beyond the pale of old gardens from Pennsylvania and Indiana northward into Canada. Slight encouragement in starting runaways would easily induce the hardy little evergreen to be as common here as it is in England.

The LANCE-LEAVED LOOSESTRIFE (Steironema lanceolatum), most common in the West and South, although it is by no means rare in the northeastern States, produces either single blossoms or few-flowered, spreading, axillary clusters on slender peduncles, each unspotted, yellow corolla half an inch across or over; the petal edges as if gnawed by the finest of teeth; the pointed calyx segments showing between them. Sterile stamens in addition to the fertile ones characterize this clan. In moist soil it blooms from June to August. It is a strange fact that female bees of the genus Macropis have never been taken on plants outside the loosestrife connection. Here there appears to be the closest interdependence between flower and insect. Even in Germany, Muller found them by far the most abundant visitors, "diligently sweeping the flowers (L. vulgaris) and piling large masses of moistened pollen on their hind legs." He inclined to believe that such blossoms in this group as have spots or streaks on their petals - pathfinders for insect visitors - are largely dependent on them, and cannot easily fertilize themselves; whereas the unmarked blossoms, growing in such situations as are less favorable to insect visits, are regularly self-fertile.

BUTTERFLY-WEED; PLEURISY-ROOT; ORANGE-ROOT; ORANGE MILKWEED
(Asclepias tuberosa) Milkweed family

Flowers - Bright reddish orange, in many-flowered, terminal clusters, each flower similar in structure to the common milkweed (q.v.). Stem: Erect, 1 to 2 ft. tall, hairy, leafy, milky juice scanty. Leaves: Usually all alternate, lance-shaped, seated on stem. Fruit: A pair of erect, hoary pods, 2 to 5 in. long, at least containing silky plumed seeds. Preferred Habitat - Dry or sandy fields, hills, roadsides. Flowering Season - June-September. Distribution - Maine and Ontario to Arizona, south to the Gulf of Mexico.

Intensely brilliant clusters of this the most ornamental of all native milkweeds set dry fields ablaze with color. Above them butterflies hover, float, alight, sip, and sail away - the great, dark, velvety, pipe-vine swallow-tail (Papilio philenor), its green-shaded hind wings marked with little white half moons; the yellow and brown, common, Eastern swallow-tail (P. asterias), that we saw about the wild parsnip and other members of the carrot family the exquisite, large, spice-bush swallow-tail, whose bugaboo caterpillar startled us when we unrolled a leaf of its favorite food supply (see spice-bush); the small, common, white, cabbage butterfly (Pieris protodice); the even more common little sulphur butterflies, inseparable from clover fields and mud puddles; the painted lady that follows thistles around the globe; the regal fritillary (Argynnis idalia), its black and fulvous wings marked with silver crescents, a gorgeous creature developed from the black and orange caterpillar that prowls at night among violet plants; the great spangled fritillary of similar habit; the bright fulvous and black pearl crescent butterfly (Phyciodes tharos), its small wings usually seen hovering about the asters; the little grayish-brown, coral hair-streak (Thecla titus), and the bronze copper (Chrysophanus thoe), whose caterpillar feeds on sorrel (Rumex); the delicate, tailed blue butterfly (Lycaena comyntas), with a wing expansion of only an inch from tip to tip; all these visitors duplicated again and again - these and several others that either escaped the net before they were named, or could not be run down, were seen one bright midsummer day along a Long Island roadside bordered with butterfly weed. Most abundant of all was still another species, the splendid monarch (Anosia plexippus), the most familiar representative of the tribe of milkweed butterflies (see common milkweed). Swarms of this enormously prolific species are believed to migrate to the Gulf States, and beyond at the approach of cold weather, as regularly as the birds, traveling in numbers so vast that the naked trees on which they pause to rest appear to be still decked with autumnal foliage. This milkweed butterfly "is a great migrant," says Dr. Holland, "and within quite recent years, with Yankee instinct, has crossed the Pacific, probably on merchant vessels, the chrysalids being possibly concealed in bales of hay, and has found lodgment in Australia where it has greatly multiplied in the warmer parts of the Island Continent, and has thence spread northward and westward, until in its migrations it has reached Java and Sumatra, and long ago took possession of the Philippines.... It has established a more or less precarious foothold for itself in southern England. It is well established at the Cape Verde Islands, and in a short time we may expect to hear of it as having taken possession of the Continent of Africa, in which the family of plants upon which the caterpillars feed is well represented."

Surely here is a butterfly flower if ever there was one, and such are rare. Very few are adapted to tongues so long and slender that the bumblebee cannot help himself to their nectar; but one almost never sees him about the butterfly-weed. While other bees, a few wasps, and even the ruby-throated hummingbird, which ever delights in flowers with a suspicion of red about them, sometimes visit these bright clusters, it is to the ever-present butterfly that their marvelous structure is manifestly adapted. Only visitors long of

limb can easily remove the pollinia, which are usually found dangling from the hairs of their legs. We may be sure that after generously feeding its guests, the flower does not allow many to depart without rendering an equivalent service. The method of compelling
visitors to withdraw pollen-masses from one blossom and deposit
them in another - an amazing process - has been already described
under the common milkweed. Lacking the quantity of sticky milky
juice which protects that plant from crawling pilferers, the butterfly-
weed suffers outrageous robberies from black ants. The hairs on its
stem, not sufficient to form a stockade against them, serve only as a
screen to reflect light lest too much may penetrate to the interior
juices. We learned, in studying the prickly pear cactus, how necessary it is for plants living in dry soil to guard against the escape of
their precious moisture.

Transplanted from Nature's garden into our own, into what Thoreau termed "that meager assemblage of curiosities, that poor apo-
logy for Nature and Art which I call my front yard," clumps of butterfly-weed give the place real splendor and interest. It is said the
Indians used the tuberous root of this plant for various maladies,
although they could scarcely have known that because of the alleged healing properties of the genus Linnaeus dedicated it to Aes-
culapius, of whose name Asclepias is a Latinized corruption.

HORSE-BALM; CITRONELLA; RICH-WEED; STONE-ROOT; HORSE-WEED
(Collinsonia Canadensis) Mint family

Flowers - Light yellowish, lemon-scented, about 1/2 in. long,
mostly opposite, in numerous spreading racemes, forming long,
loose terminal clusters. Calyx bell-shaped, 2-lipped, upper lip 3-
toothed, lower lip 2-cleft; corolla 5-lobed, 4 lobes nearly equal, the
fifth much larger, fringed; stamens protruding, 2 anther-bearing; 1
long style, the stigma forked. Preferred Habitat - Rich, moist woods.
Flowering Season - July-October. Distribution - New England, Ontario, and Wisconsin, south to Florida and Kansas.

Now that we have come to read the faces of flowers much as their insect friends must have done for countless ages, we suspect at a glance that the strong-scented horse-balm, with its profusion of lemon-colored, irregular little blossoms, is up to some ingenious trick. The lower lip, out of all proportion to the rest of the corolla, flaunting its enticing fringes; the long stamens protruding from some flowers, and only the long style from others on the same plant, excite our curiosity. Where many fragrant clumps grow in cool, shady woods at midsummer, is an excellent place to rest a while and satisfy it. Presently a bumblebee, attracted by the odor from afar, alights on the fringed platform too weak to hold him. Dropping downward, he snatches the filaments of the two long stamens to save himself; and, as he does so, pollen jarred out of their anther sacs falls on his thorax at the juncture of his wings. Hanging beneath the flower a second, he sips its nectar and is off. Many bees, large and small, go through a similar performance. Now the young, newly opened flowers have the forked stigmas of the long style only protruding at this stage, the miniature stamens being still curled within the tube. Obviously a pollen-dusted bee coming to one of these young flowers must rub off some of the vitalizing dust on the sticky fork that purposely impedes his entrance at the precise spot necessary. Notice that after a flower's stamens protrude in the second stage of its development the fork is turned far to one side to get out of harm's way - self-fertilization being an abomination. It was the lamented William Hamilton Gibson who first called attention to the horse-balm's ingenious scheme to prevent it.

VIRGINIA GROUND CHERRY
(Physalis Virginiana; P. Pennsylvanica of Gray) Potato family

Flowers - Sulphur or greenish yellow, with 5 dark purplish dots, 1 in. across or less, solitary from the leaf axils. Calyx 5-toothed, much inflated in fruit; corolla open bell-shaped, the edge 5-cleft; 5 stamens, the anthers yellow, style slender, 2-cleft. Stem: 1 1/2 to 3 ft. tall, erect, more or less hairy or glandular, branched, from a thick rootstock. Leaves: Ovate to lanceolate, tapering at both ends or wedge-shaped, often yellowish green, entire or sparingly wavy-toothed. Fruit: An inflated, 5-angled capsule, sunken at the base, loosely surrounding the edible reddish berry. Preferred Habitat - Open ground; rich, dry pastures; hillsides. Flowering Season - July-September Distribution - New York to Manitoba, south to the Gulf States.

A common plant, so variable, however, that the earlier botanists thought it must be several distinct species, lanceolata among others. A glance within shows that the open flower is not so generous as its spreading form would seem to indicate, for tufts of dense hairs at each side of grooves where nectar is secreted, conceal it from the mob, and, with the thickened filaments, almost close the throat. Doubtless these hairs also serve as footholds for the welcome bee clinging to its pendent host. The dark spots are pathfinders. One anther maturing after another, a visitor must make several trips to secure all the pollen, and if she is already dusted from another blossom, nine chances out of ten she will first leave some of the vitalizing dust on the stigma poked forward to receive it before collecting more. Professor Robertson says that all the ground cherries near his home in Illinois are remarkable for their close mutual relation with two bees of the genus Colletes. So far as is known, the insignificant little greenish or purplish bell-shaped flowers of the Alum-root (Heuchera Americana), with protruding orange anthers, are the only other ones to furnish these females with pollen for their babies' bread. Slender racemes of this species are found blooming in dry or rocky woods from the Mississippi eastward, from May to July, by which time the ground cherry is ready to provide for the bee's wants. The similar Philadelphia species was formerly cultivated for its "strawberry tomato." Many birds which feast on all this highly attractive fruit disperse the numerous kidney-shaped seeds.

GREAT MULLEIN; VELVET or FLANNEL PLANT; MULLEIN DOCK; AARON'S ROD
 (Verbascum Thapsus) Figwort family

Flowers - Yellow, 1 in. across or less, seated around a thick, dense, elongated spike. Calyx 5-parted; corolla of 5 rounded lobes; 5 anther-bearing stamens, the 3 upper ones short, woolly; 1 pistil. Stem: Stout, 2 to 7 ft. tall, densely woolly, with branched hairs. Leaves: Thick, pale green, velvety-hairy, oblong, in a rosette on the ground; others alternate, strongly clasping the stem. Preferred Habitat - Dry fields, banks, stony waste land. Flowering Season - June-September. Distribution - Minnesota and Kansas, eastward to Nova Scotia and Florida. Europe.

Leaving the fluffy thistle-down he has been kindly scattering to the four winds, the goldfinch spreads his wings for a brief undulating flight, singing in waves also as he goes to where tall, thick-set mullein stalks stand like sentinels above the stony pasture. Here companies of the exquisite little black and yellow minstrels delight to congregate with their somber families and feast on the seeds that rapidly follow the erratic flowers up the gradually lengthening spikes.

Delpino long ago pointed out that the blossom is best adapted to pollen-collecting bees, which, alighting on the two long, protruding stamens, rub off pollen on their undersides while clinging for support to the wool on the three shorter stamens, whose anthers supply their needs. As a bee settles on another flower, the stigma is calculated to touch the pollen on his under side before he gets dusted with more; thus cross-pollination is effected. Three stamens furnish a visitor with food, two others clap pollen on him. Numerous flies assist in removing the pollen, too.

"I have come three thousand miles to see the mullein cultivated in a garden, and christened the velvet plant," says John Burroughs in "An October Abroad." But even in England it grows wild, and much more abundantly in Southern Europe, while its specific name is said to have been given it because it was so common in the neighborhood of Thapsus; but whether the place of that name in Africa, or the Sicilian town mentioned by Ovid and Virgil, is not certain. Strange that Europeans should labor under the erroneous impression that this mullein is native to America, whereas here it is only an immigrant from their own land. Rapidly taking its course of empire westward from our seaports into which the seeds smuggled their passage among the ballast, it is now more common in the Eastern States, perhaps, than any native. Forty or more folk-names have been applied to it, mostly in allusion to its alleged curative powers, its use for candlewick and funeral torches in the Middle Ages. The generic title, first used by Pliny, is thought to be a corruption of Barbascum = with beards, in allusion to the hairy filaments, or, as some think, to the leaves.

Of what use is this felt-like covering to the plant? The importance of protecting the delicate, sensitive, active cells from intense light, draught, or cold, have led various plants to various practices; none more common, however, than to develop hairs on the epidermis of their leaves, sometimes only enough to give it a downy appearance, sometimes to coat it with felt, as in this case, where the hairs branch and interlace. Fierce sunlight in the exposed, dry situations where the mullein grows; prolonged drought, which often occurs at flowering season, when the perpetuation of the species is at stake; and the intense cold which the exquisite rosettes formed by year-old plants must endure through a winter before they can send up a flower-stalk the second spring - these trials the well-screened, juicy, warm plant has successfully surmounted through its coat of felt. Hummingbirds have been detected gathering the hairs to line their tiny nests. The light, strong stalk makes almost as good a cane as bamboo, especially when the root end, in running under a stone, forms a crooked handle. Pale country beauties rub their cheeks with the velvety leaves to make them rosy.

MOTH MULLEIN
(Verbascum Blattaria) Figwort family

Flowers - Yellow, or frequently white, 5-parted, about 1 in. broad, marked with brown; borne on spreading pedicles in a long, loose raceme; all the filaments with violet hairs; 1 protruding pistil. Stem: Erect, slender, simple, about 2 ft. high, sometimes less, or much taller. Leaves: Seldom present at flowering time; oblong to ovate, toothed, mostly sessile, smooth. Preferred Habitat - Dry, open wasteland; roadsides, fields. Flowering Season - June-November. Distribution - Naturalized from Europe and Asia, more or less common throughout the United States and Canada.

Quite different from its heavy and sluggish looking sister is this sprightly, slender, fragile-flowered mullein. "Said to repel the cockroach (Blatta). hence the name Blattaria; frequented by moths, hence moth mullein." (Britton and Brown's "Flora.") Are the latter frequent visitors? Surely there is nothing here to a moth's liking. New England women used to pack this plant among woolen garments in summer to keep out the tiny clothes moths. The flower, whose two long stamens and pistil protrude as from the great mullein's blossom, and whose filaments are tufted with violet wool footholds - unnecessary provisions for moths, which rarely alight on any flower, but suck with their wings in motion - are cross-fertilized by pollen-collecting bees and flies as described in the account of the great mullein.

"Of beautiful weeds quite a long list might be made without including any of the so-called wild flowers," says John Burroughs. "A favorite of mine is the little moth mullein that blooms along the highway, and about the fields, and maybe upon the edge of the lawn." Even in winter, when the slender stem, set with round brown seed-vessels, rises above the snow, the plant is pleasing to the human eye, as it is to that of hungry birds.

BUTTER-AND-EGGS; YELLOW TOAD-FLAX; EGGS-AND-BACON; FLAXWEED;
BRIDEWEED
(Linaria Linaria; L. vulgaris of Gray) Figwort family

Flowers - Light canary yellow and orange, 1 in. long or over, irregular, borne in terminal, leafy-bracted spikes. Corolla spurred at the base, 2-lipped, the upper lip erect, 2-lobed; the lower lip spreading, 3-lobed, its base an orange-colored palate closing the throat; 4 stamens in pairs within; 1 pistil. Stem: 1 to 3 ft. tall, slender, leafy. Leaves: Pale, grass-like. Preferred Habitat - Wasteland, roadsides, banks, fields. Flowering Season - June-October. Distribution - Nebraska and Manitoba, eastward to Virginia and Nova Scotia. Europe and Asia.

An immigrant from Europe, this plebeian perennial, meekly content with waste places, is rapidly inheriting the earth. Its beautiful spikes of butter-colored cornucopias, apparently holding the yolk of a diminutive Spanish egg, emit a cheesy odor, suggesting a close dairy. Perhaps half the charm of the plant consists in the pale bluish-green grass-like leaves with a bloom on the surface, which are put forth so abundantly from the sterile shoots. (See blue toad-flax.)

Guided by the orange palate pathfinder to where the curious, puzzling flower opens, the big velvety bumblebee alights, his weight depressing the lower lip until a comfortable entrance through the gaping mouth is offered him. In he goes, and his long tongue readily reaches the nectar in the deep spur, while his back brushes off pollen from the stamens in his way overhead. Then he backs out, and the gaping mouth springs shut after him - for the linaria is akin to the snapdragon in the garden. As its stamens are of two lengths, the flower is able to fertilize itself in stormy weather, insects failing to transfer its pollen. To drain ten of these spurs a minute is no difficult task for the bumblebee. But how slowly, painfully, the little lightweight hive-bees and leaf-cutters squeeze in between the tight lips. An occasional butterfly inserts its long, thin tongue, and, without transferring a grain of pollen for the flower, robs it of sweets clearly intended for the bumblebee alone. Even when ants - the worst pilferers extant - succeed in entering, they cannot reach the nectar, owing to the hairy stockade bordering the groove where it runs. Beetles, out for pollen, also occasionally steal an entrance, if nothing more. Grazing cattle let the plant alone to ripen seed in peace, for it secretes disagreeable juices in its cells - juices that were once mixed with milk by farmers' wives to poison flies.

DOWNY FALSE FOXGLOVE
(Dasystoma flava; Gerardia flava of Gray) Figwort family

Flowers - Pale yellow, 1 1/2 to 2 in. long; in showy, terminal, leafy-bracted racemes. Calyx bell-shaped, 5-toothed; corolla funnel form, the 5 lobes spreading, smooth outside, woolly within; 4 stamens in pairs, woolly; 1 pistil. Stem: Grayish, downy, erect, usually simple, 2 to 4 ft. tall. Leaves: Opposite, lower ones oblong in outline, more or less irregularly lobed and toothed; upper ones small, entire. Preferred Habitat - Gravelly or sandy soil, dry thickets, open woods. Flowering Season - July-August Distribution - "Eastern Massachusetts to Ontario and Wisconsin, south to southern New York, Georgia, and Mississippi." (Britton and Brown.)

In the vegetable kingdom, as in the spiritual, all degrees of back-sliding sinners may be found, each branded with a mark of infamy according to its deserts. We have seen how the dodder vine lost both leaf and roots after it consented to live wholly by theft of its hardworking host's juices through suckers that penetrate to the vitals; how the Indian pipe's blanched face tells the story of guilt perpetrated under cover of darkness, in the soil below; how the broom-rape and beech-drops lost their honest green color; and, finally, the foxgloves show us plants with their faces so newly turned toward the path of perdition, their larceny so petty, that only the expert in criminal botany cases condemns them. Like its cousins the gerardias (q.v.), the downy false foxglove is only a partial parasite, attaching its roots by disks or suckers to the roots of white oak or witch hazel (q.v.); not only that, but, quite as frequently, groping blindly in the dark, it fastens suckers on its own roots, actually thieving from itself! It is this piratical tendency which makes transplanting of foxgloves into our gardens so very difficult; even when lifted with plenty of their beloved vegetable mould. The term false foxglove, it should be explained, is by no means one of reproach for dishonesty; it was applied simply to distinguish this group of plants from the true foxgloves cultivated, not wild, here, which yield digitalis to the doctors.

But if these foxgloves live at others' expense, there are creatures which in turn prey upon them. Caterpillars of a peacock butterfly, known as the buckeye (Junonia coenia), with eye-like spots on its tawny, reddish-gray wings, divide their unwelcome attentions between various species of plantain, the snapdragon in the garden, gerardias, and foxgloves.

The SMOOTH FALSE FOXGLOVE (D. Virginica; G. quercifolia of Gray) - which delights in rich woods, moist or dry, bears similar, but slightly larger, blossoms on a smooth, usually branched, and taller stem, whose lower leaves especially are much cleft (pinnatifid). This species is commoner South and West, blooming from July to September. All the foxgloves elevate their sticky stigmas to the mouth of their tubes, that the pollen-dusted bumblebee may leave some of the vitalizing dust brought from another flower on its surface before she turns upside down and enters in this unusual fashion to receive a fresh supply on her way to the nectar in the base of the tube. Her pressure against the pointed anther-tips causes the light, dry pollen to sift out; on the removal of her pressure the gaping chinks close to save it from small bees and flies. It falls out, therefore, only when the bee is in the right position to receive it for export to another foxglove's stigma. Hairy footholds on anthers and filaments are provided lest the bee fall while reversed and sifting out the pollen.

The FERN-LEAVED or LOUSEWORT FALSE FOXGLOVE (D. pedicularia; G. pedicularia of Gray) - a very leafy species found in dry woods and thickets from the Mississippi and Ontario eastward to the Atlantic, north and south, has all its leaves once or twice pinnatifid, the lobes much cut and toothed. It is a rather sticky, hairy, slender, and much branched plant, growing from one to four feet tall; the broad, trumpet-shaped, yellow flower, which is sticky outside, measures an inch or an inch and a half long, and is sometimes almost as wide across. "The most abundant visitor, and the one for which the flower is most perfectly adapted," says Professor Robertson, "is Bombus Americanorum. This bee always turns head downwards on entering the flower. When it enters, or backs out, the basal joints of its legs strike the tips of the anther-cells, when the pollen falls out. I had often wondered why this bee turned upside down to enter the flower…. I discovered that the form of the flower requires it. The modification which requires the bees to reverse is associated with the peculiar mode of pollen discharge. Smaller bumblebees and some other bees which never or rarely try to suck hang under the anthers and work out the pollen by striking the trigger-like awns. They reverse of their own accord, since they are so small they are not compelled to do so on account of the form of the flower. The tube is large…so that most bumblebee workers could easily reach the nectar if the tube were not curved in the opposite direction from that of most flowers, and if the anthers did not obstruct the entrance." Sometimes small bees, despairing of getting into the tube through the mouth, suck at holes in the flower's sides, because legitimate feasting was made too difficult for the poor little things. The ruby-throated hummingbird, hovering a second above the tube, drains it with none of the clown-like performances exacted from the bumblebee. Pilfering ants find death as speedy on the sticky surfaces here as on any catchfly.

GREATER BLADDERWORT; HOODED WATER-MILFOIL; POP-WEED
(Utricularia vulgaris) Bladderwort family

Flowers - Yellow, about 1/2 in. across, 3 to 20 on short pedicels in a raceme at the top of a stout, naked scape 3 to 14 in. high. Calyx deeply 2-lobed; corolla 2-lipped, the upper lip erect, the lower lip larger, its palate prominent, the lip slightly 3-lobed, and spurred at the base; 2 stamens; 1 pistil; the stigma 2-lipped. Leaves: Very finely divided into threadlike segments, bearing little air bladders. Preferred Habitat - Floating free in ponds and slow streams, or rooting in mud. Flowering Season - June-August. Distribution - Throughout nearly the whole of North America, Cuba, and Mexico. Europe and Asia.

Here is an extraordinary little plant indeed, which, by its amazing cleverness, now overruns the globe - one of the higher order of intelligence so closely akin to the animals that the gulf which separates such from them seems not very wide after all. In studying the water-crowfoots (q.v.) and other aquatic plants, we learned why submerged leaves must be so finely cut; but what mean the little bladders tipped with bristles among the pop-weed's threadlike foliage? Formerly these were regarded as mere floats - a thoughtless theory, for branches without bladders might have been observed floating perfectly. It is now known they are traps for capturing tiny aquatic creatures: nearly every bladder you examine under a microscope contains either minute crustaceans or larvae, worms, or lower organisms, some perhaps still alive, but most of them more or less advanced toward putrefaction - a stage hastened, it is thought, by a secretion within the bladders; for the plant cannot digest fresh food; it can only absorb, through certain processes within the bladder's walls, the fluid products of decay. The little insectivorous sundew (q.v.), on the contrary, not only digests, but afterward absorbs, animal matter. Tiny aquatic creatures, ever seeking shelter from larger ones ready to devour them, enter the pop-weed bladders by bending inward the free edge of the valve, which, being strongly elastic, snaps shut again behind them instantly. "Abandon hope, all ye who enter here," might be written above the entrance. No victim ever escapes from that prison. Scientists are not agreed that the bristles draw creatures into the bladder. Whatever touches the sensitive valves is at once drawn in. "To show how closely the edge fits," says Charles Darwin, "I may mention that my son found a daphnia which had inserted one of its antennae into the slit, and it was thus held fast during a whole day. On three or four occasions I have seen long narrow larvae, both dead and alive, wedged between the corner of the valve and collar, with half their bodies within the bladder and half out. Professor Cohn of Germany tells of immersing a plant of this bladderwort one evening in clear water swarming with tiny crustaceans, and by the next morning most of the bladders contained them, entrapped and swimming around in their prisons.

So much for what is going on below the surface of the water: what above it? Several flowers on the showy spike attract numerous insects. One alighting on the lower lip must thrust his tongue beneath the upper one to reach the nectar in the spur, passing on its way the irritable stigma, which receives any pollen he has brought in. Instantly it is touched, the stigma folds up to be out of the way of the tongue when it is withdrawn from the spur now laden with fresh pollen. It is thus that self-fertilization is escaped. Many vigorous seeds follow in each capsule. This marvelous piece of mechanism is what Thoreau termed "a dirty-conditioned flower, like a sluttish woman with a gaudy yellow bonnet"!

Not through its seeds alone, however, has the little plant succeeded in firmly establishing itself. In early autumn the stems terminate in large buds which, falling off, lie dormant all winter at the bottom of the pond. In spring they root and put forth leaves bearing bladders, which at this stage of existence are filled with water to help anchor the plant. As flowering season approaches, the bladders undergo an internal change to fit them for a change of function; they now fill with air, when the buoyed plant rises toward the surface to send up its flowering scape, while the bladders proceed with their nefarious practices to nourish it more abundantly while its system is heavily taxed.

The HORNED BLADDERWORT (U. cornuta), found in sandy swamps, along the borders of ponds, marshy lake margins, and in bogs from Newfoundland to Florida, westward to Minnesota and Texas, bears from one to six deliciously fragrant yellow flowers on its leafless scape from June to August. It is "perhaps the most fragrant flower we have," says John Burroughs. "In a warm moist atmosphere its odor is almost too strong…. Its perfume is sweet and spicy in an eminent degree." The low scape, rooting in the mud, has some root-like stems and branches, sometimes with a few entire leaves and bladders. Its benefactors, bumblebees and butterflies, with their highly developed aesthetic taste, are attracted from afar by this pleasing flower, whose acute, curved spur filled with nectar may not be drained by small fry, to whom the hairy throat is an additional discouragement.

SWEET WILD HONEYSUCKLE, or WOODBINE; ITALIAN OR PERFOLIATE
HONEYSUCKLE
(Lonicera Caprifoliuin; L. grata of Gray) Honeysuckle family

Flowers - White within, the tube pinkish, soon fading yellow, 1 to 1 1/2 in. long, very fragrant; borne in terminal whorls seated in the united pair of upper leaves. Calyx small, 5-toothed; corolla slender, tubular, 2-lipped; upper lip 4-lobed; lower lip narrow, curved downward; 5 stamens and 1 style far protruding. Stem: Climbing high, smooth. Leaves: Upper pairs united around the stem into an oval disk or shallow cup; lower leaves opposite, but not united oval, entire. Fruit: Red berries, clustered. Preferred Habitat - Thickets, wayside hedges, rocky woodlands. Flowering Season - May-June. Distribution - New England and Michigan to the Southern States.

"Escaped from cultivation and naturalized." How does it happen that this vine, a native of Europe, is now so common in the Eastern United States as to be called the American woodbine? Had Columbus been a botanist and wandered about our continent in search of flowers, he would have found very few that were familiar to him at home, except such as were common both to Europe and Asia also. Where the Aleutian Islands jut far out into the Pacific, and the strongest of ocean currents flows our way, must once have been a substantial highroad for beasts, birds, and vegetables, if not for men as well; but in the wide, briny Atlantic no European seed could live long enough to germinate after drifting across to our shores, if, indeed, it ever reached here. Once the American colonies came to be peopled, with homesick Europeans, who sent home for everything portable they had loved there, enormous numbers of trees, shrubs, plants, and seeds were respectably carried across in ships; the seeds of others stole a passage, as they do this day, among the hay used in packing. This was the chance for expansion they had been waiting for for ages. While many cultivated species found it practically impossible to escape from the vigilance of gardeners here, others, with a better plan for disseminating seed, quickly ran wild. Now some of the commonest plants we have are of European origin. This honeysuckle, by bearing red berries to attract migrating birds in autumn, soon escaped the confines of gardens. Its undigested seeds, dropped in the woodland far from the parent vine, germinated quite as readily as in Europe, and pursued in peace their natural mode of existence, until here too we now have banks

"Quite over-canopied with luscious woodbine."

The HAIRY HONEYSUCKLE, or ROUGH WOODBINE (L. hirsuta), with a more northerly and westerly range, bears clusters of flowers that are yellow on the outside, and orange within the tube, the terminal clusters slightly elevated above a united pair of dull green leaves that are softly hairy underneath. The slender flower tube is sticky outside to protect it from pilfering ants, and the hairs at the base of the stamens serve to hide the nectar from unbidden guests. Berries, bright orange. Flowering season, June-July.

The deliciously fragrant CHINESE or JAPANESE HONEYSUCK- LE (L. Japonica), as commonly grown on garden trellises and fences here as the morning-glory, has freely escaped from cultivation from New York southward to West Virginia and North Carolina. Everyone must be familiar with the pairs of slender, tubular, two-lipped, white or pinkish flowers, quickly turning yellow, which are borne in the leaf axils along the sprays. The smooth, dark green, opposite leaves, pale beneath, cling almost the entire year through. The stem, in winding, follows the course taken by the hands of a clock. Were the berries red instead of black, they would, doubtless, have attracted more birds to disperse their seeds, and the vine would have traveled as fast in its wild state as the Italian honeysuckle has done. It blooms from June to August, and sparingly again in autumn. When daylight begins to fade, these long, slender-tubed buds expand to welcome their chosen benefactors, the sphinx moths, wooing them with fragrance so especially strong and sweet at this time that, long after dark, guests may be guided from afar by it alone, and entertaining them with copious draughts of deeply hidden nectar, which their long tongues alone may drain. Poised above the blossoms, they sip without pause of their whirring wings, and it is not strange that many people mistake them in the half light for hummingbirds. Indeed, they are often called hummingbird moths. Darting away suddenly and swift as thought, they have also earned the name of hawk moths. Because the caterpillars have a curious trick of raising the fore part of their bodies and remaining motionless so long (like an Egyptian sphinx), the commoner name seems most appropriate. A sphinx moth at rest curls up its exceedingly long tongue like .a watch- spring: in action only the hummingbird can penetrate to such depths; hence that honeysuckle which prefers to woo the tiny bird, whose decided preference is for red, is the TRUMPET or CORAL HONEYSUCKLE; whereas the other twiners developed deep, tubular flowers that are white or yellow, so that the moths may see them in the dark, when red blossoms are engulfed in the prevailing blackness. Moreover, the latter bloom at a season when the crepuscular and nocturnal moths are most abundant. Rough rounded pollen grains, carried on the hairs and scales on the under side of the moth's body from his head to his abdomen, including antennae, tongue, legs, and wings, cannot but be rubbed off on the protruding sticky stigma of the next honeysuckle tube entered; hence cross-fertilization is regularly effected by moths alone. The next day such interlopers as bees, flies, butterflies, and even the outwitted hummingbird, may take whatever nectar or pollen remains. If the previous evening has been calm and fine, they will find little or none; but if the night has been wild and stormy, keeping the moths under cover, the tubes will brim with sweets. After fertilization the corolla turns yellow to let visitors know the mutual benefit association has gone out of business.

BUSH HONEYSUCKLE; GRAVEL-WEED
(Diervilla Diervilla; D. trifida of Gray) Honeysuckle family

Flowers - Yellow, small, fragrant, 1 to 5 (usually 3) together on a peduncle from upper leaf-axils. Calyx tube slender, elongated; corolla narrowly funnel-form, about 3/4 in. long, its 5 lobes spreading, 3 of them somewhat united; 5 stamens; 1 pistil projecting. Stem: A smooth, branching shrub 2 to 4 ft. high. Leaves. Opposite, oval, and taper-pointed, finely saw-edged. Fruit: Slender, beaked pods crowned with the 5 calyx lobes. Preferred Habitat - Dry or rocky soil, woodlands, hills. Flowering Season - May-August. Distribution - British Possessions southward to Michigan and North Carolina.

The coral honeysuckle determined to woo the hummingbird by wearing his favorite color; the twining white and yellow honeysuckles of our porches chose for their benefactors the sphinx moths, attracting them by delicious fragrance and deeply hidden nectar in slender tubes that are visible even in the dark; whereas the small-flowered bush honeysuckles still cater to the bees which, in all probability, once sufficed for the entire family. For them a conspicuous landing place has been provided in the more highly colored lower lobe of this flower, from which the visitor cannot fail to find the pocket full of nectar that swells the base of the tube but when he alights, pollen laden from another blossom, he must pay toll by leaving some of the vitalizing dust on the projecting stigma before he feasts and dusts himself afresh. After they have been plundered, and consequently fertilized, all the honeysuckles change color, this one taking on a deeper yellow to let the bees know the larder is empty, that they may waste no precious time, but confine their visits where they are needed. "Many flowers adapted to bees show butterflies, hawk moths and hummingbirds as intruders," says Professor Robertson; "and this is important, since it enables us to understand how bee-flowers might become modified to suit them" - just as certain of the honeysuckles have done. Once the Oriental pink weigelias, grown in nearly every American garden, were thought to belong to the Diervilla clan, from which later-day systematists have banished them.

The EARLY FLY or TWIN HONEYSUCKLE (Lonicera ciliata), found in moist, cool woods from Pennsylvania and Michigan far northward, sends forth pairs of funnel-form, honey-yellow flowers, about three-quarters of an inch long, with five, regular lobes, on a slender footstalk from the leaf axils in May. It is a straggling, shrubby bush from three to five feet tall. The opposite leaves are thin, oval, bright green on both sides, the edges hairy. Two little ovoid, light red berries follow the flowers.

Another species, a shrubby SWAMP FLY-HONEYSUCKLE (L. oblongifolia), found in wet ground and bogs throughout a similar range, blooming about two weeks later, coats the under side of its young leaves with fine hairs to prevent their pores from clogging with vapors arising from its moist retreats. The little pale yellow flowers, also growing in pairs on a footstalk from the leaf axils, have their tubular corollas strongly cleft into two lips. Reddish markings within serve as pathfinders for the bumblebee, who finds so much nectar at the base that a tiny bulging pocket had to be provided to hold it. Sometimes the two flowers join below like Siamese twins, in which case the pair of crimson berries become more or less united.

> "So we grew together,
> Like to a double cherry, seeming parted."

One occasionally finds the pink and white twin-flowered TARTA-RIAN
BUSH HONEYSUCKLE (L. Tartarica) escaped from cultivation in the
Eastern States through the agency of birds which feast upon its

little round, red, translucent berries.

COMMON DANDELION; BLOWBALL; LION'S-TOOTH; PEASANT'S CLOCK
(Taraxacum Taraxacum; T. Densleonis of Gray) Chicory family

Flower-head - Solitary, golden yellow, to 2 in. across, containing 150 to 200 perfect ray florets on a flat receptacle at the top of a hollow, milky scape 2 to 18 in. tall. Leaves: From a very deep, thick, bitter root; oblong to spatulate in outline, irregularly jagged. Preferred Habitat - Lawns, fields, grassy waste places. Flowering Season - Every month in the year. Distribution - Around the civilized world.

"Dear common flower that grow'st beside the way,
    Fringing the dusty road with harmless gold.
    . . . .
Gold such as thine ne'er drew the Spanish prow
    Through the primeval hush of Indian seas,
Nor wrinkled the lean brow
    Of age, to rob the lover's heart of ease.
'Tis the spring's largess, which she scatters now
    To rich and poor alike, with lavish hand
    Though most hearts never understand
To take it at God's value, but pass by
The offered wealth with unrewarded eye."

Let the triumphant Anglo-Saxon with dreams of expansion that include the round earth, the student of sociology who wishes an insight into cooperative methods as opposed to individualism, the young man anxious to learn how to get on, parents with children to be equipped for the struggle for existence, business men and employers of labor, all sit down beside the dandelion and take its lesson to heart. How has it managed without navies and armies - for it is no imperialist - to land its peaceful legions on every part of the civilized world and take possession of the soil? How can this neglected wayside composite weed triumph over the most gorgeous hothouse individual on which the horticulturist expends all the science at his command; to flourish where others give up the struggle defeated; to send its vigorous offspring abroad prepared for similar conquest of adverse conditions wherever met to attract myriads of customers to its department store, and by consummate executive ability to make every visitor unwittingly contribute to its success? Any one who doubts the dandelion's fitness to survive, should humble himself by spending days and weeks on his knees, trying to eradicate the plant from even one small lawn with a knife, only to find the turf starred with golden blossoms, or, worse still from his point of view, hoary with seed balloons, the following spring.

Deep, very deep, the stocky bitter root penetrates where heat and drought affect it not, nor nibbling rabbits, moles, grubs of insects, and other burrowers break through and steal. Cut off the upper portion only with your knife, and not one, but several, plants will likely sprout from what remains; and, however late in the season, will economize stem and leaf to produce flowers and seeds, cuddled close within the tuft, that set all your pains at naught. "Never say die" is the dandelion's motto. An exceedingly bitter medicine is extracted from the root of this dandelion, formerly known as T. officinale. Likewise are the leaves bitter. Although they appear so early in the spring, they must be especially tempting to grazing cattle and predaceous insects, the rosettes remain untouched, while other succulent, agreeable plants are devoured wholesale. Only Italians and other thrifty Old-World immigrants, who go about then with sack and knife collecting the fresh young tufts, give the plants pause but even they leave the roots intact. When boiled like spinach or eaten with French salad dressing, the bitter juices are extracted from the leaves or disguised - mean tactics by an enemy outside the dandelion's calculation. All nations know the plant by some equivalent for the name dent de lion = lion's tooth, which the jagged edges of the leaves suggest.

Presently a hollow scape arises to display the flower above the surrounding grass. Bridge builders and constructing engineers know how yielding and economical, yet how invincibly strong, is the hollow tube. March winds may buffet and bend the dandelion's stem without harm. How children delight to split this slippery tube, and run it in and out of their mouths until curls form! At the top of the scape is a double involucre of narrow, green, leaf-like scales similar to what all composites have. Half the involucre bends downward to protect the flower from crawling pilferers, half stands erect to play the role for the community of florets within that the calyx does for individual blossoms. When it is time to close the dandelion shop, business being ended for the day, this upper-half of the involucre protects it like the heavy shutters merchants put up at their windows.

Seated on a fleshy receptacle, not one flower, but often two hundred minute, perfect florets generously cooperate. "In union there is strength" is another motto adopted, not only by the chicory clan, but by the entire horde of composites. Each floret of itself could hope for no attention from busy insects; united, how gorgeously attractive these disks of overlapping rays are! Doubtless each tiny flower was once a five-petaled blossom, for in the five teeth at the top and the five lines are indications that once distinct parts have been welded together to form a more showy and suitable corolla. Each floret insures cross-pollination from insects crawling over the head, much as the minute yellow tubes in the center of a daisy do (q.v.). Quantities of small bees, wasps, flies, butterflies, and beetles - over a hundred species of insects - come seeking the nectar that wells up in each little tube, and the abundant pollen, which are greatly appreciated in early spring, when food is so scarce. In rainy weather and at night, when its benefactors are not flying, the canny dandelion closes completely to protect its precious attractions. Because the plant, which is likely to bloom every month in the year, may not always certainly reckon on being pollinated by insects, each neglected floret will curl the two spreading, sticky branches of its style so far backward that they come in contact with any pollen that has been carried out of the tube by the sweeping brushes on their tips. Occasional self-fertilization is surely better than setting no seed at all when insects fail. Not a chance does the dandelion lose to "get on."

After flowering, it again looks like a bud, lowering its head to mature seed unobserved. Presently rising on a gradually lengthened scape to elevate it where there is no interruption for the passing breeze from surrounding rivals, the transformed head, now globular, white, airy, is even more exquisite, set as it is with scores of tiny parachutes ready to sail away. A child's breath puffing out the time of day, a vireo plucking at the fluffy ball for lining to put in its nest, the summer breeze, the scythe, rake, and mowing machines, sudden gusts of winds sweeping the country before thunderstorms - these are among the agents that set the flying vagabonds free. In the hay used for packing they travel to foreign lands in ships, and, once landed, readily adapt themselves to conditions as they find them. After soaking in the briny ocean for twenty-eight days - long enough for a current to carry them a thousand miles along the coast - they are still able to germinate.

The DWARF DANDELION, CYNTHIA, or VIRGINIA GOATS-BEARD (Adepogon Virginicum; formerly Krigia Virginica) - with from two to six long-peduncled, flat, deep yellow or reddish-orange flower heads, about an inch and a half across, on the summit of its stem from May to October, elects to grow in moist meadows, woodlands, and shady rocky places. How it glorifies them! From a tuffet of spatulate, wavy-toothed or entire leaves, the smooth, shining, branching stem arises bearing a single oblong, clasping leaf below the middle. Particularly beautiful is its silvery seed-ball, the pappus consisting of about a dozen hairlike bristles inside a ring of small oblong scales, on which the seed sails away. Range, from Massachusetts to Manitoba, south to Georgia and Kansas.

A charming little plant, the CAROLINA DWARF DANDELION or KRIGIA (A. Carolinianum), once confounded with the above, sends up several unbranched scapes from the same tuffet. It blooms in dry, sandy soil from April to August, from Maine and Minnesota to the Gulf States.

Like a small edition of Lowell's "dear common flower" is the TALL DANDELION, or AUTUMNAL HAWKBIT (Leontodon autumnale), its slender, wiry, branching scape six inches to two feet high, terminated by several flower-heads, each on a separate peduncle, which is usually a little thickened and scaly just below it. Only forty to seventy five-toothed ray florets spread in a flat golden disk from an oblong involucre. They close in rainy weather and at night. From June to November, in spite of its common name, it blooms in fields and along roadsides, its brownish seed-plumes

rapidly following; but these are produced at the frightfully extravagant cost of over two hundred thousand grains of pollen to each
head, it is estimated. The Greek generic name, meaning lion's tooth,
refers to the shape of the lobes of the narrowly oblong leaves in a
tuft at the base. Range, from New Jersey and Ohio far northward.
Naturalized from Europe and Asia.

FIELD SOW-THISTLE; MILK THISTLE
(Sonchus arvensis) Chicory family

Flower-heads - Bright yellow, very showy, to 2 in. across, several
or numerous, on rough peduncles in a spreading cluster. Involucre
nearly 1 in. high; the scales narrow, rough. Stem: 2 to 4 ft. high, leafy
below, naked, and paniculately branched above, from deep roots
and creeping rootstocks. Leaves: Long, narrow, spiny, but not
sharp-toothed; deeply cut, mostly clasping at base. Preferred Habitat - Meadows, fields, roadsides, saltwater marshes. Flowering Season - July-October. Distribution - Newfoundland to Minnesota and
Utah, south to New Jersey.

It cannot be long, at their present rate of increase, before this and
its sister immigrant become very common weeds throughout our
entire area, as they are in Europe and Asia.

The ANNUAL SOW-THISTLE or HARE'S LETTUCE (S. ole-
raceus), its smaller, pale yellow flower-heads, with smooth involucres more closely grouped, now occupies our fields and waste
places with the assurance of a native. Honeybees chiefly, but many
other bees, wasps, brilliant little flower-flies (Syrphidae), and butterflies among other winged visitors which alight on the flowers,
from May to November, are responsible for the copious, soft, fine,
white-plumed seeds that the winds waft away to fresh colonizing
ground. The leaves clasp the stem by deep ear-like or arrow-shaped
lobes, or the large lower ones are on petioles, lyrate-pinnatifid, the
terminal division commonly large and triangular; the margins all
toothed. Frugal European peasants use them as a potherb or salad.
One of the plant's common folk-names in the Old World is hare's
palace. According to the "Grete Herbale," if "the hare come under it,

he is sure no beast can touch hym!' That was the spot Brer Rabbit was looking for when Brer Fox lay low! Another early writer declares that "when hares are overcome with heat they eat of an herb called hare's-lettuce, hare's-house, hare's-palace; and there is no disease in this beast the cure whereof she does not seek for in this herb." Who has detected our cottontails nibbling the succulent leaves?

TALL or WILD LETTUCE; WILD OPIUM
(Lactuca Canadensis) Chicory family

Flower-heads - Numerous small, about 1/4 in. across, involucre cylindric, rays pale yellow; followed by abundant, soft, bright white pappus; the heads growing in loose, branching, terminal clusters. Stem: Smooth, 3 to 10 ft. high, leafy up to the flower panicle; juice milky. Leaves: Upper ones lance shaped; lower ones often 1 ft. long, wavy-lobed, often pinnatifid, taper pointed, narrowed into flat petioles. Preferred Habitat - Moist, open ground; roadsides. Flowering Season - June-November. Distribution - Georgia, westward to Arkansas, north to the British Possessions.

Few gardeners allow the table lettuce (sativa) to go to seed but as it is next of kin to this common wayside weed, it bears a strong likeness to it in the loose, narrow panicles of cream-colored flowers, followed by more charming, bright white little pompons. Where the garden varieties originated, or what they were, nobody knows. Herodotus says lettuce was eaten as a salad in 550 B.C.; in Pliny's time it was cultivated, and even blanched, so as to be had at all seasons of the year by the Romans. Among the privy-purse expenses of Henry VIII is a reward to a certain gardener for bringing "lettuze" and cherries to Hampton Court. Quaint old Parkinson, enumerating "the vertues of the lettice," says, "They all cool a hot and fainting stomache." When the milky juice has been thickened (lactucarium), it is sometimes used as a substitute for opium by regular practitioners - a fluid employed by the plants themselves, it is thought, to discourage creatures from feasting at their expense (see milkweed). Certain caterpillars, however, eat the leaves readily; but offer lettuce or poppy foliage to grazing cattle, and they will go without food rather than touch it.

"What's one man's poison, Signor,
  Is another's meat or drink."

Rabbits, for example, have been fed on the deadly nightshade for a week without injury.

The HAIRY or RED WILD LETTUCE (L. hirsuta), similar to the preceding, but often with dark reddish stem, peduncles, and tiny flower-cups, the ray florets varying from yellow to pale reddish or purplish, has longer leaves, deeply cut or lobed almost to the wide midrib. After what we learned when studying the barberry and the prickly pear cactus, for example, about plants that choose to live in dry soil, it is not surprising to find that this is a lower, less leafy, and more hairy plant than the moisture-loving tall lettuce.

An European immigrant, naturalized here but recently, the PRICKLY LETTUCE (L. Scariola) has nevertheless made itself so very much at home in a short time that it has already become a troublesome weed from New England to Pennsylvania, westward to Minnesota and Missouri. But when we calculate that every plant produces over eight thousand fluffy white-winged seeds on its narrow panicle, ready to sail away on the first breeze, no wonder so well endowed and prolific an invader marches triumphantly across continents. The long, pale green, spiny-margined, milky leaves, with stiff prickles on the midrib beneath, are doubly protected against insect borers and grazing cattle.

"Look at this delicate plant that lifts its head from the meadow;
  See how its leaves all point to the North as true as the magnet."

While Longfellow must have had the coarse-growing, yellow-flowered, daisy-like PRAIRIE ROSIN-WEED (Silphium laciniatum) in mind when he wrote this stanza of "Evangeline," his lines apply with more exactness to the delicate prickly lettuce, our eastern compass plant. Not until 1895 did Professor J. C. Arthur discover that when the garden lettuce is allowed to flower, its stem leaves also exhibit polarity. The great lower leaves of the rosin-weed, which stand nearly vertical, with their faces to the east and west, and their edges to the north and south, have directed many a traveler, not from Acadia only, across the prairie until it has earned the titles pilot-weed, compass or polar plant. Various theories have been advanced to account for the curious phenomenon, some claiming that the leaves contained sufficient iron to reader them magnetic - a theory promptly exploded by chemical analysis. Others supposed that the resinous character of the leaves made them susceptible to magnetic influence; but as rosin is a non-conductor of electricity, of course this hypothesis likewise proved untenable. At last Dr. Asa Gray brought forward the only sensible explanation: inasmuch as both surfaces of the rosin-weed leaf are essentially alike, there being very nearly as many stomata on the upper side as on the under, both surfaces are equally sensitive to sunlight; therefore the leaf twists on its petiole until both sides share it as equally as is possible. While the polarity of the prickly lettuce leaves is by no means so marked, Dr. Gray's theory about the rosin-weed may be applied to them as well.

ORANGE or TAWNY HAWKWEED; GOLDEN MOUSE-EAR HAWKWEED; DEVIL'S PAINT-BRUSH
(Hieracium aurantiacum) Chicory family

Flower-beads - Reddish orange; 1 in. across or less, the 5-toothed rays overlapping in several series; several heads on short peduncles in a terminal cluster. Stem: Usually leafless, or with 1 to 2 small sessile leaves; 6 to 20 in. high, slender, hairy, from a tuft of hairy, spatulate, or oblong leaves at the base. Preferred Habitat - Fields, woods, roadsides, dry places. Flowering Season - June-September. Distribution - Pennsylvania and Middle States northward into British Possessions.

Peculiar reddish-orange disks, similar in shade to the butterfly weed's umbels, attract our eyes no less than those of the bees, flies, and butterflies for whom such splendor was designed. After cross-fertilization has been effected, chiefly through the agency of the smaller bees, a single row of slender, brownish, persistent bristles attached to the seeds transforms the head into the "devil's paint-brush." Another popular title in England, from whence the plant originally came, is Grimm the Collier. All the plants in this genus take their name from hierax = a hawk, because people in the old country once thought that birds of prey swooped earthward to sharpen their eyesight with leaves of the hawkweed, hawkbit, or speerhawk, as they are variously called. Transplanted into the garden, the orange hawkweed forms a spreading mass of unusual, splendid color.

The RATTLESNAKE-WEED, EARLY or VEIN-LEAF HAWK-WEED, SNAKE or POOR ROBIN'S PLANTAIN (H. venosum), with flower-heads only about half an inch across, sends up a smooth, slender stem, paniculately branched above, to display the numerous dandelion-yellow disks as early as May, although October is not too late to find this generous bloomer in pine woodlands, dry thickets, and sandy soil. Purplish-veined oval leaves, more or less hairy, that spread in a tuft next the ground, are probably as efficacious in curing snakebites as those of the rattlesnake plantain (q.v.). When a credulous generation believed that the Creator had indicated with some sign on each plant the special use for which each was intended, many leaves were found to have veinings suggesting the marks on a snake's body; therefore, by simple reasoning, they must extract venom. How delightful is faith cure!

Unlike the preceding, the CANADA HAWKWEED (H. Canadense), lacks a basal tuft at flowering time, but its firm stem, that may be any height from one to five feet, is amply furnished with oblong to lance-shaped leaves seated on it, their midrib prominent, the margins sparingly but sharply toothed. In dry, open woods and thickets, and along shady roadsides, its loosely clustered heads of clear yellow, about one inch across, are displayed from July to September; and later the copious brown bristles remain for sparrows to peck at.

The ROUGH HAWKWEED (H. scabrum), with a stout, stiff stem crowned with a narrow branching cluster of small yellow flower-heads on dark bristly peduncles, also lacks a basal tuft at flowering time. Its hairy oblong leaves are seated on the rigid stem. In dry, open places, clearings, and woodlands from Nova Scotia to Georgia, and westward to Nebraska, it blooms from July to September.

More slender and sprightly is the HAIRY HAWKWEED (H. Gronovii), common in sterile soil from Massachusetts and Illinois to the Gulf States. The basal leaves and lower part of the stiff stem, especially, are hairy, not to allow too free transpiration of precious moisture.

GOLDEN ASTER
(Chrysopsis Mariana) Thistle family

Plower-heads - Composite, yellow, 1 in. wide or less, a few corymbed flowers on glandular stalks; each composed of perfect tubular disk florets surrounded by pistillate ray florets the involucre campanulate, its narrow bracts overlapping in several series. Stem: Stout, silky-hairy when young, nearly smooth later, 1 to 2 1/2 ft. tall. Leaves: Alternate, oblong to spatulate, entire. Preferred Habitat - Dry soil, or sandy, not far inland. Flowering Season - August-September. Distribution - Long Island and Pennsylvania to the Gulf States.

Whoever comes upon clumps of these handsome flowers by the dusty roadside cannot but be impressed with the appropriateness of their generic name (Chrysos = gold; opsis = aspect). Farther westward, north and south. it is the HAIRY GOLDEN ASTER (C. villosa), a pale, hoary-haired plant with similar flowers borne at midsummer, that is the common species.

GOLDENRODS
(Solidago) Thistle family

When these flowers transform whole acres into "fields of the cloth-of-gold," the slender wands swaying by every roadside, and purple asters add the final touch of imperial splendor to the autumn landscape, already glorious with gold and crimson, is any parterre of Nature's garden the world around more gorgeous than that portion of it we are pleased to call ours? Within its limits eighty-five species of goldenrod flourish, while a few have strayed into Mexico and South America, and only two or three belong to Europe, where many of ours are tenderly cultivated in gardens, as they should be here, had not Nature been so lavish. To name all these species, or the asters, the sparrows, and the warblers at sight is a feat probably no one living can perform; nevertheless, certain of the commoner goldenrods have well-defined peculiarities that a little field practice soon fixes in the novice's mind.

Along shady roadsides, and in moist woods and thickets, from August to October, the BLUE-STEMMED, WREATH or WOODLAND GOLDENROD (S. caesia) sways an unbranched stem with a bluish bloom on it. It is studded with pale golden clusters of tiny florets in the axils of lance-shaped, feather-veined leaves for nearly its entire length. Range from Maine, Ontario, and Minnesota to the Gulf States. None is prettier, more dainty, than this common species.

In rich woodlands and thicket borders we find the ZIG-ZAG or BROAD-LEAVED GOLDENROD (S. flexicaulis; S. latifolia of Gray) its prolonged, angled stem that grows as if waveringly uncertain of the proper direction to take, strung with small clusters of yellow florets, somewhat after the manner of the preceding species. But its saw-edged leaves are ovate, sharply tapering to a point, and narrowed at the base into petioles. It blooms from July to September. Range from New Brunswick to Georgia, and westward beyond the Mississippi.

During the same blooming period, and through a similar range, our only albino, with an Irish-bull name, the WHITE GOLDENROD, or more properly SILVER-ROD (S. bicolor), cannot be mistaken. Its cream-white florets also grow in little clusters from the upper axils of a usually simple and hairy gray stem six inches to four feet high. Most of the heads are crowded in a narrow, terminal pyramidal cluster. This plant approaches more nearly the idea of a rod than its relatives. The leaves; which are broadly oblong toward the base of the stem, and narrowed into long margined petioles, are frequently quite hairy, for the silver-rod elects to live in dry soil, and its juices must be protected from heat and too rapid transpiration.

In swamps and peat bogs the BOG GOLDENROD (S. uliginosa) sends up two to four feet high a densely flowered, oblong, terminal spire; its short branches so appressed that this stem also has a wand-like effect. The leaves, which are lance-shaped or oblong, gradually increase in size and length of petiole until the lowest often measure nine inches long. Season, July to September. Range, from Newfoundland to Pennsylvania and westward beyond the Mississippi.

Now we leave the narrow, unbranched, wand goldenrods strung with clusters of minute florets, which, however slender and charming, are certainly far less effective in the landscape than the following members of their clan which have their multitudes of florets arranged in large, compound, more or less widely branching, terminal, pyramidal clusters. On this latter plan the SHOWY or NOBLE GOLDENROD (S. speciosa) displays its splendid, dense, ascending branches of bloom from August to October. European gardeners object to planting goldenrods, complaining that they so quickly impoverish a rich bed that neighboring plants starve. This noble species becomes ignoble indeed, unless grown in rich soil, when it spreads in thrifty circular tufts. The stout stem, which often assumes reddish tints, rises from three to seven feet high, and the smooth, firm, broadly oval, saw-toothed lower leaves are long-petioled. Range, from Nova Scotia to the Carolinas, westward to Nebraska.

When crushed in the hand, the dotted, bright green, lance-shaped, entire leaves of the SWEET GOLDENROD or BLUE MOUNTAIN TEA (S. odora) cannot be mistaken, for they give forth a pleasant anise scent. The slender, simple, smooth stem is crowned with a graceful panicle, whose branches have the florets seated all on one side. Dry soil. New England to the Gulf States, July to September.

The WRINKLE-LEAVED or TALL, HAIRY GOLDENROD or BITTERWEED (S. rugosa), a perversely variable species, its hairy stem perhaps only a foot high, or, maybe, over seven feet, its rough leaves broadly oval to lance-shaped, sharply saw-edged, few if any furnished with footstems, lifts a large, compound, and gracefully curved panicle, whose florets are seated on one side of its spreading branches. Sometimes the stem branches at the summit. One usually finds it blooming in dry soil from July to November, throughout a range extending from Newfoundland and Ontario to the Gulf States.

Usually the ELM-LEAVED GOLDENROD (S. ulmifolia) sends up several slender, narrow spires of deep yellow bloom from about the same point at the summit of the smooth stem, like long, tapering fingers. Small, oblong, entire leaves are seated on these elongated sprays, while below the inflorescence the large leaves taper to a sharp point, and are coarsely and sharply toothed. In woods and copses from Maine and Minnesota to Georgia and Texas this common goldenrod blooms from July to September.

The unusually beautiful, spreading, recurved, branching panicle of bloom borne by the EARLY, PLUME, or SHARP-TOOTHED GOLDENROD or YELLOW-TOP (S. juncea), so often dried for winter decoration, may wave four feet high, but usually not over two, at the summit of a smooth, rigid stem. Toward the top, narrow, elliptical, uncut leaves are seated on the stalk; below, much larger leaves, their sharp teeth slanting forward, taper into a broad petiole, whose edges may be cut like fringe. In dry, rocky soil this is, perhaps, the first and last goldenrod to bloom, having been found as early as June, and sometimes lasting into November. Range, from North Carolina and Missouri very far north.

West of the Mississippi how beautiful are the dry prairies in autumn with the MISSOURI GOLDENROD (S. Missouriensis), its short, broad, spreading panicle waving at the summit of a smooth, slender stem from two to five feet tall. Its firm, rather thick leaves are lance-shaped, triple-nerved, entire, very rough-margined, or perhaps the lowest ones with a few scattered teeth.

Perhaps the commonest of all the lovely clan east of the Mississippi, or throughout a range extending from Arizona and Florida northward to British Columbia and New Brunswick, is the CANADA GOLDENROD or YELLOW-WEED (S. Canadensis). Surely everyone must be familiar with the large, spreading, dense-flowered panicle, with recurved sprays, that crowns a rough, hairy stem sometimes eight feet tall, or again only two feet. Its lance-shaped, acutely pointed, triple-nerved leaves are rough, and the lower ones saw-edged. From August to November one cannot fail to find it blooming in dry soil.

Most brilliantly colored of its tribe is the low-growing GRAY or FIELD GOLDENROD or DYER'S WEED (S. nemoralis). The rich, deep yellow of its little spreading, recurved, and usually one-sided panicles is admirably set off by the ashy gray, or often cottony, stem, and the hoary, grayish-green leaves in the open, sterile places where they arise from July to November. Quebec and the Northwest Territory to the Gulf States.

No longer classed as a true Solidago, but the type of a distinct genus, the LANCE-LEAVED, BUSHY, or FRAGRANT GOLDENROD (Euthamia graminifolia; formerly S. lanceolata) lifts its flat-topped, tansy-like, fragrant clusters of flower-heads from two to four feet above moist ground. From July to September it transforms whole riverbanks, low fields, and roadsides into a veritable El Dorado. Its numerous leaves are very narrow, lance-shaped, triple or five nerved, uncut, sometimes with a few resinous dots. Range, from New Brunswick to the Gulf, and westward to Nebraska.

"Along the roadside, like the flowers of gold
    That tawny Incas for their gardens wrought,
Heavy with sunshine droops the goldenrod."

Bewildered by the multitude of species, and wondering at the enormous number of representatives of many of them, we cannot but inquire into the cause of such triumphal conquest of a continent by a single genus. Much is explained simply in the statement that goldenrods belong to the vast order of Compositae, flowers in reality made up sometimes of hundreds of minute florets united into a far-advanced socialistic community having for its motto, "In union there is strength." (See Daisy) In the first place, such an association of florets makes a far more conspicuous advertisement than a single flower, one that can be seen by insects at a great distance; for most of the composite plants live in large colonies, each plant, as well as each floret, helping the others in attracting their benefactors' attention. The facility with which insects are enabled to collect both pollen and nectar makes the goldenrods exceedingly popular restaurants.

Finally, the visits of.insects are more likely to prove effectual, because any one that alights must touch several or many florets, and cross-pollinate them simply by crawling over a head. The disk florets mostly contain both stamens and pistil, while the ray florets in one series are all male. Immense numbers of wasps, hornets, bees, flies, beetles, and "bugs" feast without effort here indeed, the budding entomologist might form a large collection of Hymenoptera, Diptera, Coleoptera, and Hemiptera from among the. visitors to a single field of goldenrod alone. Usually to be discovered among the throng are the velvety black Lytta or Cantharis, that impostor wasp-beetle, the black and yellow wavy-banded, red-legged locust-tree borer, and the painted Clytus, banded with yellow and sable, squeaking contentedly as he gnaws the florets that feed him.

Where the slender, brown, plume-tipped wands etch their charming outline above the snow-covered fields, how the sparrows, finches, buntings, and juncos love to congregate, of course helping to scatter the seeds to the wind while satisfying their hunger on the swaying, down-curved stalks. Now that the leaves are gone, some of the goldenrod stems are seen to bulge as if a tiny ball were concealed under the bark. In spring a little winged tenant, a fly, will emerge from the gall that has been his cradle all winter.

ELECAMPANE; HORSEHEAL; YELLOW STARWORT
(Inula Helenium) Thistle family

Flower-heads - Large, yellow, solitary or a few, 2 to 4 in. across; on long, stout peduncles; the scaly green involucre nearly 1 in. high, holding disk florets surrounded by a fringe of long, very narrow, 3-toothed ray florets. Stem: Usually unbranched, 2 to 6 ft. high, hairy above. Leaves: Alternate, large, broadly oblong, pointed, saw-edged, rough above, woolly beneath some with heart-shaped, clasping bases. Preferred Habitat - Roadsides, fields, fence rows, damp pastures. Flowering Season - July-September. Distribution - Nova Scotia to the Carolinas, and westward to Minnesota and Missouri.

"September may be described as the month of tall weeds;" says John Burroughs. "Where they have been suffered to stand, along fences, by roadsides, and in forgotten corners,- redroot, ragweed, vervain, goldenrod, burdock, elecampane, thistles, teasels, nettles, asters, etc. - how they lift themselves up as if not afraid to be seen now! They are all outlaws; every man's hand is against them yet how surely they hold their own. They love the roadside, because here they are comparatively safe and ragged and dusty, like the common tramps that they are, they form one of the characteristic features of early fall."

Yet the elecampane has not always led a vagabond existence. Once it had its passage paid across the Atlantic, because special virtue was attributed to its thick, mucilaginous roots as a horse-medicine. For over two thousand years it has been employed by home doctors in Europe and Asia; and at first Old World immigrants thought they could not live here without the plant on their farms. Once given a chance to naturalize itself, no composite is slow in seizing it. The vigorous elecampane, rearing its fringy, yellow disks above lichen-covered stone walls in New England, the Virginia rail fence, and the rank weedy growth along barbed-wire barriers farther west, now bids fair to cross the continent.

CUP-PLANT; INDIAN-CUP; RAGGED CUP; ROSIN-PLANT
(Silphium perfoliatum) Thistle family

Plower-heads - Yellow, nearly flat; 2 to 3 in. across; 20 to 30 narrow, pistillate ray florets, about 1 in. long, overlapping in 2 or 3 series around the perfect but sterile disk florets. Stem: 4 to 8 ft. tall, square, smooth, usually branched above. Leaves: Opposite, ovate, upper ones united by their bases to form a cup; lower ones large, coarsely toothed, and narrowed into margined petioles; all filled with resinous juice. Preferred Habitat - Moist soil, low ground near streams. Flowering Season - July-September. Distribution - Ontario, New York, and Georgia, westward to Minnesota, Nebraska, and Louisiana.

It behooves a species related to the wonderful compass-plant (q.v.) to do something unusual with its leaves; hence this one makes cups to catch rain by uniting its upper pairs. Darwin's experiments with infinitesimal doses of ammonia in stimulating leaf activity may throw some light on this singular arrangement. So many plants provide traps to catch rain, although fourteen gallons of it contain only one grain of ammonia, that we must believe there is a wise physiological reason for calling upon the leaves to assist the roots in absorbing it, A native of Western prairies, the cup-plant has now become naturalized so far east as the neighborhood 6f New York City.

FALSE SUNFLOWER; OX-EYE
(Heliopsis helianthoides; H. laevis of Gray) Thistle family

Flower-heads - Entirely golden yellow, daisy-like, 1 1/2 to 2 1/2 in. across, the perfect disk florets inserted on a convex, chaffy receptacle, and surrounded by pistillate, fertile, 3-toothed ray florets; usually numerous solitary heads borne on long peduncles from axils of upper leaves. Stem: 3 to 5 ft. tall, branching above, smooth. Leaves: Opposite, ovate, and tapering to a sharp point, sharply and evenly toothed. Preferred Habitat - Open places; rich, low ground; beside streams. Flowering Season - July-September. Distribution - Southern Canada to Florida, westward to Illinois and Kentucky.

Along the streams the numerous flower-heads of this gorgeous sunbearer shine out from afar, brightening a long, meandering course across the low-lying meadows. Like heralds of good things to come, they march a little in advance of the brilliant pageant of wild flowers that sweeps across the country from midsummer till killing frost. Most people mistake them for true, yellow-disked sunflowers, whose ray florets are neutral, not fertile as these long persistent ones are, But no one should confuse them with the dark cone-centered ox-eye daisy. Small bees, wasps, hornets, flies, little butterflies, beetles, and lower insects come to feast on the nectar and pollen within the minute tubular disk florets. The bright fulvous and black pearl crescent butterfly, with a trifle over an inch wing expanse; the common hairstreak; the even commoner little white butterfly; and the tiny black sooty wing, among others, appear to find generous entertainment here. The last named little fellow, when in the caterpillar stage, formed a cradle for himself by folding together a leaf of the ubiquitous green-flowered pigweed or lamb's quarters (Cizenopodium album) and stitching the edges together with a few silken threads. Here it slept by day, emerging only at night to feed. Usually one has not long to wait before discovering the white-dotted sooty wing among the midsummer composites.

BLACK-EYED SUSAN; YELLOW or OX-EYE DAISY; NIGGER-HEAD; GOLDEN JERUSALEM; PURPLE CONE-FLOWER
(Rudbeckia hirta) Thistle family

Flower-heads - From 10 to 20 orange-yellow neutral rays around a conical, dark purplish-brown disk of florets containing both stamens and pistil. Stem: 1 to 3 ft. tall, hairy, rough, usually unbranched, often tufted. Leaves: Oblong to lance-shaped, thick, sparingly notched, rough. Preferred Habitat - Open sunny places; dry fields. Flowering Season - May-September. Distribution - Ontario and the Northwest Territory south to Colorado and the Gulf States.

So very many weeds having come to our Eastern shores from Europe, and marched farther and farther west year by year, it is but fair that black-eyed Susan, a native of Western clover fields, should travel toward the Atlantic in bundles of hay whenever she gets the chance, to repay Eastern farmers in their own coin. Do these gorgeous heads know that all our showy rudbeckias - some with orange red at the base of their ray florets - have become prime favorites of late years in European gardens, so offering them still another chance to overrun the Old World, to which so much American hay is shipped? Thrifty farmers may decry the importation into their mowing lots, but there is a glory to the cone-flower beside which the glitter of a gold coin fades into paltry nothingness. Having been instructed in the decorative usefulness of all this genus by European landscape gardeners, we Americans now importune the Department of Agriculture for seeds through members of Congress, even Representatives of States that have passed stringent laws against the dissemination of "weeds." Inasmuch as each black-eyed Susan puts into daily operation the business methods of the white daisy (q.v.), methods which have become a sort of creed for the entire composite horde to live by, it is plain that she may defy both farmers and legislators. Bees, wasps, flies, butterflies, and beetles could not be kept away from an entertainer so generous; for while the nectar in the deep, tubular brown florets may be drained only by long, slender tongues, pollen is accessible to all. Anyone who has had a jar of these yellow daisies standing on a polished table indoors, and tried to keep its surface free from a ring of golden dust around the flowers, knows how abundant their pollen is. There are those who vainly imagine that the slaughter of dozens of English sparrows occasionally is going to save this land of liberty from being overrun with millions of the hardy little gamins that have proved themselves so fit in the struggle for survival. As vainly may farmers try to exterminate a composite that has once taken possession of their fields.

Blazing hot sunny fields, in which black-eyed Susan feels most comfortable, suit the TALL or GREEN-HEADED CONE-FLOWER OR THIMBLEWEED (R. laciniata) not at all. Its preference is for moist thickets such as border swamps and meadow runnels. Consequently it has no need of the bristly-hairy coat that screens the yellow daisy from too tierce, sunlight, and great need of more branches and leaves. (See prickly pear.) This is a smooth, much branched plant, towering sometimes twelve feet high, though commonly not even half that height; its great lower leaves, on long petioles, have from three to seven divisions variously lobed and toothed; while the stem leaves are irregularly three to five parted or divided. The numerous showy heads, which measure from two and a half to four inches across, have from six to ten bright yellow rays drooping a trifle around a dull greenish-yellow conical disk that gradually lengthens to twice its breadth, if not more, as the seeds mature. July-September, Quebec to Montana, and southward to the Gulf of Mexico.

TALL or GIANT SUNFLOWER
(Heliainthus giganteus) Thistle family

Flower-heads - Several, on long, rough-hairy peduncles; 1 1/2 to 2 1/4 in. broad; 10 to 20 pale yellow neutral rays around a yellowish disk whose florets are perfect, fertile. Stem: 3 to 12 ft. tall, bristly-hairy, usually branching above, often reddish from a perennial, fleshy root. Leaves: Rough, firm, lance-shaped, saw-toothed, sessile. Preferred Habitat - Low ground, wet meadows, swamps. Flowering Season - August-October. Distribution - Maine to Nebraska and the Northwest Territory, south to the Gulf of Mexico.

To how many sun-shaped golden disks with outflashing rays might not the generic name of this clan (helios = the sun, anthos = a flower) be as fittingly applied: from midsummer till frost the earth seems given up to floral counterparts of his worshipful majesty. If, as we are told, one-ninth of all flowering plants in the world belong to the composite order, of which over sixteen hundred species are found in North America north of Mexico, surely over half this number are made up after the daisy pattern (q.v.), the most successful arrangement known, and the majority of these are wholly or partly yellow. Most conspicuous of the horde are the sunflowers, albeit they never reach in the wild state the gigantic dimensions and weight that cultivated, dark brown centered varieties produced from the COMMON SUNFLOWER (H. annus) have attained. For many years the origin of the latter flower, which suddenly shone forth in European gardens with unwonted splendor, was in doubt. Only lately. it was learned that when Champlain and Segur visited the Indians on Lake Huron's eastern shores about three centuries ago, they saw them cultivating this plant, which must have been brought by them from its native prairies beyond the Mississippi - a plant whose stalks furnished them with a textile fiber, its leaves fodder, its flowers a yellow dye, and its seeds, most valuable of all, food and hair oil. Early settlers in Canada were not slow in sending home to Europe so decorative and useful an acquisition. Swine, poultry, and parrots were fed on its rich seeds. Its flowers, even under Indian cultivation had already reached abnormal size. Of the sixty varied and interesting species of wild sunflowers known to scientists, all are North American. Moore's pretty statement,

"As the sunflower turns on her god when he sets
The same look which she turn'd when he rose,"

lacks only truth to make it fact. The flower does not travel daily
on its stalk from east to west. Often the top of the stem turns
sharply toward the light to give the leaves better exposure, but the
presence or absence of a terminal flower affects its action not at all.

Formerly the garden species was thought to be a native, not of
our prairies, but of Mexico and Peru, because the Spanish conquer-
ors found it employed there as a mystic and sacred symbol, much as
the Egyptians employed the lotus in their sculpture. In the temples
the handmaidens wore upon their breasts plates of gold beaten into
the likeness of the sunflower. But none of the eighteen species of
helianthus found south of our borders produces under cultivation
the great plants that stand like a golden-helmeted phalanx in every
old-fashioned garden at the North. Many birds, especially those of
the sparrow and finch tribe, come to feast on the oily seeds; and
where is there a more charming sight than when a family of goldfinches settle upon the huge, top-heavy heads, unconsciously forming a study in sepia and gold?

On prairies west of Pennsylvania to South Dakota, Missouri, and
Texas, the SAW-TOOTH SUNFLOWER (H. grosse-serratus) is
common. Deep yellow instead of pale rays around a yellowish disk
otherwise resemble the tall sunflower's heads in appearance as in
season of bloom. The smooth stalk, with a bluish-hoary bloom on its
surface, may have hairs on the branches only. Long, lance-shaped,
pointed leaves, the edges of lower ones especially sharply saw-
toothed, their upper surface rough, and underneath soft-hairy, are
on slender, short petioles, the lower ones opposite, the upper ones
alternate. Honeybees find abundant refreshment in the tubular disk
florets in which many of their tribe may be caught sucking; brilliant
little Syrphidae, the Bombilius cheat, and other flies come after pollen; butterflies feast here on nectar, too and greedy beetles, out for
pollen, often gnaw the disks with their pinchers.

Very common in dry woodlands and in roadside thickets from
Ontario to Florida, and westward to Nebraska, is the ROUGH OR
WOODLAND SUNFLOWER (H. divaricatus). Its stem, which is

smooth nearly to the summit, does not often exceed three feet in height, though it may be less, or twice as high. Usually all its widespread leaves are opposite, sessile, lance-shaped to ovate, slightly toothed, and rough on their upper surface. Few or solitary flowerheads, about two inches across, have from eight to fifteen rays round a yellow disk.

The THIN-LEAVED or TEN-PETALLED SUNFLOWER (H. decapetalus), on the contrary, chooses to dwell in moist woods and thickets, beside streams, no farther west than Michigan and Kentucky. Its smooth, branching stem may be anywhere from one foot to five feet tall; its thin, membranous, sharply saw-edged leaves, from ovate to lance-shaped, with a rounded base, roughest above and soft underneath, are commonly alternate toward the summit, while the lower ones, on slender petioles, are opposite. There are by no means always ten yellow rays around the yellow disks produced in August and September; there may be any number from eight to fifteen, although this free-flowering species, like the PALE-LEAVED WOOD SUNFLOWER (H. strumosus), an earlier bloomer, often arranges its "petals" in tens.

JERUSALEM ARTICHOKE, EARTH APPLE, CANADA POTATO, GIRASOLE (H. tuberosus), often called WILD SUNFLOWER, too, has an interesting history similar to the dark-centered, common garden sunflower's. In a musty old tome printed in 1649, and entitled "A Perfect Description of Virginia," we read that the English planters had "rootes of several kindes, Potatoes, Sparagus, Carrets and Hartichokes" - not the first mention of the artichoke by Anglo-Americans. Long before their day the Indians, who taught them its uses, had cultivated it; and wherever we see the bright yellow flowers gleaming like miniature suns above roadside thickets and fence rows in the East, we may safely infer the spot was once an aboriginal or colonial farm. White men planted it extensively for its edible tubers, which taste not unlike celery root or salsify. As early as 1617 the artichoke was introduced into Europe, and only twelve years later Parkinson records that the roots had become very plentiful and cheap in London. The Italians also cultivated it under the name Girasole Articocco (sunflower artichoke), but it did not take long for the girasole to become corrupted into Jerusalem, hence the name Jerusalem Artichoke common to this day. When the greater value of the potato came to be generally recognized, the use of artichoke roots gradually diminished. Quite different from this sunflower is the true artichoke (Cynara Scolymus), a native of Southern Europe, whose large, unopened flower-heads offer a tiny edible morsel at the base of each petal-like part.

The Jerusalem artichoke sends up from its thickened, fleshy, tuber-bearing rootstock, hairy, branching stems six to twelve feet high. Especially are the flower-stalks rough, partly to discourage pilfering crawlers. The firm, oblong leaves, taper pointed at the apex and saw-edged, are rough above, the lower leaves opposite each other on petioles, the upper alternate. The brilliant flower-heads, which are produced freely in September and October, defying frost, are about two or three inches across, and consist of from twelve to twenty lively yellow rays around a dull yellow disk. The towering prolific plant prefers moist but not wet soil from Georgia and Arkansas northward to New Brunswick and the Northwest Territory. Omnivorous small boys are not always particular about boiling, not to say washing, the roots before eating them.

LANCE-LEAVED TICKSEED; GOLDEN COREOPSIS
  (Coreopsis lanceolata) Thistle family

  Flowers-heads - Showy, bright golden yellow, the 6 to io wedge-shaped, coarsely toothed ray florets around yellowish disk florets soon turning brown; each head on a very long, smooth, slender footstalk. Stems. 1 to 2 ft. high, tufted. Leaves: A few seated on stem, lance-shaped to narrowly oblong; or lower ones crowded, spatulate, on slender petioles. Preferred Habitat - Open, sunny places, moist or dry. Flowering Season - May-September. Distribution - Western Ontario to Missouri and the Gulf States; escaped from gardens in the East.

  Glorious masses of this prolific bloomer persistently outshine all rivals in the garden beds throughout the summer. Cut as many slender-stalked flowers and buds as you will for vases indoors, cut them by armfuls, and two more soon appear for every one taken.

From seeds scattered by the wind over a dry, sandy field adjoining a Long Island garden one autumn, myriads of these flowers swarmed like yellow butterflies the next season. Very slight encouragement induces this coreopsis to run wild in the East. Grandiflora, with pinnately parted narrow leaves and similar flowers, a Southwestern species, is frequently a runaway. Bees and flies, attracted by the showy neutral rays which are borne solely for advertising purposes, unwittingly cross-fertilize the heads as they crawl over the tiny, tubular, perfect florets massed together in the central disk; for some of these florets having the pollen pushed upward by hair brushes and exposed for the visitor's benefit, while others have their sticky style branches spread to receive any vitalizing dust brought to them, it follows that quantities of vigorous seed must be set.

"There is a natural rotation of crops, as yet little understood," says Miss Going. "Where a pine forest has been cleared away, oaks come up; and a botanist can tell beforehand just what flowers will appear in the clearings of pine woods. In northern Ohio, when a piece of forestland is cleared, a particular sort of grass appears. When that is ploughed under, a growth of the golden coreopsis comes up, and the pretty yellow blossoms are followed in their turn by the plebeian rag-weed which takes possession of the entire field."

The charmingly delicate, wiry GARDEN TICKSEED, known in seedsmen's catalogues as CALLIOPSIS (Coreopsis tinctoria), which has also locally escaped to roadsides and waste places eastward, is at home in moist, rich soil from Louisiana, Arizona, and Nebraska northward into Minnesota and the British Possessions. >From May to September its fine, slender, low-growing stems are crowned with small yellow composite flowers whose rays are velvety maroon or brown at the base. (Coreopsis = like a bug, from the shape of the seeds.)

LARGER or SMOOTH BUR-MARIGOLD; BROOK SUNFLOWER
(Bidens laevis; B. chrysanthemoides of Gray) Thistle family

Flower-heads - Showy golden yellow, 1 to 2 1/2 in. across, numerous, on short peduncles; 8 to 10 neutral rays around a dingy yellowish or brown disk of tubular, perfect, fertile florets. Stem: 1 to 2 ft. high. Leaves: Opposite, sessile, lance-shaped, regularly saw-toothed. Preferred Habitat - Wet ground, swamps, ditches, meadows. Flowering Season - August-November. Distribution - Quebec and Minnesota, southward to the Gulf States and Lower California.

Next of kin to the golden coreopsis, it behooves some of the bur-marigolds to redeem their clan's reputation for ugliness and certainly the brook sunflower is a not unworthy relative. How gay the ditches and low meadows are with its bright, generous bloom in late summer, and until even the goldenrod wands turn brown! Yet all this show is expended merely for advertising purposes. The golden ray florets, sacrificing their fertility to the general welfare of the cooperative community, which each flower-head is in reality, have grown conspicuous to attract bees and wasps, butterflies, flies, and some beetles to the dingy mass of tubular florets in the center, in which nectar is concealed, while pollen is exposed for the visitors to transfer as they crawl. The rays simply make a show; within the minute, insignificant looking tubes is transacted the important business of life.

Later in the season, when the bur-marigolds are transformed into armories bristling with rusty, two-pronged, and finely-barbed pitchforks (Bidens = two teeth), our real quarrel with the tribe begins. The innocent passerby - man, woman, or child, woolly sheep, cattle with switching tails, hairy dogs or foxes, indeed, any creature within reach of the vicious grappling-hooks - must transport them on his clothing; for it is thus that these tramps have planned to get away from the parent plant in the hope of being picked off, and the seeds dropped in fresh colonizing ground; travelling in the disreputable company of their kinsmen the beggar-ticks and Spanish needles, the burdock burs, cleavers, agrimony, and tick-trefoils.

BEGGAR-TICKS, STICK-TIGHT, RAYLESS MARIGOLD, BEGGAR-LICE, PITCHFORKS, or STICK-SEED (B. frondosa) sufficiently explains its justly defamed character in its popular names. Numerous dull, dark, tawny orange flower-heads without, rays, or with insignificant ones scarcely to be detected, and surrounded by taller leaf-like bracts, add little to the beauty of the moist fields and roadsides where they rear themselves on long peduncles from July to October. The smooth, erect, branched, and often reddish, stem may be anywhere from two to nine feet tall. Usually the upper leaves are not divided, but the lower ones are pinnately compounded of three to five divisions, the segments lance-shaped or broader, and sharply toothed. As in all the bur-marigolds, we find each floret's calyx converted into a barbed implement - javelin, pitchfork, or halberd - for grappling the clothing of the first innocent victim unwittingly acting as a colonizing agent.

SNEEZEWEED; SWAMP SUNFLOWER
(Helenium autumnale) Thistle family

Flower-heads - Bright yellow, to 2 in. across, numerous, borne on long peduncles in corymb-like clusters; the rays 3 to 5 cleft, and drooping around the yellow or yellowish-brown disk. Stem: 2 to 6 ft. tall, branched above. Leaves: Alternate, firm, lance-shaped to oblong, toothed, seated on stem or the bases slightly decurrent; bitter. Preferred Habitat - Swamps, wet ground, banks of streams. Flowering Season - August-October. Distribution - Quebec to the Northwest Territory; southward to Florida and Arizona.

September, which also brings out lively masses of the swamp sunflower in the low-lying meadows, was appropriately called our golden month by an English traveler who saw for the first time the wonderful yellows in our autumn foliage, the surging seas of goldenrod; the tall, showy sunflowers, ox-eyes, rudbeckias, marigolds, and all the other glorious composites in Nature's garden, as in men's, which copy the sun's resplendent disk and rays to brighten with one final dazzling outburst the somber face of the dying year.

To the swamp sunflowers honey-bees hasten for both nectar and pollen, velvety bumblebees suck the sweets, leaf-cutter and mason bees, wasps, some butterflies, flies, and beetles visit them daily, for the round disks mature their perfect fertile florets in succession.

Since the drooping ray flowers, which are pistillate only, are fertile too, there is no scarcity of seed set, much to the farmer's dismay. Most cows know enough to respect the bitter leaves' desire to be let alone; but many a pail of milk has been spoiled by a mouthful of Helenium among the herbage. Whoever cares to learn from experience why this was called the sneezeweed, must take a whiff of snuff made of the dried and powdered leaves.

The PURPLE-HEAD SNEEZEWEED (H. nudiflorum), its yellow rays sometimes wanting, occurs in the South and West.

TANSY; BITTER-BUTTONS
(Tanacetum vulgare) Thistle family

Flower-heads - Small, round, of tubular florets only, packed within a depressed involucre, and borne, in flat-topped corymbs.
Stem: 1 1/2 to 3 ft. tall, leafy. Leaves: Deeply and pinnately cleft into narrow, toothed divisions; strong scented.
Preferred Habitat - Roadsides; commonly escaped from gardens.
Flowering Season - July-September.
Distribution - Nova Scotia, westward to Minnesota, south to Missouri and North Carolina. Naturalized from Europe.

"In the spring time, are made with the leaves hereof newly sprung up, and with eggs, cakes or Tansies which be pleasant in taste and goode for the Stomache," wrote quaint old Gerarde. That these were popular dainties in the seventeenth century we further know through Pepys, who made a "pretty dinner" for some guests, to wit: "A brace of stewed carps, six roasted chickens, and a jowl of salmon, hot, for the first course; a tansy, and two neat's tongues, and cheese, the second." Cole's "Art of Simpling," published in 1656, assures maidens that tansy leaves laid to soak in buttermilk for nine days "maketh the complexion very fair." Tansy tea, in short, cured every ill that flesh is heir to, according to the simple faith of mediaeval herbalists - a faith surviving in some old women even to this day. The name is said to be a corruption of athanasia, derived from two Greek words meaning immortality. When some monks in reading Lucian came across the passage where Jove, speaking of Ganymede to Mercury, says, "Take him hence, and when he has tasted immortality let him return to us," their literal minds inferred that this plant must have been what Ganymede tasted, hence they named it athanasia! So great credence having been given to its medicinal powers in Europe, it is not strange the colonists felt they could not live in the New World without tansy. Strong-scented pungent tufts topped with bright yellow buttons - runaways from old gardens - are a conspicuous feature along many a roadside leading to colonial homesteads.

GOLDEN RAGWORT; GROUNDSEL; SQUAW-WEED
(Senecio aureus) Thistle family

Flower-heads - Golden yellow, about 3/4 in. across, borne on slender peduncles in a loose, leafless cluster; rays 8 to 12 around minute disk florets. Stem: Slender, 1 to 2 1/2 ft. high, solitary or tufted, from a strong-scented root. Leaves: From the root, on long petioles, rounded or heart-shaped, scalloped-edged, often purplish; stem leaves variable, lance-shaped or lyrate, deeply cut, sessile. Preferred Habitat - Swamps, wet ground, meadows. Flowering Season - May-July. Distribution - Gulf States northward to Missouri, Ontario, and Newfoundland.

While the aster clan is the largest we have in North America, this genus Senecio is really the most numerous branch of the great composite tribe, numbering as it does nearly a thousand species, represented in all quarters of the earth. It is said to take its name from senex = an old man, in reference to the white hairs on many species; or, more likely, to the silky pappus that soon makes the fertile disks hoary headed. "I see the downy heads of the senecio gone to seed, thistle like but small," wrote Thoreau in his journal under date of July 2nd, when only the pussy-toes everlasting could have plumed its seeds for flight over the dry uplands in a similar fashion. Innumerable as the yellow, daisy-like composites are, most of them appear in late summer or autumn, and so the novice should have little difficulty in naming these loosely clustered, bright, early blooming small heads.

RED AND INDEFINITES

"I want the inner meaning and the understanding of the wildflowers in the meadow. Why are they? What end? What purpose? The plant knows, and sees, and feels; where is its mind when the petal falls? Absorbed in the universal dynamic force, or what? They make no shadow of pretence, these beautiful flowers, of being beautiful for my sake; of bearing honey for me; in short, there does not seem to be any kind of relationship understood between us, and yet . . . language does not express the dumb feelings of the mind any more than the flower can speak. I want to know the soul of the flowers! . . . All these life-laboured monographs, these classifications, works of Linnaeus, and our own classic Darwin, microscope, physiology - and the flower has not given us its message yet.' ' - Richard Jeffries.

JACK-IN-THE-PULPIT; INDIAN TURNIP
(Arisaema triphyllum) Arum family

Flowers - Minute, greenish yellow, clustered on the lower part of a smooth, club-shaped, slender spadix within a green and maroon or whitish-striped spathe that curves in a broad-pointed flap above it. Leaves: 3-foliate, usually overtopping the spathe, their slender petioles 9 to 30 in. high, or as tall as the scape that rises from an acrid corm. Fruit: Smooth, shining red berries clustered on the thickened club. Preferred Habitat - Moist woodland and thickets. Flowering Season - April-June. Distribution - Nova Scotia westward to Minnesota, and southward to the Gulf States.

A jolly looking preacher is Jack, standing erect in his particolored pulpit with a sounding-board over his head; but he is a gay deceiver, a wolf in sheep's clothing,, literally a "brother to dragons," an arrant upstart, an ingrate, a murderer of innocent benefactors! "Female botanizing classes pounce upon it as they would upon a pious young clergyman," complains Mr. Ellwanger. A poor relation of the stately calla lily one knows Jack to be at a glance, her lovely white robe corresponding to his striped pulpit, her bright yellow spadix to his sleek reverence. In the damp woodlands where his pulpit is erected beneath leafy cathedral arches, minute flies or gnats, recently emerged from maggots in mushrooms, toadstools, or decaying logs, form the main part of his congregation.

Now, to drop the clerical simile, let us peep within the sheathing spathe, or, better still, strip it off altogether. Dr. Torrey states that the dark-striped spathes are the fertile plants, those with green and whitish lines, sterile. Within are smooth, glossy columns, and near the base of each we shall find the true flowers, minute affairs, some staminate; others, on distinct plants, pistillate, the berry bearers; or rarely both male and female florets seated on the same club, as if Jack's elaborate plan to prevent self-fertilization were not yet complete. Plants may be detected in process of evolution toward their ideals: just as nations and men are. Doubtless, when Jack's mechanism is perfected, his guilt will disappear. A little way above the florets the club enlarges abruptly, forming a projecting ledge that effectually closes the avenue of escape for many a guileless victim. A fungus gnat, enticed perhaps by the striped house of refuge from cold spring winds, and with a prospect of food below, enters and slides down the inside walls or the slippery colored column: in either case descent is very easy; it is the return that is made so difficult, if not impossible, for the tiny visitors. Squeezing past the projecting ledge, the gnat finds himself in a roomy apartment whose floor - the bottom of the pulpit - is dusted over with fine pollen; that is, if he is among staminate flowers already mature. To get some of that pollen, with which the gnat presently covers himself, transferred to the minute pistillate florets waiting for it in a distant chamber is, of course, Jack's whole aim in enticing visitors within his polished walls; but what means are provided for their escape? Their efforts to crawl upward over the slippery surface only land them weak and discouraged where they started. The projecting ledge overhead prevents them from using their wings; the passage between the ledge and the spathe is far too narrow to permit flight. Now, if a gnat be persevering, he will presently discover a gap in the flap where the spathe folds together in front, and through this tiny opening he makes his escape, only to enter another pulpit, like the trusted, but too trusting, messenger he is, and leave some of the vitalizing pollen on the fertile florets awaiting his coming.

But suppose the fly, small as he is, is too large to work his way out through the flap, or too bewildered or stupid to find the opening, or too exhausted after his futile efforts to get out through the overhead route to persevere, or too weak with hunger in case of long detention in a pistillate trap where no pollen is, what then? Open a dozen of Jack's pulpits, and in several, at least, dead victims will be found - pathetic little corpses sacrificed to the imperfection of his executive system. Had the flies entered mature spathes, whose walls had spread outward and away from the polished column, flight through the overhead route might have been possible. However glad we may be to make every due allowance for this sacrifice of the higher life to the lower, as only a temporary imperfection of mechanism incidental to the plant's higher development, Jacks present cruelty shocks us no less. Or, it may be, he will become insectivorous like the pitcher plant in time. He comes from a rascally family, anyhow. (See cuckoo pint.)

In June and July the thick-set club, studded over with bright berries, becomes conspicuous, to attract hungry woodland rovers in the hope that the seeds will be dropped far from the parent plant. The Indians used to boil the berries for food. The farinaceous root (corm) they likewise boiled or dried to extract the stinging, blistering juice, leaving an edible little "turnip," however insipid and starchy.

The GREEN DRAGON, or DRAGON-ROOT (A. Dracontium), to which Jack is brother, is found in similar situations or beside streams in wet, shady ground, and sends up a narrow greenish or whitish tapering spathe, one or two inches long, enwrapping a slender, pointed spadix, that projects sometimes seven inches beyond its tip. Within, tiny pistillate florets are seated around the base, while on the staminate plants the inflorescence extends higher. A large, solitary, dark green leaf, divided into from five to seventeen oblong, pointed segments, spreads above. Large ovoid heads of reddish-orange berries are the plant's most conspicuous feature.

SKUNK OR SWAMP CABBAGE
(Spathyema fetida; Symplocarpus fetidus of Gray) Arum family

Flowers - Minute, perfect, fetid; many scattered over a thick, rounded, fleshy spadix, and hidden within a swollen, shell-shaped, purplish-brown to greenish-yellow, usually mottled, spathe, close to the ground, that appears before the leaves. Spadix much enlarged and spongy in fruit, the bulb-like berries imbedded in its surface. Leaves: In large crowns like cabbages, broadly ovate, often 1 ft. across, strongly nerved, their petioles with deep grooves, malodorous. Preferred Habitat - Swamps, wet ground. Flowering Season - February-April. Distribution - Nova Scotia to Florida, and westward to Minnesota and Iowa.

This despised relative of the stately calla lily proclaims spring in the very teeth of winter, being the first bold adventurer above ground. When the lovely hepatica, the first flower worthy the name to appear, is still wrapped in her fuzzy furs, the skunk cabbage's dark incurved horn shelters within its hollow, tiny, malodorous florets. Why is the entire plant so fetid that one flees the neighborhood, pervaded as it is with an odor that combines a suspicion of skunk, putrid meat, and garlic? After investigating the carrion-flower (q.v.) and the purple trillium, among others, we learned that certain flies delight in foul odors loathsome to higher organisms; that plants dependent on these pollen carriers woo them from long distances with a stench, and in addition sometimes try to charm them with color resembling the sort of meat it is their special mission, with the help of beetles and other scavengers of Nature, to remove from the face of the earth. In such marshy ground as the skunk cabbage lives in, many small flies and gnats live in embryo under the fallen leaves during the winter. But even before they are warmed into active life, the hive-bees, natives of Europe, and with habits not perfectly adapted as yet to our flora (nor our flora's habits to theirs - see milkweed), are out after pollen. Where would they find any so early, if not within the skunk cabbage's livid horn of plenty? Not even an alder catkin or a pussy willow has expanded yet. In spite of the bee's refined taste in the matter of perfume and color, she has no choice, now, but to enter so generous an entertainer. At the top of the thick rounded spadix within, the skunk cabbage florets there first mature their stigmas, and pollen must therefore be carried to them on the bodies of visitors. Later these stigmas wither, and abundant pollen is shed from the now ripe anthers.

Meantime the lower, younger florets having matured their stigmas, some pollen may fall directly on them from the older flowers above. A bee crawling back and forth over the spadix gets thoroughly dusted, and flying off to another cluster of florets cross-fertilizes them - that is, if all goes well. But because the honeybee never entered the skunk cabbage's calculations, useful as the immigrant proved to be, the horn that was manifestly designed for smaller flies often proves a fatal trap. Occasionally a bee finds the entrance she has managed to squeeze through too narrow and slippery for an exit, and she perishes miserably.

"A couple of weeks after finding the first bee," says Mr. William Trelease in the "American Naturalist," "the spathes will be found swarming with the minute black flies that were sought in vain earlier in the season, and their number is attested not only by the hundreds of them which can be seen, but also by the many small but very fat spiders whose webs bar the entrance to three-fourths of the spathes. During the present spring a few specimens of a small scavenger beetle have been captured within the spathes of this plant…. Finally, other and more attractive flowers opening, the bees appear to cease visiting those of this species, and countless small flies take their place, compensating for their small size by their great numbers." These, of course, are the benefactors the skunk cabbage catered to ages before the honeybee reached our shores.

After the flowering time come the vivid green crowns of leaves that at least please the eye. Lizards make their home beneath them, and many a yellowthroat, taking advantage of the plant's foul odor, gladly puts up with it herself and builds her nest in the hollow of the cabbage as a protection for her eggs and young from four-footed enemies. Cattle let the plant alone because of the stinging, acrid juices secreted by it, although such tender, fresh, bright foliage must be especially tempting, like the hellebore's, after a dry winter diet. Sometimes tiny insects are found drowned in the wells of rain water that accumulate at the base of the grooved leafstalks.

RED, WOOD, FLAME, or PHILADELPHIA LILY
  (Lilium Philadelphicum) Lily family

Flowers - Erect, tawny or red-tinted outside; vermilion, or sometimes reddish orange, and spotted with madder brown within; 1 to 5, on separate peduncles, borne at the summit. Perianth of 6 distinct, spreading, spatulate segments, each narrowed into a claw, and with a nectar groove at its base; 6 stamens; 1 style, the club-shaped stigma 3-lobed. Stem: 1 to 3 ft. tall, from a bulb composed of narrow, jointed, fleshy scales. Leaves: In whorls of 3's to 8's, lance-shaped, seated at intervals on the stem. Preferred Habitat - Dry woods, sandy soil, borders, and thickets. Flowering Season - June-July. Distribution - Northern border of United States, westward to Ontario, south to the Carolinas and West Virginia.

Erect, as if conscious of its striking beauty, this vivid lily lifts a chalice that suggests a trap for catching sunbeams from fiery old Sol. Defiant of his scorching rays in its dry habitat, it neither nods nor droops even during prolonged drought; and yet many people confuse it with the gracefully pendent, swaying bells of the yellow Canada lily, which will grow in a swamp rather than forego moisture. Li, the Celtic for white, from which the family derived its name, makes this bright-hued flower blush to own it. Seedmen, who export quantities of our superb native lilies to Europe, supply bulbs so cheap that no one should wait four years for flowers from seed, or go without their splendor in our over-conventional gardens. Why this early lily is radiantly colored and speckled is told in the description of the Canada lily (q.v.).

The WESTERN RED LILY (L. umbellatum), that takes the place of the Philadelphia species from Ohio, Minnesota, and the Northwest Territory, southward to Missouri, Arkansas, and Colorado, lifts similar but smaller red, orange, or yellow flowers on a more slender stem, two feet high or less, set with narrow, linear, alternate leaves, or perhaps the upper ones in whorls. It blooms in June or July, in dry soil, preferably in open, sandy situations.

LARGE CORAL-ROOT
 (Corallorhiza multiflora) Orchid family

Flowers - Dull brownish purple, about 1/2 in. high; 10 to 30 borne in a raceme 2 to 8 in. long. Petals about the length of sepals, and somewhat united at the base; spur yellowish, the oval lip white, spotted and lined with purplish; 3-lobed, wavy edged. Scape, 8 to 20 in. tall, colored, furnished with several flat scales. Leaves: None. Root: A branching, coral-like mass. Preferred Habitat - Dry woods. Flowering Season - July-September. Distribution - Nova Scotia, westward to British Columbia; south to Florida, Missouri, and California.

To the majority of people the very word orchid suggests a millionaire's hothouse, or some fashionable florist's show window, where tropical air plants send forth gorgeous blossoms, exquisite in color, marvelous in form; so that when this insignificant little stalk pokes its way through the soil at midsummer and produces some dull flowers of indefinite shades and no leaves at all to help make them attractive, one feels that the coral-root is a very poor relation of theirs indeed. The prettily marked lower lip, at once a platform and nectar guide to the insect alighting on it, is all that suggests ambition worthy of an orchid.

If poverty of men and nations can be traced to certain radical causes by the social economist, just as surely can the botanist account for loss of leaves - riches - by closely examining the poverty-stricken plant. Every phenomenon has its explanation. A glance at the extraordinary formation under ground reveals the fact that the coralroots, although related to the most aristocratic and highly organized plants in existence, have stooped to become ghoulish saprophytes. An honest herb abounds in good green coloring matter (chlorophyll), that serves as a light screen to the cellular juices of leaf and stem. It also forms part of its digestive apparatus, aiding a plant in the manufacture of its own food out of the soil, water, and gases; whereas a plant that lives by piracy - a parasite - or a saprophyte, that sucks up the already assimilated products of another's decay, loses its useless chlorophyll as surely as if it had been kept in a cellar. In time its equally useless leaves dwindle to bracts, or disappear. Nature wastes no energy. Fungi, for example, are both parasites and saprophytes; and so when plants far higher up in the evolutionary scale than they lose leaves and green color too, we may know they are degenerates belonging to that disreputable gang of branded sinners which includes the Indian-pipe, broom-rape, dodder, pine-sap, and beech-drops. Others, like the gerardias and foxgloves, may even now be detected on the brink of a fall from grace.

The EARLY CORAL-ROOT (C. Corallorhiza; C. innata of Gray) - a similar but smaller species, whose loose spike of dull purplish flowers likewise terminates a scaly purplish or yellowish scape arising from a mass of short, thick, whitish, fleshy, blunt fibers, may be found in the moist woods blooming in May or June. It has a more northerly range, however, extending from the mountains of Georgia, it is true, but chiefly from the northern boundary of the United States, from New England westward to the State of Washington, and northward to Nova Scotia and Alaska.

ADAM AND EVE; PUTTY-ROOT
(Aplectrum spicatum; A. hyemale of Gray)) Orchid family

Flowers - Dingy yellowish brown and purplish, about 1 in. long, each on a short pedicel, in a few-flowered, loose, bracted raceme 2 to 4 in. long. No spur; sepals and petals similar, small and narrow, the lip wavy-edged. Scape: to 2 ft. high, smooth, with about 3 sheathing scales. Leaf: Solitary, rising from the corm in autumn, elliptic, broad, plaited-nerved, 4 to 6 in. long. Root: A corm usually attached to one of the preceding season. Preferred Habitat - Moist woods or swamps. Flowering Season - May-June. Distribution - Georgia, Missouri, and California northward, into British Possessions.

More curious than beautiful is this small orchid whose dingy flowers of indefinite color and without spurs interest us far less than the two corms barely hidden below ground. These singular solid bulbs, about an inch thick, are connected by a slender stalk, suggesting to the imaginative person who named the plant our first parents standing hand in hand in the Garden of Eden.

But usually several old corms - not always two, by any means - remain attached to the nearest one, a bulb being produced each year until Cain and Abel often join Adam and Eve to make up quite a family group. A strong, glutinous matter within the corms has been used as a cement, hence the plant's other popular name. From the newest bulb added, a solitary large leaf arises in late summer or autumn, to remain all winter. The flower stalk comes up at one side of it the following spring. Meantime the old corms retain their life, apparently to help nourish the young one still joined to them, while its system is taxed with flowering.

WILD GINGER; CANADA SNAKEROOT; ASARABACCA
(Asarum Canadense) Birthwort family

Flower - Solitary, dull purplish brown, creamy white within, about 1 in. broad when expanded, borne on a short peduncle close to or upon the ground. Calyx cup-shaped, deeply cleft, its 3 acutely pointed lobes spreading, curved; corolla wanting; 12 short, stout stamens inserted on ovary; the thick style 6-lobed, its stigmas radiating on the lobes. Leaves: A single pair, dark green, reniform, 4 to 7 in. broad, on downy petioles 6 to 12 in. high, from a creeping, thick, aromatic, pungent rootstock. Preferred Habitat - Rich, moist woods; hillsides. Flowering Season - March-May. Distribution - North Carolina, Missouri, and Kansas, northward, to New Brunswick and Manitoba.

Like the wicked servant who buried the one talent entrusted to his care, the wild ginger hides its solitary flower if not actually under the dry leaves that clothe the ground in the still leafless woodlands, then not far above them. Why? When most plants flaunt their showy blossoms aloft, where they may be seen of all, why should this one bear only one dull, firm cup, inconspicuous in color as in situation? In early spring - and it is one of the earliest flowers - gnats and small flies are warming into active life from the maggots that have lain under dead leaves and the bark of decaying logs all winter. To such guests a flower need offer few attractions to secure them in swarms. Bright, beautiful colors, sweet fragrance, luscious nectar, with which the highly specialized bees, butterflies, and moths are wooed, would all be lost on them, lacking as they do esthetic taste. For flies, a snug shelter from cold spring winds such as Jack-in-the-pulpit, the marsh calla, the pitcher-plant, or the skunk cabbage offers; sometimes a fetid odor like the latter's, or dull purplish red or brownish color resembling stale meat, which the purple trillium likewise wears as an additional attraction, are necessary when certain carrion flies must be catered to; and, above all, an abundance of pollen for food - with any or all of these seductions a flower dependent on flies has nothing to fear from neglect. Therefore the wild ginger does not even attempt to fertilize itself. Within the cozy cup one can usually find a contented fly seeking shelter or food. Close to the ground it is warm and less windy. When the cup first opens, only the stigmas are mature and sticky to receive any pollen the visitors may bring in on their bodies from other asylums where they have been hiding. These stigmas presently withering, up rise the twelve stamens beside them to dust with pollen the flies coming in search of it. Only one flower from a root compels cross-fertilizing between flowers of distinct plants - a means to insure the most vigorous seed, as Darwin proved. Evidently the ginger is striving to attain some day the ambitious mechanism for temporarily imprisoning its guests that its cousin the Dutchman's pipe has perfected. After fertilization the cup nods, inverted, and the leathery capsule following it bursts irregularly, discharging many seeds.

No ruminant will touch the leaves, owing to their bitter juices, nor will a grub or nibbling rodent molest the root, which bites like ginger; nevertheless credulous mankind once utilized the plant as a tonic medicine.

DUTCHMAN'S PIPE; PIPE-VINE
(Aristolochia macrophylla; A. Sipho of Gray))

Flower - An inflated, curved, yellowish-green, veiny tube (calyx), pipe-shaped, except that it abruptly broadens beyond the contracted throat into 3 flat, spreading, dark purplish or reddish-brown lobes; pipe 1 to 1 1/2 in. long, borne on a long, drooping peduncle, either solitary or 2 or 3 together, from the bracted leaf-axils; 6 anthers, without filaments, in united pairs under the 3 lobes of the short, thick stigma. Stem: A very long, twining vine, the branches smooth and green. Leaves: Thin, reniform to heart-shaped, slender petioled, downy underneath when young; 6 to 15 in. broad when mature. Fruit: An oblong, cylindric capsule, containing quantities of seeds within its six sections. Preferred Habitat - Rich, moist woods. Flowering Season - May-June. Distribution - Pennsylvania, westward to Minnesota, south to Georgia and Kansas. Escaped from cultivation further north.

After learning why the pitcher plant, Jack-in-the-pulpit, and skunk cabbage are colored and shaped as they are, no one will be surprised on opening this curious flower to find numbers of little flies within the pipe. Certain relatives of this vine produce flowers that are not only colored like livid, putrid meat around the entrance, but also emit a fetid odor to attract carrion flies especially. (See purple trillium.)

In May, when the pipe-vine blooms, gauzy-winged small flies and gnats gladly seek food and shelter from the wind within so attractive an asylum as the curving tube offers. They enter easily enough through the narrow throat, around which fine hairs point downward - an entrance resembling an eel trap's. Any pollen they may bring in on their bodies now rubs off on the sticky stigma lobes, already matured at the bottom of a newly opened flower, in which they buzz, crawl, slide, and slip, seeking an avenue of escape. None presents itself: they are imprisoned. The hairs at the entrance, approached from within, form an impenetrable stockade. Must the poor little creatures perish? Is the flower heartless enough to murder its benefactors, on which the continuance of its species depends? By no means is it so shortsighted! A few tiny drops of nectar exuding from the center table prevent the visitors from starving. Presently the fertilized stigmas wither, and when they have safely escaped the danger of self-fertilization, the pollen hidden under their lobes ripens and dusts afresh the little flies so impatiently awaiting the feast. Now, and not till now, it is to the advantage of the species that the prisoners be released, that they may carry the vitalizing dust to stigmas waiting for it in younger flowers. Accordingly, the slippery pipe begins to shrivel, thus offering a foothold; the once stiff hairs that guarded its exit grow limp, and the happy gnats, after a generous entertainment and snug protection, escape uninjured, and by no means unwilling to repeat the experience. Evidently the wild ginger, belonging to a genus next of kin, is striving to perfect a similar prison. In the language of the street, the ginger flower does not yet "work" its.visitors "for all they are worth."

Later, when we see the exquisite dark, velvety, blue-green, pipe-vine, swallow-tail butterfly (Papilio philenor) hovering about verandas or woodland bowers that are shaded with the pipe-vine's large leaves, we may know she is there only to lay eggs that her caterpillar descendants may find themselves on their favorite food store.

The VIRGINIA SNAKEROOT or SERPENTARY (A. serpentaria), found in dry woods, chiefly in the Middle States and South, although its range extends northward to Connecticut, New York, and Michigan, is the species whose aromatic root is used in medicine. It is a low-growing herb, not a vine; its heart-shaped leaves, which are narrow and tapering to a point, are green on both sides, and the curious, greenish, S-shaped flower, which grows alone at the tip of a scaly footstalk from the root, appears in June or July. Sometimes the flowers are cleistogamous (see violet wood-sorrel).

FIRE PINK; VIRGINIA CATCHFLY
(Silene Virginica) Pink family

Flowers - Scarlet or crimson, 1 1/2 in. broad or less, a few on slender pedicels from the upper leaf-axils. Calyx sticky, tubular, bell-shaped, 5-cleft, enlarged in fruit; corolla of 5 wide-spread, narrow, notched petals, sometimes deeply 2-cleft; 10 stamens; 3 styles. Stem: 1 to 2 ft. high; erect, slender, sticky. Leaves: Thin, spatulate, 3 to 5 in. long; or upper ones oblong to lance-shaped. Preferred Habitat - Dry, open woodland. Flowering Season - May-September. Distribution - Southern New Jersey to Minnesota, south to Georgia and Missouri.

The rich, glowing scarlet of these pinks that fleck the Southern woodland as with fire, will light up our Northern rock gardens too, if we but sow the seed under glass in earliest spring, and set out the young plants in well-drained, open ground in May. Division of old perennial roots causes the plants to sulk; dampness destroys them.

To the brilliant blossoms butterflies chiefly come to sip (see wild pink), and an occasional hummingbird, fascinated by the color that seems ever irresistible to him, hovers above them on whirring wings. Hapless ants, starting to crawl up the stem, become more and more discouraged by its stickiness, and if they persevere in their attempts to steal from the butterfly's legitimate preserves, death overtakes their erring feet as speedily as if they ventured on sticky fly paper. How humane is the way to protect flowers from crawling thieves that has been adopted by the high-bush cranberry and the partridge pea (q.v.), among other plants! These provide a free lunch of sweets in the glands of their leaves to satisfy pilferers, which then seek no farther, leaving the flowers to winged insects that are at once despoilers and benefactors.

WILD COLUMBINE
(Aquilegia Canadensis) Crowfoot family

Flower - Red outside, yellow within, irregular, 1 to 2 in. long, solitary, nodding from a curved footstalk from the upper leaf-axils. Petals 5, funnel-shaped, but quickly narrowing into long, erect, very slender hollow spurs, rounded at the tip and united below by the 5 spreading red sepals, between which the straight spurs ascend; numerous stamens and 5 pistils projecting. Stem: 1 to 2 ft. high; branching, soft-hairy or smooth. Leaves: More or less divided, the lobes with rounded teeth; large lower compound leaves on long petioles. Fruit: An erect pod, each of the 5 divisions tipped with a long, sharp beak. Preferred Habitat - Rocky places, rich woodland. Flowering Season - April-July. Distribution - Nova Scotia to the Northwest Territory; southward to the Gulf States. Rocky Mountains.

Although under cultivation the columbine nearly doubles its size, it never has the elfin charm in a conventional garden that it possesses wild in Nature's. Dancing in red and yellow petticoats to the rhythm of the breeze, along the ledge of overhanging rocks, it coquettes with some Punchinello as if daring him to reach her at his peril. Who is he? Let us sit a while on the rocky ledge and watch for her lovers.

Presently a big muscular bumblebee booms along. Owing to his great strength, an inverted, pendent blossom, from which he must cling upside down, has no more terrors for him than a trapeze for the trained acrobat. His long tongue - if he is one of the largest of our sixty-two species of Bombus - can suck almost any flower unless it is especially adapted to night-flying sphinx moths, but can he drain this? He is the truest benefactor of the European columbine (q.v.), whose spurs suggested the talons of an eagle (aquila) to imaginative Linnaeus when he gave this group of plants its generic name. Smaller bumblebees, unable through the shortness of their tongues to feast in a legitimate manner, may be detected nipping holes in the tips of all columbines, where the nectar is secreted, just as they do in larkspurs, Dutchman's breeches, squirrel corn, butter and eggs, and other flowers whose deeply hidden nectaries make dining too difficult for the little rogues. Fragile butterflies, absolutely dependent on nectar, hover near our showy wild columbine with its five tempting horns of plenty, but sail away again, knowing as they do that their weak legs are not calculated to stand the strain of an inverted position from a pendent flower, nor are their tongues adapted to slender tubes unless these may be entered from above. The tongues of both butterflies and moths bend readily only when directed beneath their bodies. It will be noticed that our columbine's funnel-shaped tubes contract just below the point where the nectar is secreted - doubtless to protect it from small bees. When we see the honeybee or the little wild bees - Haliclus chiefly - on the flower, we may know they get pollen only.

Finally a ruby-throated hummingbird whirs into sight. Poising before a columbine, and moving around it to drain one spur after another until the five are emptied, he flashes like thought to another group of inverted red cornucopias, visits in turn every flower in the colony, then whirs away quite as suddenly as he came. Probably to him, and no longer to the outgrown bumblebee, has the flower adapted itself. The European species wears blue, the bee's favorite color according to Sir John Lubbock; the nectar hidden in its spurs, which are shorter, stouter, and curved, is accessible only to the largest humblebees. There are no hummingbirds in Europe. (See jewel-weed.) Our native columbine, on the contrary, has longer, contracted, straight, erect spurs, most easily drained by the ruby-throat which, like Eugene Field, ever delights in "any color at all so long as it's red."

To help make the columbine conspicuous, even the sepals become red; but the flower is yellow within, it is thought to guide visitors to the nectaries. The stamens protrude like a golden tassel. After the anthers pass the still immature stigmas, the pollen of the outer row ripens, ready for removal, while the inner row of undeveloped stamens still acts as a sheath for the stigmas. Owing to the pendent position of the flower, no pollen could fall on the latter in any case. The columbine is too highly organized to tolerate self-fertilization. When all the stamens have discharged their pollen, the styles then elongate; and the feathery stigmas, opening and curving sidewise, bring themselves at the entrance of each of the five cornucopias, just the position the anthers previously occupied. Probably even the small bees, collecting pollen only, help carry some from flower to flower but perhaps the largest bumblebees, and certainly the hummingbird, must be regarded as the columbine's legitimate benefactors. Caterpillars of one of the dusky wings (Papilio lucilius) feed on the leaves.

Very rarely is the columbine white, and then its name, derived from words meaning two doves, does not seem wholly misapplied.

"O Columbine, open your folded wrapper
Where two twin turtle-doves dwell,"

lisp thousands of children speaking the "Songs of Seven" as a first "piece" at school. How Emerson loved the columbine! Dr. Prior says the flower was given its name because "of the resemblance of the nectaries to the heads of pigeons in a ring around a dish - a favorite device of ancient artists."

This exquisite plant was forwarded from the Virginia colony to England for the gardens of Hampton Court by a young kinsman of Tradescant, gardener and herbalist to Charles I.

PITCHER-PLANT; SIDE-SADDLE FLOWER; HUNTSMAN'S CUP; INDIAN DIPPER
(Sarracenea purpurea) Pitcher-plant family

Flower - Deep reddish purple, sometimes partly greenish, pink, or red, 2 in. or more across, globose; solitary, nodding from scape 1 to 2 ft. tall. Calyx of 5 sepals, with 3 or 4 bracts at base; 5 overlapping petals, enclosing a yellowish, umbrella-shaped dilation of the style, with 5 rays terminating in 5-hooked stigmas; stamens indefinite. Leaves: Hollow, pitcher-shaped through the folding together of their margins, leaving a broad wing; much inflated, hooded, yellowish green with dark maroon or purple lines and veinings, 4 to 12 in. long, curved, in a tuft from the root. Preferred Habitat - Peat bogs; spongy, mossy swamps. Flowering Season - May-June. Distribution - Labrador to the Rocky Mountains, south to Florida, Kentucky, and Minnesota.

> "What's this I hear
> About the new carnivora?
>   Can little piants
>   Eat bugs and ants
>   And gnats and flies? -
> A sort of retrograding:
>   Surely the fare
>   Of flowers is air
>   Or sunshine sweet
>   They shouldn't eat
> Or do aught so degrading!"

There must always be something shocking in the sacrifice of the higher life to the lower, of the sensate to what we are pleased to call the insensate, although no one who has studied the marvelously intelligent motives that impel a plant's activities can any longer consider the vegetable creation as lacking sensibility. Science is at length giving us a glimmering of the meaning of the word universe, teaching, as it does, that all creatures in sharing the One Life share in many of its powers, and differ from one another only in degree of possession, not in kind. The transition from one so-called kingdom into another presumably higher one is a purely arbitrary line marked by man, and often impossible to define. The animalcule and the insectivorous plant know no boundaries between the animal and the vegetable. And who shall say that the sun-dew or the bladderwort is not a higher organism than the amoeba? Animated plants, and vegetating. animals parallel each other. Several hundred carnivorous plants in all parts of the world have now been named by scientists.

It is well worth a journey to some spongy, sphagnum bog to gather clumps of pitcher-plants which will furnish an interesting study to an entire household throughout the summer while they pursue their nefarious business in a shallow bowl on the veranda. A modification of the petiole forms a deep hollow pitcher having for its spout a modification of the blade of the leaf. Usually the pitchers are half filled with water and tiny drowned victims when we gather them. Some of this fluid must be rain, but the open pitcher secretes much juice too. Certain relatives, whose pitchers have hooded lids that keep out rain, are nevertheless filled with fluid. On the Pacific Coast the golden jars of Darlingtonia Californica, with their overarching hoods, are often so large and watery as to drown small birds and field mice. Note in passing that these otherwise dark prisons have translucent spots at the top, whereas our pitcher-plant is lighted through its open transom.

A sweet secretion within the pitcher's rim, which some say is intoxicating, others, that it is an anaesthetic, invites insects to a fatal feast. It is a simple enough matter for them to walk into the pitcher over the band of stiff hairs, pointing downward like the withes of a lobster pot, that form an inner covering, or to slip into the well if they attempt crawling over its polished upper surface. To fly upward in a perpendicular line once their wings are wet is additionally hopeless, because of the hairs that guard the mouth of the trap; and so, after vain attempts to fly or crawl out of the prison, they usually sink exhausted into a watery grave.

When certain plants live in soil that is so poor in nitrogen compounds that protein formation is interfered with, they have come to depend more or less on a carnivorous diet. The sundew (q.v.) actually digests its prey with the help of a gastric juice similar to what is found in the stomach of animals; but the bladderwort (q.v.) and pitcher-plants can only absorb in the form of soup the products of their victims' decay. Flies and gnats drowned in these pitchers quickly yield their poor little bodies; but owing to the beetle's hard-shell covering, many a rare specimen may be rescued intact to add to a collection.

A similar ogre plant is the YELLOW-FLOWERED TRUMPET-LEAF (S. flava) found in bogs in the Southern States.

GROUND-NUT
(Apios Apios; A. tuberosa of Gray) Pea family

Flowers - Fragrant, chocolate brown and reddish purple, numerous, about 1/2 in. long, clustered in racemes from the leaf-axils. Calyx 2-lipped, corolla papilionaceous, the broad standard petal turned backward, the keel sickle-shaped; stamens within it 9 and 1. Stem: From tuberous, edible rootstock; climbing, slender, several feet long, the juice milky. Leaves: Compounded of 5 to 7 ovate leaflets. Fruit: A leathery, slightly curved pod, 2 to 4 in. long. Preferred Habitat - Twining about undergrowth and thickets in moist or wet ground. Flowering Season - July-September. Distribution - New Brunswick to Ontario, south to the Gulf States and Kansas.

No one knows better than the omnivorous "barefoot boy" that

"where the ground-nut trails its vine"

there is hidden something really good to eat under the soft, moist soil where legions of royal fern, usually standing guard above it, must be crushed before he digs up the coveted tubers. He would be the last to confuse it with the WILD KIDNEY BEAN or BEAN VINE (Phaseolus polystachyus; P. perennis of Gray). The latter has loose racemes of smaller purple flowers and leaflets in threes; nevertheless it is often confounded with the ground-nut vine by older naturalists whose knowledge was "learned of schools."

Usually a bee, simply by alighting on the wings of a blossom belonging to the pea family, releases the stamens and pistil from the keel; not so here. The sickle-shaped keel of the ground-nut's flower rests its tip firmly in a notch of the standard petal, nor will any jar or pressure from outside release it. A bee, guided to the nectary by the darker color of the underside of the curved keel which spans the open cavity of the flower, enters, at least partially, and so releases by his pressure, applied from underneath, the tip of the sickle from its notch in the standard. Now the released keel curves all the more, and splits open to release the stigmatic tip of the style that touches any pollen the bee may have brought from another blossom. Continuing to curve and coil while the bee sucks, it presently dusts him afresh with pollen from the now released anthers. A mass of pulp between anthers and stigma prevents any of the flower's own pollen from self-fertilizing it. These little blossoms, barely half an inch long, with their ingenious mechanism to compel cross-fertilization, repay the closest study.

At midnight the leaves of the ground-nut.and wild bean "are hardly to be recognized in their queer antics," says William Hamilton Gibson. "The garden beans too play similar pranks. Those lima bean poles of the garden hold a sleepy crowd."

PINE SAP; FALSE BEECH-DROPS; YELLOW BIRD'S-NEST (Hypopitis Hypopitis; Monolropa Hypopitis of Gray) Indian-pipe family

Flowers - Tawny, yellow,ecru, brownish pink, reddish, or bright crimson, fragrant, about 1/2 in. long; oblong bell-shaped; borne in a one-sided, terminal, slightly drooping raceme, becoming erect after maturity. Scapes: Clustered from a dense mass of fleshy, fibrous roots; 4 to 12 in. tall, scaly bracted, the bractlets resembling the sepals. Leaves: None. Preferred Habitat - Dry woods, especially under fir, beech, and oak trees. Flowering Season - June-October. Distribution - Florida and Arizona, far northward into British Possessions. Europe and Asia.

Branded a sinner, through its loss of leaves and honest green coloring matter (chlorophyll), the pine sap stands among the disreputable 'gang' of thieves that includes its next of kin the Indian-pipe, the broom-rape, dodder, coral-root, and beech-drops (q.v.). Degenerates like these, although members of highly respectable, industrious, virtuous families, would appear to be as low in the vegetable kingdom as any fungus, were it not for the flowers they still bear. Petty larceny, no greater than the foxglove's at first, then greater and greater thefts, finally lead to ruin, until the pine-sap parasite either sucks its food from the roots of the trees under which it takes up its abode, or absorbs, like a ghoulish saprophyte, the products of vegetable decay. A plant that does not manufacture its own dinner has no need of chlorophyll and leaves, for assimilation of crude food can take place only in those cells which contain the vital green. This substance, universally found in plants that grub in the soil and literally sweat for their daily bread, acts also as a moderator of respiration by its absorptive influence on light, and hence allows the elimination of carbon dioxide to go on in the cells which contain it. Fungi and these degenerates which lack chlorophyll usually grow in dark, shady woods.

Within each little fragrant pine-sap blossom a fringe of hairs, radiating from the style, forms a stockade against short-tongued insects that fain would pilfer from the bees. As the plant grows old, whatever charm it had in youth disappears, when an unwholesome mold overspreads its features.

SCARLET PIMPERNEL; POOR MAN'S or SHEPHERD'S WEATHER-GLASS; RED
CHICKWEED; BURNET ROSE; SHEPHERD'S CLOCK
 (Anagallis arvensis) Primrose family

 Flower - Variable, scarlet, deep salmon, copper red, flesh colored, or rarely white; usually darker in the center; about 1/4 in. across; wheel-shaped; 5-parted; solitary, on thread-like peduncles from the leaf-axils. Stem: Delicate; 4-sided, 4 to 12 in. long, much branched, the sprays weak and long. Leaves: Oval, opposite, sessile, black dotted beneath. Preferred Habitat - Waste places, dry fields and roadsides, sandy soil. Flowering Season - May-August. Distribution - Newfoundland to Florida, westward to Minnesota and Mexico.

Tiny pimpernel flowers of a reddish copper or terra cotta color have only to be seen to be named, for no other blossoms on our continent are of the same peculiar shade. Thrifty patches of the delicate little annuals have spread themselves around the civilized globe; dying down every autumn, and depending on seeds alone to keep the foothold once gained here, in Mexico and South America,

Europe, Egypt, Abyssinia, Cape of Good Hope, Mauritius, New Holland, Nepal, Persia, and China. What amazing travelers plants are! The blue-flowered plants are now believed to be a distinct species (A. coerulea).

Notwithstanding the fact that many birds delight to feast on the seeds, or perhaps because of it, for many must be dropped undigested, the scarlet pimpernel is one of the most widely distributed species known.

Before a storm, when the sun goes under a cloud, or on a dull day, each little weather prophet closes. A score of pretty folk names given it in every land it adopts testifies to its sensitiveness as a barometer. Under bright skies the flower may be said to open out flat at about nine in the morning and to begin to close at three in the afternoon. No nectar is secreted unless there may be some in the colored hairs which clothe the filaments. As if it knew perfectly well that however.desirable insect visitors are - and it has an excellent device for compelling them to transfer pollen - it is likewise independent of them, it takes no risk in exposing the precious vitalizing dust to wind and rain, but closes up tight, thereby bringing its pollen-laden stamens in contact with its stigma. Manifestly, it is better for a plant having aspirations to colonize the globe to set even self-fertilized seed than none at all.

HOUND'S TONGUE; GYPSY FLOWER
(Cynoglossum officinale) Borage family

Flowers - Dull purplish red, about 1/3 in. across, borne in a curved raceme or panicle that straightens as the bloom advances upward. Calyx 5-parted; corolla salverform, its 5 lobes spreading; 5 stamens; 1 pistil. Stem: Erect, stout, hairy, leafy, usually branched, 1 1/2 to 3 ft. high. Leaves: Rather pale, lower ones large, oblong, slender petioled; upper ones lance-shaped, sessile, or clasping. (Thought to resemble a dog's tongue.) Preferred Habitat - Dry fields, waste places. Flowering Season - May-September. Distribution - Quebec to Minnesota, south to the Carolinas and. Kansas.

This is still another weed "naturalized from Europe" which, by contenting itself with waste land, has been able in an incredibly short time to overrun half our continent. How easy conquest of our vast unoccupied area is for weeds that have proved fittest for survival in the overcultivated Old World! Protected from the ravages of cattle by a disagreeable odor suggesting a nest of mice, and foliage that tastes even worse than it smells; by hairs on its stem that act as a light screen as well as a stockade against pilfering ants; by humps on the petals that hide the nectar from winged trespassers on the bees' and butterflies' preserves, the hound's tongue goes into the battle of life further armed with barbed seeds that sheep must carry in their fleece, and other animals, including most unwilling humans, transport to fresh colonizing ground. For a plant to shower its seeds beside itself is almost fatal; so many offspring impoverish the soil and soon choke each other to death, if, indeed, ants and such crawlers have not devoured the seeds where they lie on the ground. Some plants like the violet, jewelweed, and witch-hazel forcibly eject theirs a few inches, feet or yards. The wind blows millions about with every gust. Streams and currents of water carry others; ships and railroads give free transportation to quantities among the hay used in packing; birds and animals lift many on their feet - Darwin raised 537 plants from a ball of mud carried between the toes of a snipe! - and such feathered and furred agents as feed on berries and other fruits sometimes drop the seeds a thousand miles from the parent. but it will be noticed that such vagabonds as travel by the hook or by crook method, getting a lift in the world frpm every passer-by -.burdocks, beggar-ticks, cleavers, pitchforks, Spanish needles, and scores of similar tramps that we pick off our clothing after every walk in autumn - make, perhaps, the most successful travelers on the globe. The hound's tongue's four nutlets, grouped in a pyramid, and with barbed spears as grappling-hooks, imbed themselves in our garments until they pucker the cloth. Wool growers hurl anathemas at this whole tribe of plants.

A near relative, the common VIRGINIA STICKSEED (Lappula Virginiana; C. Morisoni of Gray) produces similar little barbed nutlets, following insignificant, tiny, palest blue or white flowers up the spike. These bristling seeds, shaped like sad-irons, reflect in their title the ire of the persecuted man who named them Beggar's Lice. If as Emerson said, a weed, is a plant whose virtues have not yet been discovered, the hound's tongue, the similar but blue-flowered WILD COMFREY (C. Virginicum), next of kin, and the stickseed are no weeds; for ages ago the caterpillars of certain tiger moths learned to depend on their foliage as a food store,

OSWEGO TEA; BEE BALM; INDIAN'S PLUME; FRAGRANT BALM; MOUNTAIN MINT
(Monarda didyma) Mint family

Flowers - Scarlet, clustered in a solitary, terminal, rounded head of dark-red calices, with leafy bracts below it. Calyx narrow, tubular, sharply 5-toothed; corolla tubular, widest at the mouth, 2-lipped, 1 1/2 to 2 inches long; 2 long, anther-bearing stamens ascending, protruding; 1 pistil; the style 2-cleft. Stem: 2 to 3 ft. tall. Leaves: Aromatic, opposite, dark green, oval to oblong lance-shaped, sharply saw-edged, often hairy beneath, petioled; upper leaves and bracts often red. Preferred Habitat - Moist soil, especially near streams, in hilly or mountainous regions. Flowering Season - July-September. Distribution - Canada to Georgia, west to Michigan.

Gorgeous, glowing scarlet heads of bee balm arrest the dullest eye, bracts and upper leaves often taking on blood-red color, too, as if it had dripped from the lacerated flowers. Where their vivid doubles are reflected in a shadowy mountain stream, not even the cardinal flower is more strikingly beautiful. Thrifty clumps transplanted from Nature's garden will spread about ours and add a splendor like the flowers of salvia, next of kin, if only the roots get a frequent soaking.

With even longer flower tubes than the wild bergamot's (q.v.), the bee balm belies its name, for, however frequently bees may come about for nectar when it rises high, only long-tongued bumblebees could get enough to compensate for their trouble. Butterflies, which suck with their wings in motion plumb the depths. The ruby-throated hummingbird - to which the Brazilian salvia of our gardens has adapted itself - flashes about these whorls of Indian plumes just as frequently - of course transferring pollen on his needle-like bill as he darts from flower to flower. Even the protruding stamens and pistil take on the prevailing hue. Most of the small, blue or purple flowered members of the mint family cater to bees by wearing their favorite color; the bergamot charms butterflies with magenta, and tubes so deep the short-tongued mob cannot pilfer their sweets; and from the frequency of the hummingbird's visits, from the greater depth of the bee balm's tubes and their brilliant, flaring red - an irresistibly attractive color to the ruby-throat - it would appear that this is a bird flower. Certainly its adaptation is quite as perfect as the salvia's. Mischievous bees and wasps steal nectar they cannot reach legitimately through bungholes of their own making in the bottom of the slender casks.

"This species," says Mr. Ellwanger, "is said to give a decoction but little inferior to the true tea, and was largely used as a substitute" by the Indians and the colonists, who learned from them how to brew it.

SCARLET PAINTED CUP; INDIAN PAINT-BRUSH
 (Castilleja coccinea) Figwort family

Flowers - Greenish yellow, enclosed by broad, vermilion, 3-cleft floral bracts; borne in a terminal spike. Calyx flattened, tubular, cleft above and below into 2 lobes; usually green, sometimes scarlet; corolla very irregular, the upper lip long and arched, the short lower lip 3-lobed; 4 unequal stamens; pistil. Stem: 1 to 2 ft. high, usually unbranched, hairy. Leaves: Lower ones tufted, oblong, mostly uncut; stem leaves deeply cleft into 3 to 5 segments, sessile. Preferred Habitat - Meadows; prairies; moist, sandy soil; thickets. Flowering Season - May-July. Distribution - Maine to Manitoba, south to Virginia, Kansas, and Texas.

Here and there the fresh green meadows show a touch of as vivid a red as that in which Vibert delighted to dip his brush.

        "Scarlet tufts
   Are glowing in the green like flakes of fire;

The wanderers of the prairie know them well,
And call that brilliant flower the 'painted cup.'"

Thoreau, who objected to this name, thought flame flower a better one, the name the Indians gave to Oswego tea; but here the floral bracts, not the flowers themselves, are on fire. Lacking good, honest, deep green, one suspects from the yellowish tone of calices, stem, and leaves, that this plant is something of a thief. That it still possesses foliage, proves only petty larceny against it, similar to the foxglove's (q.v.). Caterpillars of certain checker-spot butterflies in turn prey upon Castilleja. Under cover of darkness, in the soil below, the roots of our painted cup occasionally break in and steal from the roots of its neighbors such juices as the plant must work over into vegetable tissue. Therefore it still needs leaves, indispensable parts of a digestive apparatus. Were it wholly given up to piracy, like the dodder, or as parasitic as the Indian pipe, even the green and the leaf that it hath would be taken away from this slothful servant.

But even without honest leaf green (chlorophyll), we know that plants as low in the scale as fungi often take on the most brilliant of yellows and reds. In the painted cup the bracts, which enfold the insignificant yellowish cloistered flowers like a cape, render them great service in attracting the ruby-throated hummingbird by donning his favorite color. No lip landing place is provided for insects, as in other members of the figwort family dependent on bees; although bumblebees, which desire one, and butterflies, which suck with their wings in motion, may be rarely caught robbing the short tubes. Among the wild flowers, only the columbine, with an almost parallel blooming season, rivals the painted cup for the bird's beneficent attentions. The latter flowers at about the time the ruby-throat flashes northward out of the tropics to spend the summer. Professor Robertson of Illinois says, "In 1886 the first hummingbird seen was on May 5, visiting the Castilleja."

WOOD BETONY; LOUSEWORT; BEEFSTEAK PLANT; HIGH
HEAL-ALL

{Pedicularis Canadensis) Figwort family

Flowers - Greenish yellow and purplish red, in a short dense spike. Calyx oblique, tubular, cleft on lower side, and with 2 or 3 scallops on upper; corolla about 3/4 in. long, 2-lipped, the upper lip arched, concave, the lower 3-lobed; 4 stamens in pairs; 1 pistil. Stems: Clustered, simple, hairy, 6 to 18 in. high. Leaves: Mostly tufted, oblong lance-shaped in outline, and pinnately lobed. Preferred Habitat - Dry, open woods and thickets. Flowering Season - April-June. Distribution - Nova Scotia to Florida, westward to Manitoba, Colorado, and Kansas.

When the Italians wish to extol someone they say, "He has more virtues than betony," alluding, of course, to the European species, Betonica officinalis, a plant that was worn about the neck and cultivated in cemeteries during the Middle Ages as a charm against evil spirits; and prepared into plasters, ointments, syrups, and oils, was supposed to cure every ill that flesh is heir to. Our commonest American species fulfils its mission in beautifying roadside banks and dry, open woods and copses with thick, short spikes of bright flowers, that rise above large rosettes of coarse, hairy, fern-like foliage. At first, these flowers, beloved of bumblebees, are all greenish yellow; but as the spike lengthens with increased bloom, the arched, upper lip of the blossom becomes dark purplish red, the lower one remains pale yellow, and the throat turns reddish, while some of the beefsteak color often creeps into stems and leaves as well.

Farmers once believed that after their sheep fed on the foliage of this group of plants a skin disease, produced by a certain tiny louse (pediculus), would attack them - hence our innocent betony's repellent name.

BEECH-DROPS
 (Septamnium Virginianum; Epifegus Virginiana of Gray)
 Broom-rape family

Flowers - Small, dull purple and white, tawny, or brownish striped; scattered along loose, tiny bracted, ascending branches. Stem:

Brownish or reddish tinged, slender, tough, branching above, 6 in. to 2 ft. tall, from brittle, fibrous roots. Preferred Habitat - Under beech, oak, and chestnut trees. Flowering Season - August-October. Distribution - New Brunswick, westward to Ontario and Missouri, south to the Gulf States.

Nearly related to the broom-rape is this less attractive pirate, a taller, brownish-purple plant, with a disagreeable odor, whose erect, branching stem without leaves is still furnished with brownish scales, the remains of what were once green leaves in virtuous ancestors, no doubt. But perhaps even these relics of honesty may one day disappear. Nature brands every sinner somehow; and the loss of green from a plant's leaves may be taken as a certain indication that theft of another's food stamps it with this outward and visible sign of guilt. The grains of green to which foliage owes its color are among the most essential of products to honest vegetables that have to grub in the soil for a living, since it is only in such cells as contain it that assimilation of food can take place. As chlorophyll, or leaf-green, acts only under the influence of light and air, most plants expose all the leaf surface possible; but a parasite, which absorbs from others juices already assimilated, certainly has no use for chlorophyll, nor for leaves either; and in the broom-rape, beechdrops, and Indian pipe, among other thieves, we see leaves degenerated into bracts more or less without color, according to the extent of their crime. Now they cannot manufacture carbohydrates, even if they would, any more than fungi can.

On the beech-drop's slender branches two kinds of flowers are seated: below are the minute fertile ones, which never open, but, without imported pollen, ripen an abundance of seed with literally the closest economy. Nevertheless, to save the species from still deeper degeneracy through perpetual self-fertilization, small purplish-striped flowers above them mature stigmas and anthers on different days, and invite insect visits to help them produce a few cross-fertilized seeds. Even a few will save it. Every plant which bears cleistogamous or blind flowers - violets, wood-sorrel, jewelweed, among others - must also display some showy ones.

TRUMPET-FLOWER; TRUMPET-CREEPER
(Tecoma radicans) Trumpet-creeper family

Flowers - Red and veined within, paler and inclined toward tawny without, trumpet-shaped, about 2 1/2 in. long, the limb with 5 rounded lobes; 2 to 9 flowers in the terminal clusters; anther-bearing stamens 4, in pairs, under upper part of tube; 1 pistil. Stem: A woody vine 20 to 40 ft. long, prstrate or climbing. Leaves: Opposite, pinnately compounded of 7 to 11 ovate, saw-edged leaflets. Preferred Habitat - Moist, rich woods and thickets. Flowering Season - August-September. Distribution - New Jersey and Pennsylvania, westward to Illinois, and soutb to the Gulf States. Occasionally escaped from gardens farther north.

>From early May untll the middle of October, the ruby-throated hummingbird forsakes the tropics to spend the flowery months with us. Which wild flowers undertake to feed him? Years before showy flowers were brought from all corners of the earth to adorn our gardens, about half a dozen natives in that parterre of Nature's east of the Mississippi catered to him in orderly succeswsion. In feasting at their board he could not choose but reciprocate the favor by transferring their pollen as they took pains to arrange matters. Nectar and tiny insects he is ever seeking. Of course hundreds of flowers secrete nectar which taxes them little; and while the vast majority of these are avowedly adapted to insect benefactors; what is to prevent the bird's needle-like bill from probing the sweets from most of them? Certain flowers dependent on him, finding that the mere offering of nectar was not enough to insure his fidelity, that he was constantly lured away, had to offer some especially strong attractions to make his regular visits sure. How did these learn that red is irresistibly fascinating to him, and orange scarcely less so, perhaps for the sake of the red that is mixed with the yellow? Today we find such flowers as need him sorely, wearing his favorite colors. But even this delicate attention is not enough. He demands that his refreshments shall be reserved for him in a tube so deep or inaccessible that, when he calls, he will find all he desires, notwithstanding the occasional intrusion of such long-tongued insects as bumblebees, butterflies, and moths. First the long-spurred red and yellow columbine and the painted cup, then the coral honeysuckle,

jewelweed, trumpet-creeper, Oswego tea, and cardinal flower have the honor of catering to the exacting little sprite from spring to autumn. His sojourn in our gardens is prolonged until his beloved gladioli, cannas, honeysuckles, nasturtiums, and salvia succumb to frost.

Where a trumpet vine climbs with the help of its aerial roots, like an ivy's, and sends forth clusters of brilliant tubes at the tips of long, wiry branches, there one is sure to see sooner or later, the ruby-throat flashing, whirring, darting from flower to flower. Eight birds at once were counted about a vine one sunny morning. The next, a pair of tame pigeons walked over the roof of the summer-house where the creeper grew luxuriantly, and punctured, with a pop that was distinctly heard fifty feet away, the base of every newly opened nectar-filled trumpet on it! That afternoon all the corollas discolored, and no hummers came near.

CORAL or TRUMPET HONEYSUCKLE
(Lonicera sempervirens) Honeysuckle family

Flowers - Red outside, orange yellow within; whorled round terminal spikes. Calyx insignificant; corolla tubular, slender, 1 1/2 in. long or less, slightly spread below the 5-lobed limb; 5 stamens; 1 pistil. Stem: A high, twining vine. Leaves: Evergreen in the South only; opposite, rounded oval, dark, shining green above, the upper leaves united around the stem by their bases to form a cup. Fruit: An interrupted spike of deep orange-red berries. Preferred Habitat - Rich, light, warm soil; hillsides, thickets. Flowering Season - April-September. Distribution - Connecticut, westward to Nebraska, and south to the Gulf States. Occasionally escaped from cultivation farther north,.

Small-flowered bush honeysuckles elected to serve and be served by bees; those with longer tubes welcomed bumblebees; the white and yellow flowered twining honeysuckles, deep of tube and deliciously fragrant, especially after dark, when they are still visible, cater to the sphinx moths (see sweet wild honeysuckle); but surely the longest-tongued bumblebee could not plumb the depths of this slender-tubed trumpet honeysuckle, nor the night-flying moth discover a flower that has melted into the prevailing darkness when he begins his rounds, and takes no pains to guide him with perfume. What creature, then, does it cater to? After reading of the aims of the trumpet-flower on the preceding page, no one will be surprised to hear that the ruby-throated hummingbird's visits are responsible for most of the berries that follow these charming, generous, abundant flowers, so eminently to his liking. Larger migrants than he, in search of fare so attractive, distribute the seeds far and wide. Is any other species more wholly dependent on birds?

CARDINAL FLOWER; RED LOBELIA
(Lobelia cardinalis) Bellflower family

Flowers - Rich vermilion, very rarely rose or white, 1 to 1 1/2 in, long, numerous, growing in terminal, erect, green-bracted, more or less 1-sided racemes. Calyx 5-cleft; corolla tubular, split down one side, 2-lipped; the lower lip with 3 spreading lobes, the upper lip 2-lobed, erect; 5 stamens united into a tube around the style; 2 anthers with hairy tufts. Stem: 2 to 4 1/2 ft. high, rarely branched. Leaves: Oblong to lance-shaped, slightly toothed, mostly sessile. Preferred Habitat - Wet or low ground, beside streams, ditches, and meadow runnels. Flowering Season - July-September. Distribution - New Brunswick to the Gulf States, westward to the Northwest Territory and Kansas.

By the depth and brilliancy of its incomparable hue, the shade with which Vibert delighted to illumine his rich canvases, the color of the famous hat worn by seventy ecclesiastical princes of the Roman Church, but a richer red than the bird which shares the name can boast, the cardinal flower proclaims its title to all beholders. Because its vivid beauty cannot be hid, and few withstand the temptation to pick it, its extermination goes on as rapidly as its bird namesake's.

"Hast thou named all the birds without a gun?
Loved the wood rose and left it on its stalk?"

The easy cultivation from seed of this peerless wildflower - and it is offered in many trade catalogues - might save it to those regions in Nature's wide garden that now know it no more. The ranks of floral missionaries need recruits.

Curious that the great blue lobelia should be the cardinal flower's twin sister! Why this difference of color? Sir John Lubbock proved by tireless experiment that the bees' favorite color is blue, and the shorter-tubed blue lobelia elected to woo them as her benefactors. Whoever has made a study of the ruby-throated hummingbird's habits must have noticed how red flowers entice him - columbines, painted cups, coral honeysuckle, Oswego tea, trumpet flower, and cardinal in Nature's garden; cannas, salvia, gladioli, pelargoniums, fuchsias, phloxes, verbenas, and nasturtiums among others in ours. How the cardinal flower's wonderful mechanism works to utilize his visits has already been told under great lobelia, in the description of the blue lobelia of similar construction. But with a bird so much greater than the ruby-throat that the jeweled-feathered atom could be concealed under one of its talons is the red lobelia forever associated:

"The cardinal, and the blood-red spots,
  Its double in the stream
As if some wounded eagle's breast,
  Slow throbbing o'er the plain,
Had left its airy path impressed
  In drops of scarlet rain."

APPENDICES

FRAGRANT FLOWERS OR LEAVES.

Baby's Breath. Large Purple-fringed Orchis. Smaller Purple-fringed Orchis. Hepatica (occasionally). Purple Marsh Clematis. English Violet. Wild Phlox. Catnip. Pennyroyal. Wild Thyme. Peppermint. Spear Mint. Wild Mint. Pasture Thistle. Pink Moccasin Flower. Showy Orchis. Rose Pogonia. Arethusa. Calopogon.
Night-flowering Catchfly. Bouncing Bet. Purple-flowering Raspberry. Queen-of-the-Prairie. Wild Rose. Red Clover. Musk Mallow. Prince's Pine. Bog Wintergreen. Pink Azalea. White Azalea. Trailing Arbutus. Sabbatia. Fly-trap Dogbane. Four-leaved Milkweed. Field Bindweed. Wild Bergamot. Twin-flower. Joe-Pye Weed (slightly). Wild Spikenard (slightly). White-fringed Orchis. Ladies' Tresses. Lizard's Tail. Bladder Campion. White Water Lily. Laurel Magnolia. Squirrel Corn. White Sweet Clover. Wild Grape. Sweet White Violet. Canada Violet. Sweet-Cicely. Sweet Pepperbush. Pyrola. Shin-leaf. Wintergreen. Button-bush. Partridge Vine. Elder. Clammy Everlasting. Bellwort. Adders Tongue. Small Yellow Lady's Slipper. Spice-bush. Yellow Sweet Clover. Yellow Wood-sorrel. Evening Primrose. Horse-balm. Horned Bladderwort. Honeysuckles. Fragrant Goldenrod. Ground-nut. Pine Sap. Oswego Tea.

UNPLEASANTLY SCENTED

Purple Trillium. Black Cohosh. Mandrake. Jamestown Weed. Salt-marsh Fleabane. Camomile. Carrion-flower. Barberry. Skunk Cabbage. Hound's Tongue. Beech-drops.

PLANTS AND SHRUBS CONSPICUOUS IN FRUIT

RED AND REDDISH: Nightshade. Twisted-stalk. American Cranberry.

Marsh Calla. Wild Spikenard (pale red speckled berries).
Two-leaved Solomon's Seal (pale red speckled). Wake-robins. Red
Baneberry. Red Raspberry. Strawberries. Red Choke-berry.
June-berry. Shad-bush. Hawthorns. Harmless Sumacs. Hollies.
Bittersweet. Winterberry (Black Alder). American Spikenard.
Flowering Dogwood. Dwarf Cornel or Bunebberry. Wintergreen.
Red
Bearberry. Partridge Vine. Hobble-bush. Red-berried Elder. High
Bush Cranberry. Barberry. Spice-bush. Ground Cherry. Wild
Honeysuckies. Jack-in-the-Pulpit.

BLUISH AND BLACK: Deadly Nightshade. Star-flowered Solomon's
Seal. True Solomon's Seal. Large-flowered Wake-robin. Black
Raspberry. Bush Blackberry. Dewberry. Black Choke-berry. Wild
Grapes. Virginia Creeper. Cornels. Pokeweed. Huckleberry.
Blueberries. Elder. Arrow-woods. Viburnums. Nanny-berry.
Blackberry Lily.

WHITE: White Baneberry (black eye). Poison Sumac. Poison Ivy.
Panicled Dogwood. Snowberry.

FLUFFY: Thistles. Virginia Clematis. Milkweeds. White lettuce
(cinnamon). Groundsel-bush. Spring Everlasting. Dandelions. Sow-
thistle. Lettuces. Hawkweeds (brown).

PLANT FAMILIES REPRESENTED

WATER-PLANTAIN FAMILY (Alismaceae)
Water-plantain. Arrow-head.

ARUM FAMILY (Araceae) Jack-in-the-pulpit. Green dragon. Arrow-arum. Water-arum. Skunk cabbage. Golden-club. Calamus-root.

SPIDERWORT FAMILY (Commelinaceae)
Day-flowers. Spiderwort. Wandering Jew.

PICKEREL-WEED FAMILY (Pontederiaceae)
Pickerel-weed.

BUNCH-FLOWER FAMILY (Melanthaceae)
White hellebore. Bellworts.

LILY FAMILY (Liliaceae)
Lilies. Adder's tongue. Hyacinths. Star-of-Bethlehem.
Colic-root.

LILY-OF-THE-VALLEY FAMILY (Convallariaceae)
Clintonia. Wild spikenard. Solomon's seals. False
lily-of-the-valley. Twisted-stalks. Indian cucumber-root.
Wake-robins. Carrion-flower. Cat-brier.

AMARYLLIS FAMILY (Amaryilidaceae)
Yellow star-grass.

IRIS FAMILY (Iridaceae)
Irises. Blackberry lily. Blue-eyed grass.

ORCHID FAMILY (Orchidaceae)
Ladies' slippers. Orchises. Rose pogonia. Arethusa. Ladies'
tresses. Rattlesnake plantains. Twayblades. Calypso. Coral-roots.

Calopogon. Adam and Eve.

LIZARD'S-TAIL FAMILY (Saururaceae)
Lizard's-tail.

BIRTHWORT FAMILY (Aristoltochiaceae)
Wild ginger. Dutchman's pipe. Serpentary.

BUCKWHEAT FAMILY (Polygonaceae) Persicarias. Smartweed.
Water pepper. Lady's thumb. Pink knotweed. Climbing false buckwheat. Tear-thumb. Coast jointweed.

POKEWEED FAMILY (Phytolaccaceae)
Pokeweed.

PURSLANE FAMILY (Portulacaceae)
Spring beauty. Pussley. Portulaca.

PINK FAMILY (Caryophyllaceae)
Corn cockle. Campions. Catchflies. Pinks. Bouncing Bet.
Chickweed.

WATER-LILY FAMILY (Nymphaeaceae)
Water-shield. Pond-lilies. Lotus.

MAGNOLIA FAMILY (Magnoliaceae)
Laurel magnolia.

CROWFOOT FAMILY (Ranunculaceae)
Marsh-marigold. Gold-thread. Bane-berries. Black Cohosh.
Columbines. Larkspurs. Anemones. Hepatica. Virgin's bower.
Clematis. Water-crowfoots. Spearworts. Buttercups. Meadow-rues.

BARBERRY FAMILY (Berberidaceae)
Barberries. Twin-leaf. Wild mandrake.

LAUREL FAMILY (Lauraceae)
Spice-bush.

POPPY FAMILY (Papaveraceae) Bloodroot. Celandine poppies. California poppy. Dutchman's breeches. Squirrel corn. Bleeding-heart. Climbing fumitory. Pink and Golden corydalis.

MUSTARD FAMILY (Cruciferae)
Mustards. Charlock. Cresses. Rocket. Radish. Ladies' smock. Toothworts. Shepherd's purse. Vernal whitlow grass.

PITCHER-PLANT FAMILY (Sarraceniaceae)
Pitcher-plant. Sundew.

ORPINE FAMILY (Crassulaceae)
Live-forever.

SAXIFRAGE FAMILY (Saxifragaceae)
Early saxifrage. Foam-flower. Mitrewort. Grass-of-Parnassus. Hydrangea.

WITCH-HAZEL FAMILY (Hamamelidaceae)
Witch-hazel.

ROSE FAMILY (Rosaceae)
Ninebark. Meadow-sweet. Steeplebush. Goat's beard. Indian physic. Ipecac. Raspberries. Blackberries. Dalibarda. Strawberries. Cinquefoils. Avens. Queen-of-the-prairie. Agrimony. Roses.

APPLE FAMILY (Pomaceae)
Chokeberries. June-berry. Shadbush. Hawthorns.

SENNA FAMILY (Caesalpinaceae)
Sensitive Pea. Partridge pea. Wild senna.

PEA FAMILY (Papilionaceae)
Wild indigo. Rattle-box. Wild lupine. Clovers. Sweet clovers. Goat's rue. Tick-trefoils. Bush-clovers. Blue vetches. Pea vine. Seaside pea. Butterfly-pea. Hog peanut. Milk-pea. Wild bean.

GERANIUM FAMILY (Geraniaceae)
Wild geranium. Herb Robert. Cranesbill.

WOOD-SORREL FAMILY (Oxalidaceae)
Wood-sorrels.

FLAX FAMILY (Linaceae)
Flax. Slender yellow and Ridged flax.

MILKWORT FAMILY (Polygalaceae)
Milkworts. Fringed polygala.

SPURGE FAMILY (Euphorbiaceae)
Flowering spurge.

SUMAC FAMILY (Anacardiaceae)
Sumacs. Poison ivy. Smoke bush.

HOLLY FAMILY (Ilicaceae)
Hollies. Winter-berry (black alder).

STAFF-TREE FAMILY (Celastraceae)
Climbing bitter-sweet.

JEWEL-WEED FAMILY (Balsaminaceae)
Jewel-weed. Pale touch-me-not.

BUCKTHORN FAMILY (Rhamnaceae)
New Jersey tea.

GRAPE FAMILY (Vitaceae)
Wild grapes. Virginia creeper. Ampelopsis.

MALLOW FAMILY (Malvaceae)
Mallows. Velvet leaf. Althaea.

ST. JOHNS-WORT FAMILY (Hypericaceae)
St. Peter's-wort. St. Andrew's cross. St. John's-worts.

ROCK-ROSE FAMILY (Cistaceae)
Frost-flowers. Poverty grass.

VIOLET FAMILY (Violaceae)
Violets. Pansies.

CACTUS FAMILY (Cactaceae)
Prickly pears.

LOOSESTRIFE FAMILY (Lythraceae)
Purple loosestrife. Blue wax-weed.

MEADOW-BEAUTY FAMILY (Melastomaceae)
Meadow-beauty. Deer-grass.

EVENING-PRIMROSE FAMILY (Onagraceae)
Fire-weed. Willow-herbs. Evening-primrose. Sundrops.
Enchanter's nightshade.

GINSENG FAMILY (Ariliaceae)
American spikenard. Wild sarsaparilla. Ginsings.

CARROT FAMILY (Umbelliferae)
Wild carrot. Cowbane. Parsnips. Parsley. Sanicle. Fennel.
Pimpernel. Water-hemlock. Sweet-Cicely. Poison hemlock.
Water-parsnip.

DOGWOOD FAMILY (Cornaceae)
Cornels or Dogwoods.

WHITE-ALDER FAMILY (Clethraceae)
Sweet pepperbush.

WINTERGREEN FAMILY (Pyrolaceae)
Wintergreens. Shin-leaf. Prince's pine.

INDIAN-PIPE FAMILY (Monotrotaceae)
Indian-pipe. Pine sap.

HEATH FAMILY (Ericaceae)
Labrador tea. Azaleas. Laurels. Rhodora. Rhododendrons.
Leucothoe. Wild rosemary. Fetter-bush, Stagger-bush. Andromeda.
Cassandra. Sourwood. Trailing arbutus. Creeping wintergreen.
Bearberries.

HUCKLEBERRY FAMILY (Vacciniaceae) Huckleberries. Blueberries. Squaw huckleberry. Creeping snowberry. Cranberry.

DIAPENSIA FAMILY (Diapensiaceae)
Pyxie.

PRIMROSE FAMILY (Primulaceae)
Loosestrifes. Moneywort. Star-flower. Scarlet pimpernel.
Shooting star.

PLUMBAGO FAMILY (Plumbaginaceae)
Marsh rosemary.

GENTIAN FAMILY (Gentianaceae)
Sabbatia. Sea-pink. Marsh pink. Gentians.

DOGBANE FAMILY (Apocynaceae)
Dogbane. Indian hemp.

MILKWEED FAMILY (Asclepiadaceae)
Miikweeds. Butterfly weed.

MORNING-GLORY FAMILY (Convolvulaceae)
Wild potato vine. Bindweeds.

DODDER FAMILY (Cascutaceae)
Gronovius' dodder.

PHLOX FAMILY (Polemoniaceae)
Phloxes. Moss pink.

WATER-LEAF FAMILY (Hydrophyllaceae)
Virginia water-leaf.

BORAGE FAMILY (Boraginaceae)
Hound's tongue. Comfrey. Stick-seeds. Virginia cowslip.
Lungwort. Forget-me-not. Viper's bugloss. Vervains. Verbena.

MINT FAMILY (Labiatae)
Blue curls. Skullcaps. Catnip. Gill-over-the-ground. Self-heal.
Obedient plant. Motherwort. Oswego tea. Wild bergamot.
Pennyroyal. Sweet basil. Hyssop. Mints. Wild thyme. Dittany.
Peppermint. Citronella.

POTATO FAMILY (Solanaceae)
Ground cherry. Nightshades. Thorn apples. (Jamestown weed.)

FIGWORT FAMILY (Scrophulariaceae)
Mulleins. Butter-and-eggs. Blue toad-flax. Figwort.
Turtle-head. Beard tongues. Blue-eyed Mary. Monkey-flower.
Speedwells. Brooklime. Culver's-root. False foxgloves. Gerardias.
Scarlet painted cup. Wood betony.

BLADDERWORT FAMILY (Lentibulariaceae)
Bladderworts.

BROOM-RAPE FAMILY (Orobanchaceae)
Broom-rape. Beech-drops.

TRUMPET-CREEPER FAMILY (Bignoniaceae)
Trumpet-flower.

ACANTHUS FAMILY (Acanthaceae)
Hairy ruellia.

MADDER FAMILY (Rubiaceae)
Bluets. Button-bush. Partridge-vine. Cleavers. Bedstraw.

HONEYSUCKLE FAMILY (Caprifoliaceae)
Elder bushes. Hobble-bush. Bush cranberry. Arrow-woods.

Withe-rod. Sweet viburnum. Black haw. Twin-flower. Snowberry. Honeysuckles. Fly-honeysuckles. Bush-honeysuckles.

TEASEL FAMILY (Dipsacaceae)
Card teasel.

GOURD FAMILY (Cucurbitaceae)
Star-cucumber.

BELL-FLOWER FAMILY (Campanulaceae)
Harebell. Bellflowers. Venus' looking-glass. Cardinal flower. Lobelias. Indian tobacco.

CHICORY FAMILY (Cichoriaceae)
Chicory. Cynthia. Dwarf goat's beard. Fall dandelion. Dandelions. Sow-thistles. Wild lettuces. Hawk-weeds. Rattlesnake-weed. White lettuce.

THISTLE FAMILY (Compositae)
Iron-weed. Joe-Pye weed. Boneset or Thoroughwort. White sanicle. Climbing hempweed. Blazing-star. Button snake-root. Golden aster. Goldenrods. Asters. Robin's plantain. Flea-banes. Sweet scabious. Groundsel-bush. Everlastings. Elecampane. Cup-plant. Compass-plant. Ox-eyes. Cone-flowers. Black-eyed Susan. Sunflowers. Jerusalem artichoke. Tickseeds. Bur-marigolds. Beggar-ticks. Sneezeweed. Yarrow. Camomiles. Daisy. Tansy. Ragwort. Burdock. Thistles.